HEAT TRANSFER - MODELS, METHODS AND APPLICATIONS

Edited by **Konstantin Volkov**

Heat Transfer - Models, Methods and Applications
http://dx.doi.org/10.5772/intechopen.71737
Edited by Konstantin Volkov

Contributors

Franciszek Kluza, Dariusz Góral, Katarzyna Kozłowicz, Guoxi He, Yansong Li, Baoying Wang, Mohan Lin, Yongtu Liang, Mohamed El-Sayed Ali, Shereef Sadek, Syaiful Syaiful, Lei-Yong Jiang, Yinghua Han, Prakash Patnaik, Anatoly Nikolaevich Makarov, R. C. Mehta, Irakli Shekriladze, Jianyao Yao, Kun Zhang, Jianqiang Xin, Ning Hu, Hafiz Muhammad Ali, Arslan Qasim, Sullahuddin Malik, Ghulam Murtaza, Boštjan Končar, Luka Klobučar, Yi Zheng, Alok Ghanekar, Yanpei Tian, Hyo Tae Kim, Veysel Ozceyhan, Orhan Keklikcioglu, Guangqin Huang

First published in London, United Kingdom, 2018 by IntechOpen
IntechOpen is the global imprint of INTECHOPEN LIMITED, registered in England and Wales, registration number: 11086078, The Shard, 25th floor, 32 London Bridge Street
London, SE19SG – United Kingdom
Printed in Croatia

British Library Cataloguing-in-Publication Data
A catalogue record for this book is available from the British Library

Additional hard copies can be obtained from orders@intechopen.com

Heat Transfer - Models, Methods and Applications, Edited by Konstantin Volkov
p. cm.
Print ISBN 978-1-78923-264-6
Online ISBN 978-1-78923-265-3

For EU product safety concerns:
IN TECH d.o.o., Prolaz Marije Krucifikse Kozulić 3, 51000 Rijeka, Croatia,
info@intechopen.com or visit our website at intechopen.com.

Meet the editor

Dr. Volkov is a senior lecturer in Thermofluids at the Kingston University (London, UK). He holds his PhD degree in Fluid Mechanics. After completing his PhD degree, he worked at the Baltic State Technical University (Russia), University of Central Lancashire (UK), and University of Surrey (UK). His expertise is in multidisciplinary areas from design and optimization of energy systems to fundamental problems focused on modeling and simulation of turbulent multiphase flows. He is the chartered engineer and a member of Institute of Physics, Institution of Mechanical Engineers, and Combustion Institute. He is an author of more than 120 scientific papers and a member of editorial board and scientific committee of numerous journals and conferences.

Contents

Preface

Heat transfer is a discipline of thermal engineering that concerns the generation, use, conversion, and exchange of thermal energy between physical systems. Heat transfer is classified into various mechanisms, such as conduction, convection, radiation, and transfer of energy by phase changes. Designers and engineers also consider the transfer of mass, differing chemical species to achieve heat transfer. While these mechanisms have distinct characteristics, they often occur simultaneously in the same system. Heat transfer has a broad application to the functioning of numerous devices and systems.

The book offers a general overview of the concepts, theories, and models underlying the heat and mass transfer. This book intends to provide the reader a comprehensive overview of the current state of the art in heat and mass transfer. The book also focuses on new analytical, experimental, and computational developments in the field of research of heat and mass transfer phenomena. The generation, conversion, use, and exchange of thermal energy between physical systems are considered. Various mechanisms of heat transfer such as conduction, convection, radiation, and transfer of energy by phase changes are presented. Theory and fundamental research in heat and mass transfer, numerical simulations and algorithms, experimental techniques, and measurements as they applied to all kinds of applied and emerging problems are covered.

The book covers some topics in depth while offering introductions and summaries of others. There are 4 parts and 16 chapters in the book covering different aspects related to heat and mass transfer and various practical applications.

The book is divided into four parts: principles and methods of analysis, heat transfer in complex systems, enhancement of heat transfer, and engineering applications. The first part provides a series of chapters concerned with the introductory topics that are required when solving different engineering problems in heat transfer. The second part of the book is related to heat transfer in complex systems including conjugate heat transfer and heat transfer in two-phase media. Part 3 addresses the enhancement of heat transfer in engineering applications. In addition to providing a detailed treatment of various methods and types of heat exchangers, this part also examines the impact of design and operation of industrial devices on heat transfer characteristics. The concluding part of the book covers the industrial applications.

In Part 1, the theory and models underlying various heat transfer mechanisms are described, and the principal results and formulae are presented. Available techniques for obtaining analytical, approximate, and numerical solutions to complicated problems are also considered.

In many industrial applications, convection radiation and conduction contribute simultaneously to the heat transfers. Heat transfer is complicated by the phase transition and complex geometrical configurations of industrial devices. Part 2 covers some methods of analysis of complex heat transfer including gas-liquid stratified flows in pipeline with phase change, boiling heat transfer, calculations of heat transfer in the furnaces of steam boilers, conjugate heat transfer, and nanoscale radiative thermal transport.

The heat transfer enhancement techniques are widely used in many practical applications to make possible reduction in weight and size or enhance the performance of engineering devices. These techniques are classified as active and passive techniques. The active technique required external power, while the passive technique does not need any external power. An effort has been made in Part 3 to carry out an extensive review of various turbulators, extended surfaces, rough surfaces, and others for enhancing heat transfer in heat exchangers.

Enhancement and control of forced convection heat transfer are important in many engineering applications including turbomachinery, nuclear, electronic industry, and others. Some of these applications are covered in Part 4 of the book.

The book promotes an open discussion among research institutions, academia, and industry from around the globe on research and development of enabling technologies. The book covers many aspects of theory and practice, which deliver essential contributions and provide their input and support to the cooperative efforts.

Dr. Konstantin Volkov
MEng, MSc, PhD, DSc, CEng, MIMechE, MInstP, FHEA
Department of Mechanical and Automotive Engineering
School of Mechanical and Aerospace Engineering
Faculty of Science, Engineering and Computing
Kingston University
London, United Kingdom

Section 1

Principles and Methods of Analysis

Chapter 1

Free Convection Heat Transfer from Different Objects

Mohamed Ali and Shereef Sadek

Additional information is available at the end of the chapter

http://dx.doi.org/10.5772/intechopen.75427

Abstract

Heat transfer by natural convection phenomenon has been studied by diverse researchers for several geometries and different boundary conditions. Different approaches were used such as numerical, analytical, and experimental approaches. In this chapter, we are going to review the most common published research on object shapes in natural convection heat transfer with their obtained numerical, semiempirical, and/or experimental correlations for the last two decades. These correlations are very important for engineering applications. This chapter will be stressed on cylinders in different orientations with cross sections such as circular, square, triangular, vertical, and horizontal coils and array of cylinders of different cross sections.

Keywords: natural convection, experimental heat transfer, empirical correlations, numerical heat transfer, helical coils, cylinders with different cross section, array of cylinders

1. Introduction

Many investigations have been conducted on natural convection heat transfer from single object with different cross sections such as elliptic, rectangular or square, triangular, and circular cylinders in free space. Furthermore, correlations of flat plate (vertical, horizontal, or inclined) and coils were also developed. Such correlations are those for vertical plate as reported by McAdams [1] and Churchill and Chu [2], horizontal surface by Goldstein et al. [3] and Lloyd and Moran [4], long horizontal cylinder by Morgan [5] and Churchill and Chu [6], and spheres by Churchill [7]. In Section 2, cylinders with different cross sections and orientations are presented, Section 3 reports correlations obtained from helical coils, and arrays of cylinders of different cross sections in free convection is reviewed in Section 4.

2. Cylinders of different cross sections in free convection

Atmane et al. [8] investigated experimentally the effect of vertical confinement on a single heated horizontal cylinder for a range of Rayleigh number between 3×10^4 and 3×10^6 in water. They concluded that the heat transfer nearby the cylinder and the natural convection hydrodynamics was not affected by the water surface if the distance was four times the cylinder diameter or more.

Free convection heat transfer from vertical slender cylinders has been reviewed by Popiel [9]. The classical analysis of the laminar free convection heat transfer from vertical cylinders was shown. Numerical calculations for a turbulent boundary layer flow on a vertical cylinder using a modified integral method were presented. Experimental data for laminar-turbulent transition suggested that the critical Grashof number for a vertical flat plate is $Gr_{cr} \approx 10^9$, while for a vertical cylinder is $Gr_{cr} \approx 4 \times 10^9$.

Theoretical, numerical, and experimental data for free convection heat transfer from vertical slender circular cylinders were surveyed by Popiel [9], who developed the following 3% criterion for transversal curvature effect for the average heat transfer from isothermal vertical cylinders valid in the range of Prandtl number from $Pr = 0.01$ to 100:

$$Gr_H^{0.25} \frac{D}{H} \leq a + \frac{b}{Pr^{0.25}} + \frac{c}{Pr^{0.5}} \quad (1)$$

where a = 11.474, b = 48.92, and c = 0.006085. Elsayed et al. [10] studied experimentally free convection from a constant heat flux elliptic tube cross section. They studied the effect of tube orientation on the local Nusselt number distributions as well as the average Nusselt number variation versus tube inclination angle. Results showed that the maximum average Nusselt number was achieved when the tube's major axis is vertical. An empirical correlation (Eq. (2)) has been derived to evaluate the average Nusselt number in terms of Rayleigh number based on the input heat flux:

$$Nu_{av} = 0.47(Ra^*)^{0.2}, \quad 1.1 \times 10^7 \leq Ra^* < 8 \times 10^7 \quad (2)$$

The maximum deviation between the experimental data and the correlation is about 5%. From the comparison between free convection around isothermal and constant heat flux elliptic tubes, it was found that at the steady state, the heat flux tube correlates well with $Ra^{0.25}$, similar to the isothermal tube. Ali and Al-Ansary [11] have reported experimentally steady-state natural convection heat transfer from the outer surface of vertical triangular cylinder in air. The experimental results showed the existence of laminar and transition to turbulence regimes. They observed a decrease in the local axial (perimeter-averaged) heat transfer coefficient at the lower half of the cylinder which indicated a laminar regime. On the other hand, an increase in the heat transfer coefficient along the cylinder axis was related to transition to turbulent regime at any constant heat flux. The local axial Nusselt and the modified Rayleigh numbers were used to develop two correlations for laminar (Eq. (3)) and transition regime (Eq. (4)):

$$Nu_x = 3.034(Ra^*)^{0.156}, \quad 10^7 \le Ra_x^* < 10^{12} \tag{3}$$

with correlation coefficient $R = 91\%$:

$$Nu_x = 0.359\left(Ra_x^*\right)^{0.244}, \quad 1.0 \times 10^{10} \le Ra_x^* < 2.0 \times 10^{12} \tag{4}$$

with a correlation coefficient $R = 91.6\%$. Another overall averaged correlation was obtained using the averaged data and the equilateral triangular side length as the characteristic length:

$$\overline{Nu}_L = 0.373\left(Ra_L^*\right)^{0.24}, \quad 4 \times 10^5 \le Ra_L^* \le 6.0 \times 10^8 \tag{5}$$

with a correlation coefficient $R = 93.4\%$. Critical values of the modified Rayleigh numbers were obtained and correlated for transition to turbulent:

$$(Nu_x)_{cr} = 0.457\left(Ra_x^*\right)_{cr}^{0.22}, \quad 4 \times 10^9 \le Ra_x^* \le 5.0 \times 10^{11} \tag{6}$$

with a correlation coefficient $R = 96.5\%$.

Ali and Al-Ansary [12] carried out an experimental study to investigate the steady-state natural convection from horizontal ducts with triangular cross sections in air. Two different horizontal duct orientations were considered; in the first position, one surface was facing downward and a vertex facing upward, while in the other position, the vertex faced downward and a flat surface faced upward. All ducts had an equilateral triangle cross section with side lengths of 0.044, 0.06, 0.08, 0.10, and 0.13m. A constant-heat-flux heating element was placed inside each duct to provide the heat flux. General correlations are obtained for two duct orientations using the local perimeter-averaged heat transfer data given by Eqs. (7) and (8):

$$Nu_x = 0.429\left(Ra_x^*\right)^{0.241}, \quad 2.0 \times 10^8 \le Ra_x^* \le 1.0 \times 10^{12} \tag{7}$$

with a correlation coefficient of $R = 97.5\%$:

$$Nu_x = 0.688\left(Ra_x^*\right)^{0.222}, \quad 9.0 \times 10^7 \le Ra_x^* \le 1.0 \times 10^{12} \tag{8}$$

with a correlation coefficient of $R = 97.3\%$. A critical correlation is obtained to segregate the laminar and transition regimes when the duct vertex faces upward [Eq. (9)]:

$$Nu_x = 0.325\left(Ra_x^*\right)^{0.241}, \quad 1.0 \times 10^7 \le Ra_x^* \le 1.0 \times 10^{11} \tag{9}$$

with a correlation coefficient of $R = 98.2\%$.

Zeitoun and Ali [13] have studied numerically two-dimensional laminar natural convection heat transfer in air around horizontal ducts with rectangular and square cross sections. Temperature and velocity profiles were obtained close to every surface of the ducts. Different aspect ratios were used for wide ranges of Rayleigh numbers. The process procedure was based on the finite element technique. Temperature and velocity profiles are obtained near each surface of the ducts (correlation 10) covering wide ranges of Rayleigh number for various aspect ratios (Γ), which was obtained for ducts with horizontal square and rectangular cross sections:

$$Nu = \left(0.9\Gamma^{-0.061} + 0.371\Gamma^{-0.114}Ra^{0.1445}\right)^2, \quad 700 \le Ra \le 1.3 \times 10^8 \tag{10}$$

with a correlation coefficient of $R = 99.8\%$.

Al-Ansary et al. [14] have presented a numerical study of laminar natural convection heat transfer from uniformly heated horizontal cylinders of a triangular cross section in air. Flow streamlines as well as temperature contours were presented around the cylinder's perimeter. Thermal boundary layer profiles were also shown for different modified Rayleigh numbers. General correlations of Nusselt numbers versus modified Rayleigh numbers were obtained for symmetric, transition, and asymmetric plumes for both positions of the cylinders. For triangular cylinders facing upward, correlations (11–12) were obtained for symmetric and asymmetric plumes around the cylinders, respectively. On the other hand, for triangular cylinders facing downward, correlations (13–14) were obtained for symmetric and asymmetric plumes around the cylinders, respectively:

$$Nu = 0.619(Ra^*)^{0.194}, \quad Ra^* < 7.0 \times 10^5 \tag{11}$$

$$Nu = 0.293(Ra^*)^{0.249}, \quad Ra^* \ge 7.0 \times 10^5 \tag{12}$$

$$Nu = 0.707(Ra^*)^{0.176}, \quad Ra^* \le 1.0 \times 10^6 \tag{13}$$

$$Nu = 0.133(Ra^*)^{0.301}, \quad Ra^* > 2.0 \times 10^7 \tag{14}$$

Steady-state free convection heat transfer from outer surface of horizontal cylinders having rectangular cross section in air was investigated by Ali [15]. Five cylinders have been used with aspect ratios Γ (Γ = cylinder height/cylinder width) of 2, 1, and 0.5. The cylinders had a constant heat flux boundary conditions through inserting internal heating element along the centerline of each cylinder. Temperature measurements were taken along the axial and peripheral directions of the cylinder surface. Two distinct flow regimes were observed, namely, laminar and transition to turbulence. A laminar regime was obtained for low values of convection heat flux and characterized by a decrease in Nusselt numbers at any fixed longitudinal direction x on the cylinder's surface. However, Nusselt numbers increased as x increased along the cylinder's surface for any value of the heat flux, and the regime was characterized as transition to turbulence. General correlation using the local circumference average transition to turbulence heat transfer data was obtained where Nusselt numbers were correlated using the modified Rayleigh numbers (Eq. (15)):

$$Nu_x = 0.355\left(Ra_x^*\right)^{0.297}, \quad 1.9 \times 10^8 \le Ra_x^* \le 7 \times 10^{11} \tag{15}$$

On the overall averaged basis, correlation covering the locus of critical points segregating the laminar and transition regimes was obtained using the modified Rayleigh number (Eq. (16)):

$$\overline{Nu_L} = 0.168\left(Ra_L^*\right)^{0.287} \tag{16}$$

with a correlation coefficient of $R = 95.93\%$. In addition, laminar heat transfer data were aspect ratio and area ratio dependent, as correlated using the modified Rayleigh numbers over averaged basis (Eqs. (17) and (18)):

$$\overline{Nu_L} = 12.41\left(Ra_L^*\right)^{-0.237}\kappa^{1.259}\Gamma^{-1.226} \tag{17}$$

(for) $\kappa < 100, \quad 8 \times 10^6 \le Ra_L^* \le 6 \times 10^7$

$$\overline{Nu_L} = 33,633.84\left(Ra_L^*\right)^{-0.241}\kappa^{0.86}\Gamma^{0.0779} \tag{18}$$

for $\kappa \ge 100, 1.0 \times 10^6 \le Ra_L^* \le 7 \times 10^6$ where κ is the area ratio (κ = surface area of the cylinder/cross section area of the cylinder). Ali [16] has investigated experimentally the steady-state free convection from vertical and horizontal cylinders with square or rectangular cross section. Four and three cylinders had square and rectangular cross sections, respectively. The cylinders had constant heat flux boundary conditions where the heating elements were inserted internally along the center of the cross section. Temperature measurements were taken on the cylinder's peripheral direction and on the axial direction of the cylinders. General correlation using the local perimeter-averaged transition to turbulence heat transfer data was obtained where Nusselt numbers were correlated using the modified Rayleigh numbers (Eq. (19)). Furthermore, the laminar local perimeter-averaged data were correlated using the axial distance as a parameter (Eqs. (20) and (21)):

$$Nu_x = 0.14\left(Ra_x^*\right)^{0.269}, \quad 5.0 \times 10^9 \le Ra_x^* \le 1.5 \times 10^{11} \tag{19}$$

with a correlation coefficient $R = 94\%$:

$$Nu_x = 1.7\left(Ra_x^*\right)^{0.164} \tag{20}$$

For area ratio $\kappa < 0.01, 1.0 \times 10^7 \le Ra_x^* \le 4.0 \times 10^{10}$, with a correlation coefficient $R = 97.3\%$:

$$Nu_x = 6.02\left(Ra_x^*\right)^{0.113} \tag{21}$$

For area ratio $\kappa = 0.01, 1.0 \times 10^7 \le Ra_x^* \le 1.0 \times 10^{11}$, with a correlation coefficient $R = 93.4\%$. In addition, on the overall averaged basis for laminar heat transfer, data were correlated using the modified Rayleigh numbers as seen by Eqs. (22) and (23):

$$\overline{Nu_L} = 1.52E + 11\left(Ra_L^*\right)^{-0.283}\kappa^{4.077}\Gamma^{0.392} \tag{22}$$

which is valid for

$$\kappa > 0.01, 4.0 \times 10^6 \le Ra_L^* \le 4.0 \times 10^8 \tag{23}$$

$$\overline{Nu_L} = 4.283E + 8\left(Ra_L^*\right)^{-0.0365}\kappa^{2.42} \tag{24}$$

which is valid for

$$\kappa \leq 0.01, 1.0 \times 10^6 \leq Ra_L^* \leq 1.5 \times 10^8 \quad (25)$$

Natural convection from inclined circular cylinders to air was reported experimentally by Oosthuizen [17] using a transient cooling technique. Oosthuizen pointed out that the flow over the inclined cylinder was in general three-dimensional and obtained a correlation for Nusselt numbers as a function of Grashof number and the cylinder's aspect ratio:

$$\frac{Nu_D}{(G_D \cos\phi)^{0.25}} = function\left(\frac{l}{D \tan\phi}\right) = function(l^*) \quad (26)$$

where D, l, and ϕ are the diameter, length, and angle of inclination of the circular cylinder. He observed that for l^* greater than about 10, the flow over the cylinder is the same as that over a horizontal cylinder, while for l^* less than about 1, the flow is the same as that over a vertical cylinder. Similar experiment was done by Stewart [18], and using the same technique followed by Oosthuizen [17], a constant heat flux was applied by inserting a heating element along the centerline of the cylinder. Stewart reported a general correlation valid for horizontal, vertical, or inclined cylinders at any angles:

$$\frac{Nu_D}{(Ra_D \cos\phi)^{0.25}} = 0.53 + 0.555\left[\left(\frac{D}{l \cos\phi}\right)^{0.25} - \left(\frac{D}{l}\right)^{0.25}\right] \quad (27)$$

Local heat transfer coefficient profiles were experimentally obtained by Al-Arabi and Salman [19], for inclined circular cylinder subject to boundary condition of constant heat flux. In their experiment, the laminar regime was characterized by a decrease in the local heat transfer coefficient up to the middle of the cylinder, and the transition regime was indicated by an increase of the local heat transfer coefficient in the other half of the cylinder. Experimental correlations were obtained for each angle of inclination. Another experiment was done for isothermal inclined cylinders. Al-Arabi and Khamis [20] reported that the average heat transfer coefficient decreased with increasing the cylinder diameter for constant inclination and cylinder length. General local (Eqs. (26) and (27)) and average (Eqs. (28) and (29)) Nusselt number correlations were suggested for laminar (Eqs. (26) and (28)) and turbulent (Eqs. (27) and (29)) convection heat transfer:

$$Nu_x = \left[2.3 - 1.72(\sin\phi)^{0.8}\right](Gr_D)^{-1/12}[Gr_x Pr]^{0.25+1/\left(12\sqrt{\sin\phi}\right)} \quad (28)$$

$$Nu_x = \left[0.42 - 0.16(\sin\phi)^{0.8}\right](Gr_D)^{-1/12}[Gr_x Pr]^{1/3} \quad (29)$$

$$Nu_l = \left[2.9 - 2.32(\sin\phi)^{0.8}\right](Gr_D)^{-1/12}[Gr_l Pr]^{0.25+1/\left(12\sqrt{\sin\phi}\right)} \quad (30)$$

$$Nu_l = \left[0.47 + 0.11(\sin\phi)^{0.8}\right](Gr_D)^{-1/12}[Gr_l Pr]^{1/3} \quad (31)$$

Li and Tarasuk [21] have presented a universal correlation in terms of Nusselt and Rayleigh numbers for circular cross-sectional inclined cylinder from 0 to 90°. Their results show a direct proportion of the average Nusselt numbers with the Rayleigh numbers and inversely proportional to the inclination angle of the cylinder to the horizontal for the Rayleigh number range (104 to 106). A copper electroplating system was used to simulate the free convection heat transfer using a mass transfer system based on the analogy between heat and mass transfer by Heo and Chung [22]. They have used that method to determine the heat transfer from inclined cylinders. Numerical and experimental investigation on free convection heat transfer was reported by Kalendar and Oosthuizen [23] using inclined isothermal square cylinders. Those cylinders had been exposed to the top surface. Their results for the average Nusselt number and for inclination angle range from 0 to 180° were correlated as

$$\frac{\overline{Nu}}{Ra^{0.28}} = 0.27 + \frac{0.65}{\left(W \cdot Ra^{0.25}\right)^{0.95}} \tag{32}$$

where W is the width ratio (width/height). Experimental study has been made on local and averaged natural convection heat transfer from inclined square cylinders in air at 30°, 45°, and 60° inclination angle to the horizontal by Ali [24]. His results showed that the correlation obtained for each inclination angle indicated that the natural convection has a weak dependence on the inclination angle; therefore, overall correlations were obtained to cover all cylinders at all inclination angles for laminar regime: one for local laminar profiles (Eq. (31)) and the other for averaged profiles (Eq. (32)):

$$Nu_x = 1.109\left(Ra_x^*\right)^{0.193}, \quad 1.0 \times 10^7 \le Ra_x^* \le 2.0 \times 10^{12} \tag{33}$$

$$Nu = 0.452(Ra^*)^{0.224}, \quad 4.0 \times 10^4 \le Ra^* \le 1.0 \times 10^8 \tag{34}$$

The corresponding local correlation for transition regime was obtained as

$$Nu_x = 0.842\left(Ra_x^*\right)^{0.209}, \quad 9.0 \times 10^9 \le Ra_x^* \le 1.0 \times 10^{12} \tag{35}$$

3. Helical coils in free convection

Natural convection heat transfer from helical coils is very important for its large surface area per unit volume. Helical coils are used in many engineering applications such as heating, ventilation, and air-conditioning systems. The first reference to deal with the subject of helical coils that tried to get empirical correlation is an experimental study by Ali [25]. In this study, a correlation is developed for natural convection from vertical helical coils in water for a range of Prandtl number, $3.44 \le Pr \le 5.30$. In his experiment, ten coils were used corresponding to four various coil diameters to tube diameter ratios and to five different ratios of pitch to tube diameter. It was observed that the average heat transfer coefficient is inversely proportional with the coil length for tube diameter of $d = 0.012$m but directly proportional with the coil length for $d = 0.008$m. The correlation for $d = 0.012$m, using the coil length as a characteristic length, is

$$Nu_L = 0.685 Ra_L^{0.295}, \quad 3 \times 10^{12} \le Ra_L \le 8 \times 10^{14} \tag{36}$$

where L is the length of the helical coil. Free laminar convection heat transfer from coils in air was studied by Xin and Ebadian [26] for vertical and horizontal positions. Three test coils were used in their experiment. The coils had a constant heat flux boundary condition. Averaged Nusselt numbers were obtained in correlation form for the two positions of the coils.

For the horizontal coils, the study was concentrated on the middle turn of each coil only, and they conclude that the peripheral average Nusselt number distribution around the middle turn is almost periodic. Their correlation for the three middle turns of the three test coils (did not include the coil end effect) was given by

$$Nu_d = 0.318Ra_d^{0.293}, \quad 5 \times 10^3 \le Ra_d \le 1 \times 10^5 \tag{37}$$

where d is the tube diameter of the helical coil. Ali [27] has studied experimentally the effect of laminar natural convection heat transfer from horizontal helical coils in air. Four coils of different diameters were used and heated using constant heat flux in the range of $500 - 5000\,\mathrm{W/m^2}$. In his results, the transition regime was characterized by a wavy variation in Nusselt numbers as a function of the number of turns. His laminar-averaged correlation obtained for the heat transfer coefficient indicated a decrease in the heat transfer coefficient with increasing the number of coil turns.

His overall averaged Nusselt numbers vs. Rayleigh numbers for all turns of the four test coils for $q'' = 500$ and $1000\,\mathrm{W \cdot m^{-2}}$ were shown. Least square power law fit through his data set obtained the following correlations for a heat transfer flux q'' of 500 and $1000\,\mathrm{W \cdot m^{-2}}$, respectively.

$$Nu_d = 10824.2Ra_d^{-1.196}, \quad 340 \le Ra_d \le 645 \tag{38}$$

$$Nu_d = 187508Ra_d^{-1.526}, \quad 728 \le Ra_d \le 938 \tag{39}$$

Using the horizontal coil axial distance x as a characteristic length, Ali [27] showed that an overall correlation for all heat fluxes for all coils in air can be obtained as

$$Nu_x = 0.913Ra_x^{0.301}, \quad 3 \times 10^3 \le Ra_d \le 7 \times 10^6 \tag{40}$$

Another constant heat flux experimental work using vertical helical coils in air for two different groups of coils was studied by Ali [28]. He used the coil tube diameter and the coil axial length as two different characteristic lengths for analyzing his data. His obtained Nusselt number correlation, using the tube diameter as a characteristic length, showed heat flux dependence when Rayleigh number was used in the correlation. However, using the axial distance of the coil as a characteristic length in both Nusselt and Rayleigh numbers correlation made all the data to collapse on one unique curve independent of the heat flux. Prabhanjan et al. [29] have presented an experimental investigation using three vertical coils in water using various characteristic lengths. The correlation they got was not accurate enough to be considered for a wide rage and different orientations. Therefore, they recommend that more study should be made to develop an accurate correlation for Nusselt numbers. On the other hand, their experiment was good in predicting the outer surface temperature of the coil. In order to obtain correlations for natural convection heat transfer from coils in high Prandtl number medium, Ali [30] investigated experimentally that case for laminar and transition natural convection heat transfer using vertical helical coils with various pitches in 57% glycerol-water solution by mass. That experiment was for a range of Prandtl number of 28–36. For both diameter ratio and Rayleigh numbers fixed, he showed increases of the overall heat transfer coefficient with decreasing the number of coil turns. Accordingly, keeping the diameter ratio constant and increasing the Rayleigh numbers led to increase in heat transfer coefficients. On the other hand, enhancement was obtained in heat transfer coefficients corresponding to lower diameter ratio of the coils for both constant number of turns and Rayleigh number. The overall averaged empirical correlations for coils with five and ten turns were reported, respectively, as

$$Nu_L = 2.53 \times 10^{-5} Ra_L^{0.739} \left(\frac{D}{d_0}\right)^{-1.313}, \quad 10^{12} \le Ra_L \le 10^{14} \tag{41}$$

$$Nu_L = 1.535 \times 10^{-5} Ra_L^{0.671} \left(\frac{D}{d_0}\right)^{-0.702} \tag{42}$$

$$for \quad 7 \times 10^{12} \le Ra_L \le 8 \times 10^{14}$$

where D is the coil diameter and d_0 is the outer tube diameter. The overall average Nusselt numbers for all coils used was correlated using Rayleigh number as

$$Nu_L = 0.106 Ra_L^{0.335}, \quad 2 \times 10^{12} \le Ra_L \le 8 \times 10^{14} \tag{43}$$

$$Nu_L = 0.555 Gr_L^{0.301} Pr^{0.314} \tag{44}$$

For $10^8 \le Gr_L \le 5 \times 10^{14}$ and $4.4 \le \overline{Pr} \le 345$.

An experimental study has been made on steady-state natural convection heat transfer from vertical helical coil tubes in heat transfer oil of Prandtl number range 250 − 400 by Ali [31]. He used five sets of coils; each set consisted of three coils with constant coil diameter to tube diameter ratio. Each set had two, five, and ten turns. The helical coil-to-tube diameter ratios were 30, 20.83, 17.5, 13.33, 10, and 10. The coil length was used as a characteristic length in the dimensionless groups where correlation obtained Eq. (42) for high Prandtl number and compared to that of low Prandtl number fluids of [25].

An alternative correlation was obtained using the Rayleigh number as

$$Nu_L = 0.714 Ra_L^{0.294}, \quad 5 \times 10^{10} \le Ra_L \le 8 \times 10^{14} \tag{45}$$

Natural convection heat transfer phenomena for an inclined helical coil were investigated experimentally by Moon et al. [32] for $Ra_D = 4.55 \times 10^6$. An electroplating system was used to measure mass transfer instead of heat transfer, based on the analogy concept.

4. Arrays of cylinders in free convection

To increase the surface area of the heat transfer, thermally interacting multiple horizontal and vertical cylinders have been investigated by different researchers. Most of the analyses for those types of geometries are circular. A pair of vertically aligned horizontal cylinders in water is investigated by Reymond et al. [33] for a Rayleigh numbers 2×10^6, 4×10^6, and 6×10^6 and for a range of cylinder spacing of 1.5, 2, and 3 diameters. They observed that the heat transfer from the bottom cylinder in the array was unaffected by the upper cylinder for the used spacing between cylinders. It was also found that the presence of unheated lower cylinder had no effect on the upper cylinder heat transfer. For that reason they did the experiment for both heated cylinders with different spacings. Their experiment showed that the rising plume from the lower cylinder had large effect on the upper cylinder's surface heat transfer. Their results showed that increasing the Rayleigh number increases the area-averaged heat transfer in line and the following correlation is suggested:

$$\overline{Nu} = 0.48(Ra_L)^{0.25}, \quad 10^4 < Ra_L < 10^7 \tag{46}$$

Experimental investigation on natural convection at high Rayleigh numbers from a pair of evenly spaced cylinders in water for five different center-to-center separation distances was reported by Grafsrønningen and Jensen [34]. Their results showed that at small separation distance of the upper cylinder, the increase in average Nusselt number was about 6% and for large one was about 15 to 40%. The effect of dissimilar cylinder spacing between horizontal cylinders in vertical arrays of two or three cylinders was investigated by Grafsrønningen and Jensen [35]. Their results have shown that the Nusselt number on the middle cylinder in a three-cylinder array increased compared to the lower cylinder and that of the upper cylinder increased too with no much difference than that of the middle cylinder.

Experimental investigation was performed on laminar natural convection from two cylinders arranged in vertical array with a spacing range 2 and 6 of the tube diameter by Chouikh et al. [36]. The cylinders were kept isothermal with Rayleigh numbers range 10^2 and 10^4. Their results were consistent with those in the literature. They found that the bottom cylinder's heat transfer was unaffected by the downstream cylinders and was equal to that of a single cylinder. It was also observed that the upper cylinder in the array of large spacing had an enhanced numbers, but at close distances, it had a reduced Nusselt numbers. On the other hand, for the same spacing, a direct enhancement of heat transfer from the upper cylinder with Rayleigh number was obtained. Their observation about the heat transfer from cylinders in the array agreed with the flow field around the cylinders.

Numerical work was reported on natural steady-state laminar convection heat transfer from horizontal cylinders arranged in vertical array by Corcione [37] in air. Cylinders were isothermal, and numerical simulations were done for arrays of $2-6$ with separation distance between cylinders of 2 up to more than 50 cylinder diameters. The simulation was done for Rayleigh number range 5×10^2 and 5×10^5. He has reported numerical correlations valid for the whole array of cylinders and also for any individual cylinder in the array. He compared his numerical results with those in the literature, and it was found satisfactory. In addition, it was found that at any investigated Rayleigh number, degradation was generally the rule at the smaller tube spacing, while enhancement predominated at the larger ones. Two distinct numerical correlations were obtained, as shown below, for the average Nusselt number Nui of any ith cylinder in the array in terms of Rayleigh number Ra, the cylinder location measured from the center of the bottom cylinder x/H, and the number of cylinders in the array Ni (where $H = S \times (N-1)$):

$$Nu_{ith} = Ra^{0.25}\left\{0.364 \ln\left[(x/D)^{0.4}/N_i^{0.9}\right] + 0.508\right\} \tag{47}$$

for $2 \le N_i \le 6$, $2(N_i - 1) < x/D \le 8 + N_i$, and $5 \times 10^2 \le Ra \le 5 \times 10^5$, with percent standard deviation of error $Esd = 3.19\%$ and range of error from −5.07 to +7.97%. N_i cylinder's index in an array

$$Nu_i = Ra^{0.25}\left\{0.587 \ln\left[(x/D)^{0.33}/N_i^{0.5}\right] + 0.35\right\} \tag{48}$$

for $2 \le N_i \le 6$, $(8 + N_i) < x/D \le \left(10^9/Ra\right)^{0.333}$, and $5 \times 10^2 \le Ra \le 5 \times 10^5$ with percent standard deviation of error $Esd = 3.27\%$ and range of error from −5.93 to +7.96%, and S is the center-to-center distance and D is the cylinder diameter. Another average numerical correlation for Nusselt numbers Nua for all cylinders in the array was obtained in terms of the cylinder spacing ratio S/D, the Rayleigh numbers Ra, and the total number of cylinders N in the array as

$$Nu_a = Ra^{0.235}\left\{0.292 \ln\left[(S/D)^{0.4} \times N^{-0.2}\right] + 0.447\right\} \tag{49}$$

which is valid for

$$2 \le N \le 6, \quad 5 \times 10^2 \le Ra \le 5 \times 10^5, \quad S/D \le 10 - \log(Ra)$$

with percent standard deviation of error $Esd = 2.25\%$ and range of error from −4.79 to +5.27%:

$$Nu_a = Ra^{0.235}\left\{0.277 \ln\left[(S/D)^{0.4} \times N^{0.2}\right] + 0.335\right\} \tag{50}$$

which is valid for

$$2 \le N \le 6, \quad 5 \times 10^2 \le Ra \le 5 \times 10^5, \quad S/D \le 10 - \log(Ra)$$

with percent standard deviation of error $Esd = 2.72\%$ and range of error from −6.40 to +6.09%. Sadeghipour and Asheghi [38] have reported an experimental study for natural convection heat transfer from horizontal isothermal cylinders in vertical arrays of 2–8 at a low Rayleigh number of 500, 600, and 700. Their results showed that the heat transfer coefficient for the lowest cylinder is the same as that of a single horizontal cylinder. They found that the Nusselt numbers reached a maximum enhancement of 24% for the second cylinder, in an array of two cylinders, at a cylinder spacing of $S/D > 20$. For arrays of more cylinders, they observed a maximum enhancement at distance of $S/D > 15$ from the upper cylinder in the array, but that enhancement is a little bit higher as the number of cylinders in the array increased. Therefore, they found a direct proportional between the number of cylinders in the array and the Nusselt numbers of the array. The following correlation presents their array's Nusselt numbers as a function of *Ra*, S/D and N:

$$\overline{Nu}_N = Ra^{1/4}\left[0.823 + \exp\left(-1.5(S/D)^{0.05N}\right)\right] \quad (51)$$

which is valid for

$$2 \le N \le 8, \quad 500 \le Ra \le 700, \quad 3.5 \le S/D \le 27.5$$

Park and Chang [39] have studied numerically the interactive natural convection heat transfer from a pair of vertically separated horizontal circular cylinders of equal diameter. The flow was assumed to be laminar, and each cylinder was kept isothermal. In their study, zone-dependent grid systems were successfully employed with grid overlapping in between. Results were presented, the velocity profiles and the local and overall Nusselt numbers for the Rayleigh number ranged $10^4 \sim 10^5$, Prandtl number fixed at 0.7, and for different cylinder spacings.

Corcione [40] studied steady-state laminar free convection from a pair of vertical arrays of equally spaced, horizontal isothermal cylinders set in free air, numerically. Simulations were obtained for $1 - 4$ circular cylinders arranged in tube arrays for vertical and horizontal center-to-center distance of the range 2–12 and 1.4–24 of cylinder diameters, respectively. The Rayleigh numbers used in his simulation was in the range 10^2–10^4, where he used the cylinder diameter as a characteristic length. He has concluded that the Nusselt numbers of any cylinder in the array may be reduced or enhanced compared to that of a single tube depending on the location of the cylinder in the array, the array geometry, and the investigated Rayleigh numbers. He also proposed some dimensionless correlations for the heat transfer.

Persoons et al. [41] have discussed the near interaction between local fluid dynamics and natural convection heat transfer from a couple of isothermally heated horizontal cylinders plunged in water. The presence of the next heated cylinder induced heat transfer improvements of up to 10% and strong variations in local heat transfer rate. Therefore, specific attention was motivated on how the local heat transfer characteristics of the upper cylinder were affected by buoyancy-induced fluid flow from the lower cylinder. The paper studied a variety of Rayleigh numbers between 1.8×10^6 and 5.5×10^6 and a vertical cylinder gaps between $2D$ and $4D$. A joint temporal analysis of the data has provided new insights into the governing mechanisms, which enables further optimization of the heat transfer performance. They concluded that there was no significant heat transfer fluctuations associated with the single cylinder for the range of investigated Rayleigh number $\left(1.8 \times 10^6 \le Ra \le 5.5 \times 10^6\right)$. This result was consistent with the average heat transfer rate correlation. On the other hand, for a vertical array of two horizontal cylinders with center spacing from $2D$ to $4D$, a strong periodicity associated with the local heat transfer rate was found. It was observed that fluctuations were highest around the bottom of the upper cylinder where the lower plume impinged and also at the top of it where the plume detached. Furthermore, an increase up to 10% of the average heat transfer rate was observed at the upper cylinder which depends on the separation distance and the Rayleigh number. However, a reduction of -5% in heat transfer was also obtained at low Rayleigh numbers and small separation distance $\left(Ra = 1.8 \times 10^6, S/D = 2\right)$, respectively.

Marstrrs [42] has studied experimentally the heat transfer properties of a vertical array of heated cylinders in steady-state natural convection. The results for a variety of combinations of spacing and number of cylinders were reported. The cylinder temperature in a vertical array is a function of spacing and position for a given dissipation rate. For close spacing, the temperature rises, generally, but for wide spacing, the temperature decreases monotonically. For closely spaced arrays, individual tube Nusselt numbers were found to be smaller than for a single cylinder (as much as 50% smaller). However, for wide spacing, individual tube Nusselt numbers were higher (up to about 30%) than for a single cylinder. A similar experiment was done by Sparrow and Nlethammer [43] for two-cylinder array. The distance between the two cylinders was changed between 2–9 the cylinder diameter, and it was found that degradation of the Nusselt number of the upper cylinder occurred at small separations and enhancement at large ones if compared to the lower cylinder. It was also found that the maximum enhancement observed at a separation distance range of 7–9 cylinder diameters.

Basit et al. [44] numerically investigated natural convection heat transfer from an assembly of vertical cylinders of Pakistan Atomic Research Reactor. They solved the two-dimensional axisymmetric case and presented the temperature, velocity profiles, and Nusselt number variations at different heat flux values. The normalized Nusselt number decreased along the axial distance. They observed an increase of maximum temperature along the length of the fuel pin but still far below the melting point. On the other hand, a decrease in local heat transfer coefficient was obtained along the length of the fuel pin.

Yousefi and Ashjaee [45] have studied experimentally laminar natural convection heat transfer from vertical array of horizontal isothermal elliptic cylinder. In their experiment, they studied two to five cylinder major axis at Rayleigh number between 10^3 and 2.5×10^3. They concluded that:

a. At any Rayleigh number and for each spacing between cylinders, the heat transfer rate from the bottom cylinder in the array was unaffected by the rest of the cylinders, and it was equal to that of a single cylinder.

b. For the spacing between cylinders used in their study, they observed a direct proportion between the heat transfer from the topmost cylinder in the array and the spacing between cylinders.

c. At large separation between cylinders, the buoyant flow effect was very limited, and Nusselt numbers of the downstream cylinders in the array were almost the same unaffected by the large separation between cylinders. [Where S and a are the vertical center-to-center separation distance and the minor axis, respectively].

d. The average Nusselt number of the arrays increased with increasing the Rayleigh number and the cylinder spacing. Also, for large cylinder spacing, the difference between the average Nusselt number and that for single cylinder decreased by increasing the Rayleigh number.

The experimental data for the average Nusselt number of any individual ith cylinder in the array $\overline{Nu_i}$, excluding the lowest one, were correlated to the Ra, Y/a, and N_i, by the following equation:

$$\overline{Nu_i} = Ra^{0.25}\left\{0.321 \ln\left[(Y/a)^{0.695}/N_i^{1.4}\right] + 0.575\right\} \tag{52}$$

which is valid for

$$2 \le N_i \le 8, \quad 10^3 \le Ra \le 2.5 \times 10^3, \quad N_i \le Y/a \le 4N_i$$

with percentage standard deviation of error $Esd = 2.91\%$ and range of error from -6.01 to $+6.91\%$. The experimental data for the average Nusselt number of all the cylinders in the vertical array may be correlated to the Ra and S/a, by the following equation:

$$\overline{Nu}_a = Ra^{0.25}[0.183 \ln (S/a) + 0.28] \tag{53}$$

which is valid for

$$2 \le N \le 5, \quad 10^3 \le Ra \le 2.5 \times 10^3$$

with percentage standard deviation of error $Esd = 1.34\%$ and range of error E from -2.6 to 2%. Ashjaee and Yousefi [46] have experimentally studied steady-state two-dimensional natural convection heat transfer from the vertical and inclined array of five horizontal isothermal elliptic cylinders with vertical major axis which confined between two adiabatic walls. Their correlation for any individual ith cylinders in the vertical array using the average Nusselt numbers $\overline{Nu}_{iv}$ excluding the lowest one can be correlated to the Ra, S_y/d, and Ni by the following equation:

$$\overline{Nu}_{iv} = Ra^{0.25}\left\{0.45 \ln \left[\left(S_y/d\right)^{0.45}/N_i^{0.92}\right] + 0.51\right\} \tag{54}$$

which is valid for

$$2 \le N_i \le 5, \quad 10^3 \le Ra \le 3 \times 10^3, \quad N_i \le S_y/d \le 4N_i$$

with percent standard deviation of error equal to 4.1% and range of error from -7.01 to 6.72%, where S_y, d, and the $\overline{Nu}_{iv}$ stand for vertical center-to-center distance of the ith cylinder from the bottom cylinder, diameter of the cylinders, and average Nusselt number of the ith cylinder in vertical array, respectively. On the other hand, their correlation for the vertical array presented by Nusselt number $\overline{Nu}_{av}$ was correlated to the Ra and P_y/d, by the following equation:

$$\overline{Nu}_{av} = Ra^{0.235}\left[0.47 \ln \left(P_y/d\right)^{0.4} + 0.27\right] \tag{55}$$

which is valid for

$$N = 5, \quad 10^3 \le Ra \le 3 \times 10^3, \quad 2 \le P_y/d \le 5$$

where P_y and N stand for vertical center-to-center separation distance and number of cylinders in the array, respectively. Eq. 53 had a percent standard deviation of error equal to 2.37% and range of error from -2.86 to 1.58%. Furthermore, for inclined array, the following correlation was obtained for individual cylinder where the variation of Nusselt number with Ra, S_y/d, N_i, P_x/d, and P_y/d are as follows:

$$\overline{Nu}_{ii} = Ra^{0.25}\left\{0.67 \ln \left[\left(S_y/d\right)^{0.45}/N_i^{0.83}\right] + 0.7\right\} \times \left\{0.95(P_x N_i/d)^{0.41}/\left(P_y/d\right)^{1.88} + 0.65\right\} \tag{56}$$

which is valid for

$$2 \leq N_i \leq 5, \quad 10^3 \leq Ra \leq 3 \times 10^3, \quad N_i \leq S_y/d \leq 4N_i \quad 0 \leq P_x/d \leq 2$$

and with percent standard deviation of error equal to 4.19% and range of error from −9.6 to 10.8%, and P_x stands for horizontal center-to-center separation distance. Moreover, their correlation of the average Nusselt number of the inclined array was obtained as

$$\overline{Nu}_{ai} = 0.572Ra^{0.235}\left[(P_x/d) + 0.0256\exp\left(0.24/\eta^{1.07}\right)\right]^{\eta} \tag{57}$$

where η is defined as $\eta = \left(3.115P_y/d\right)^{-0.934}$. Ali et al. [47] have investigated experimentally natural convection heat transfer from the outer surface of a vertical array of horizontal square cylinders in air. Five cylinders equally spaced were used with cross section of $0.02 \times 0.02m^2$. The cylinders were subject to constant heat flux boundary condition using internal constant heat flux heating elements. Experiment was done for arrays of 2–5 square cylinders and for four center-to-center separation distances to equivalent diameter ratios (S/D). Their study concentrated on the effect of cylinder location in the array and on the geometry of the array, where D was the cross-sectional side length of the square cylinder and Y was the vertical distance measured from the lower cylinder. Local and averaged results showed that the lowermost cylinder was uninfluenced by the downstream cylinders and its Nusselt number is identical to that of a single cylinder in free space. It is also shown that any cylinder in the array was influenced only by the upstream cylinders and unaffected by the downstream ones. Furthermore, the top (downstream) cylinder in each array always degraded the Nusselt number the most than that of a single cylinder followed by other cylinders upstream toward the lower cylinder at small $S/D = 2.5$. This degradation is generally depended on the center-to-center distance ratio S/D of the cylinders in the array. As S/D increases the degradation changed to enhancement in Nusselt numbers, and critical S/D was obtained for the upper (downstream) cylinder in each array. This enhancement depends on the number of cylinders in the array as well as on the S/D ratio. Local and averaged correlations are obtained for each cylinder in different arrays in terms of Nusselt numbers and the modified Rayleigh numbers for $S/D = 2.5$ where the degradation occurred (Eq. (56) for local and Eq. (57) for averaged). **Figure 1** shows the setup used in their experiment holding the array of square cylinders and the cross section of the cylinders filled with sand to insure uniform conduction from the heating element to the surface of the cylinders. Furthermore, general local and averaged correlations, where enhancement in heat transfer occurred, are obtained by Eqs. (58) and (59), respectively, for $S/D = 5$, 7.5, and 10 as shown below:

$$Nu_x = 0.253\left(Ra_x^*\right)^{0.254}N^{-0.099}M^{0.048}, \qquad R^2 = 95.20\% \tag{58}$$

$$Nu_D = 0.425\left(Ra_D^*\right)^{0.213}N^{-0.98}M^{0.049}, \qquad R^2 = 94.20\% \tag{59}$$

$$Nu_x = 0.247\left(Ra_x^*\right)^{0.246}N^{0.147}(S/D)^{0.084}M^{-0.058}, \qquad R^2 = 94.66\% \tag{60}$$

$$Nu_D = 0.379\left(Ra_D^*\right)^{0.205}N^{0.147}(S/D)^{0.084}M^{0.069}, \qquad R^2 = 90.40\% \tag{61}$$

where N and M stand for the number of cylinders in the array (2, 3, 4, or 5) and the sequence of the cylinder in the array (1, 2, 3, or 4), respectively.

M Sequence of cylinder in an array.

Figure 1. Setup used to hold the arrays of square cylinders and cross sections at A-A showing the center-to-center distance S and the side length D of the square cylinders.

5. Conclusion

Natural convection from individual cylinder of different cross sections is reviewed in this chapter. Circular, square, triangular, and elliptic cylinders are considered. Empirical, numerical, or experimental correlations are mentioned for such cross-sectional cylinders. Helical coils are also reviewed, and experimental correlations for vertical and horizontal coils are presented by different authors in various environments. Correlations for horizontal, vertical, and inclined array of cylinders with different cross sections are also shown either for individual cylinder in the array or as the average Nusselt numbers for the whole array of cylinders.

Acknowledgements

The author extends his appreciation to the Deanship of Scientific Research at King Saud University for funding this work through the research group project No. RGP-080.

Nomenclature

Γ cylinder aspect ratio

κ area ratio ($As = Ac$)

ϕ cylinder inclination angle

a minor axis of the elliptic

D cylinder diameter

d_0 coil outer tube diameter

G_D Grashof number based on D

http://dx.doi.org/10.5772/intechopen.75427

Gr_{cr} critical Grashof number based on H

Gr_H Grashof number

H cylinder height

L cylinder length

M sequence of cylinder in an array

N number of cylinders in an array

N_i cylinder's index in an array

N_u Nusselt number

P_y vertical center-to-center distance

P_r Prandtl number

q heat transfer flux

R correlation coefficient

R_a Rayleigh number

Ra^* modified Rayleigh number

S center-to-center distance in an array

x distance along cylinder

Y distance of any cylinder from the lower cylinder

l cylinder length

W width-to-height ratio

Author details

Mohamed Ali* and Shereef Sadek

*Address all correspondence to: mali@ksu.edu.sa

Mechanical Engineering Department, College of Engineering, King Saud University, Riyadh, Saudi Arabia

References

[1] McAdams WH. Heat Transmission, Chapter 7. 3rd ed. New York: McGraw-Hill; 1945

[2] Churchill SW, Chu HH. Correlating equations for laminar and turbulent free convection from a vertical plate. International Journal of Hear and Mass Transfer. 1975;**18**:1323-1329

[3] Goldstein R, Sparrow EM, Jones DC. Natural convection mass transfer adjacent to horizontal plates. International Journal of Heat Mass Transfer. 1973;**16**:1025-1035

[4] Lloyd JR, Moran WR. Natural Convection Adjacent to Horizontal Surfaces to Various Platforms. Paper 74-WA/HT-66, ASME; 1974

[5] Morgan VT. The overall convective heat transfer from smooth circular cylinders, vol. 11. In: Irvine TF, Hartnett JP, editors. Advances in Heat Transfer. New York: Academic Press; 1975. pp. 199-264

[6] Churchill SW, Chu HH. Correlating equations for laminar and turbulent free convection from a horizontal cylinder. International Journal of Heat and Mass Transfer. 1975;**18**:1049-1053

[7] Churchill SW. Free convection around immersed bodies. Section 2.5.7, New York: Hemisphere, 1983. In: Schlender EU, editor. Chief Heat Exchanger Design Handbook

[8] Atmane MA, Chan VS, Murray DB. Natural convection around a horizontal heated cylinder the effects of vertical confinement. International Journal of Heat and Mass Transfer. 2003;**46**:3661-3672

[9] Popiel CO. Free convection heat transfer from vertical slender cylinders: A review. Heat Transfer Engineering. 2008;**29**(6):521-536

[10] Elsayed AO, Ibrahim EZ, Elsayed SA. Free from a constant heat flux elliptic tube. Energy Conversion and Management. 2003;**44**:2445-2453

[11] Ali ME, Al-Ansary H. General correlations for laminar and transition natural convection heat transfer from vertical triangular cylinders in air. Experimental Heat Transfer. 2011;**24**:133-150

[12] Ali ME, Al-Ansary H. Experimental investigations on natural convection heat transfer around horizontal triangular duct. Heat Transfer Engineering. 2010;**31**:350-361

[13] Zeitoun O, Ali M. Numerical investigation of natural convection around isothermal horizontal rectangular ducts. Numerical Heat Transfer Part A. 2006;**50**:189-204

[14] Al-Ansary H, Zeitoun O, Ali M. Numerical modeling of natural convection heat transfer around horizontal triangular cylinders. Numerical Heat Transfer, Part A Applications. 2012;**61**(3):201-219

[15] Ali ME. Natural convection heat transfer from horizontal rectangular ducts. ASME Journal of Heat Transfer. 2007;**129**(9):1195-1202

[16] Ali ME. Natural convection heat transfer from vertical rectangular ducts. Heat and Mass Transfer. 2009;**46**:225-266

[17] Oosthuizen PH. Experimental study of free convection heat transfer from inclined cylinders. ASME, Journal of Heat Transfer. 1976;**98**(4):672-674

[18] Stewart WE. Experimental free convection from an inclined cylinder. Journal of Heat Transfer. 1981;**103**(4):817-819

[19] Al-Arabi M, Salman YK. Laminar natural convection heat transfer from an inclined cylinders. International Journal of Heat and Mass Transfer. 1980;**23**(1):45-51

[20] Al-Arabi M, Khamis M. Natural convection heat transfer from inclined cylinders. International Journal of Heat and Mass Transfer. 1982;**25**(1):3-15

[21] Li J, Tarasuk JD. Local free convection around inclined cylinders in air: An interferometric study. Experimental Thermal and Fluid Science. 1992;**5**(2):235-242

[22] Heo JH, Chung BJ. Natural convection heat transfer on the outer surface of inclined cylinders. Chemical Engineering Science. 2012;**73**:366-372

[23] Kalendar A, Oothuizen PH. A numerical and experimental study of natural convection heat transfer from an inclined isothermal square cylinder with an exposed top surface. Heat and Mass Transfer. 2013;**49**(5):601-616

[24] Ali M. Experimental free convection heat transfer from inclined square cylinders. Heat and Mass Transfer. 2017;**53**(5):1643-1655

[25] Ali ME. Experimental investigation of natural convection from vertical helical coiled tubes. International Journal of Heat and Mass Transfer. 1994;**37**(4):665-671

[26] Xin RC, Ebadian MA. Natural convection heat transfer from helicoidal pipes. Journal of Thermophysics and Heat Transfer. 1996;**12**(2):297-302

[27] Ali ME. Laminar natural convection from constant heat flux helical coiled tubes. International Journal of Heat and Mass Transfer. 1998;**41**(14):2175-2182

[28] Ali ME. Natural convection from vertical helical coiled tubes in air. In: 33rd National Heat Transfer Conference (ASME, ed.); 15–17 August 1999. Albuquerque, Paper # NHTC99–114

[29] Prabhanjan DG, Rennie TJ, Vijaya Raghavan GS. Natural convection heat transfer from helical coiled tubes. International Journal of Thermal Sciences. 2004;**43**:359-365

[30] Ali ME. Free convection heat transfer from the outer surface of vertically oriented helical coils in glycerol-water solution. Heat and Mass Transfer. 2004;**40**(8):615-620

[31] Ali ME. Natural convection heat transfer from vertical helical coils in oil. Heat Transfer Engineering. 2006;**27**(3):79-85

[32] Moon J, Heo J, Chung B. Natural convection experiments on the outer surface of an inclined helical coil. Heat and Mass Transfer. 2015;**51**:1229-1236

[33] Reymond O, Darina BM, O'Donovan TS. Natural convection heat transfer from two horizontal cylinder. Experimental Thermal and Fluid Science. 2003;**32**:1702-1702

[34] Grafsrønningen S, Jensen A. Natural convection heat transfer from two horizontal cylinders at high rayleigh numbers. International Journal of Heat and Mass Transfer. 2012;**55**: 5552-5564

[35] Grafsrønningen S, Jensen A. Natural convection heat transfer from three vertically arranged horizontal cylinders with dissimilar separation distance at moderately high rayleigh numbers. International Journal of Heat and Mass Transfer. 2013;**57**:519-527

[36] Chouikh R, Guizani A, Maalej A, Belghith A. Experimental study of the natural convection flow around an array of heated cylinders. Renewable Energy. 2000;**21**:3660-3673

[37] Corcione M. Correlation equation for free convection heat transfer from horizontal isothermal cylinders set in a vertical array. International Journal of Heat and Mass Transfer. 2005;**48**:3660-3673

[38] Sadeghipour SM, Asheghi M. Free convection heat transfer from arrays of vertically separates horizontal cylinders at low rayleigh number. International Journal of Heat and Mass Transfer. 1994;**37**:103-109

[39] Park KS, Chang SK. Numerical study on interactive laminar natural convection from a pair of vertically separated horizontal cylinders. Numerical Heat Transfer. 1988;**14**(1):61-74

[40] Corcione M. Interactive free convection from a pair of vertical tube-arrays at moderate rayleigh numbers. International Journal of Heat and Mass Transfer. 2007;**50**(1):1061-1074

[41] Persoons T, O'Gorman M, Donoghue D, Byrne G, Murray B. Natural convection heat transfer and fluid dynamics for a pair of vertically aligned isothermal horizontal cylinders. International Journal of Heat and Mass Transfer. 2011;**54**(1):5163-5172

[42] Marstrrs FG. Arrays of heated horizontal cylinders in natural convection. International Journal of Heat and Mass Transfer. 1972;**15**(1):921-933

[43] Sparrow EM, Nlethammer JE. Effect of vertical separation distance and cylinder-to-cylinder temperature imbalance on natural convection for a pair of horizontal cylinders. ASME, Journal of Heat Transfer. 1981;**103**:638-644

[44] Basit AM, Rafique M, Chughtai RI, Inayat HM. Computer simulation of natural convection heat transfer from an assembly of vertical cylinders of parr-2. Applied Thermal Engineering. 2007;**27**(1):194-201

[45] Yousefi T, Ashjaee M. Experimental study of natural convection heat transfer from vertical array of isothermal horizontal elliptic cylinders. Experimental Thermal and Fluid Science. 2005;**32**:240-248

[46] Ashjaee M, Yousefi T. Experimental study of free convection heat transfer from horizontal isothermal cylinders arranged in vertical and inclined arrays. Heat Transfer Engineering. 2007;**28**:460-471

[47] Ali M, Nuhait A, Almuzaiqer R. The Effect of Square Tube Location in a Vertical Array of Square Tubes on Natural Convection Heat Transfer. Heat Transfer Engineering. 2017 http://dx.doi.org/10.1080/01457632.2017.1358485

Chapter 2

Use of CFD Codes for Calculation of Radiation Heat Transfer: From Validation to Application

Boštjan Končar and Luka Klobučar

Additional information is available at the end of the chapter

http://dx.doi.org/10.5772/intechopen.74420

Abstract

In complex geometries, computational fluid dynamic (CFD) codes are commonly used to predict the heat and fluid transfer. To justify their use for the applications with dominant radiation heat transfer conditions, the implemented models need to be first appropriately validated on simple benchmark examples where the analytical solutions exist. The practical application discussed in this chapter considers the thermal radiation inside the vacuum vessel of the fusion reactor. Two representative benchmark examples are used to obtain the analytical solution and assess the accuracy of the real case simulations performed by CFD codes. The analytical solutions use the view factor method to calculate the net radiation heat flux on each radiating surface. Several numerical methods are available in the CFD codes to solve the thermal radiation problems. The discrete transfer method (DTM) is considered as one of the most efficient for solving the radiation fluxes between the surfaces in the case of radiatively non-participating fluid. Discussion includes description of fundamentals of analytical and numerical thermal radiation methods, validation of radiative heat exchange in simple enclosure problems, estimation of numerical errors and application to the practical case.

Keywords: radiation heat transfer, discrete transfer method, CFD code validation, analytical example, practical application

1. Introduction

There are many applications where thermal radiation plays an important role. To name just a few of them, atmospheric physics, astrophysics, astronautics, remote sensing, nuclear engineering and many other applications may be added on the list. The engineering example in this study considers the thermal radiation exchange in the vacuum vessel of the fusion reactor.

In all cases where the thermal radiation is significant, the proper choice of the thermal radiation model affects the accuracy of the solution and also the amount of time required for the computation. Accurate solutions are computationally very expensive; therefore, a special care must be taken to select an appropriate calculation method that corresponds to the physics of the radiative heat transfer under consideration. For example, different methods can be applied in the situations where the medium between the enclosing solid surfaces behaves as optically dense or it is practically transparent to the heat transferred by radiation. In the first situation the radiative heat can be scattered, absorbed or re-emitted in the medium and at the solid surfaces. The radiation models used for participating medium are usually based on the solution of general thermal radiation equation [1]. In the second situation the medium is either not present (vacuum) or it is transparent to the wavelengths at which the thermal radiation occurs. In this case the radiative heat is transferred solely between the boundary surfaces and depends on the surface properties and geometrical orientation that each surface has to the others. This type of thermal radiation is relevant for most of the engineering applications. For simple geometries analytical solutions based on the view factor evaluation exist [2, 3]. In the real cases with complex surface geometries, the exact analytical expressions are not available and approximate numerical methods must be used, such as Monte Carlo [4] or discrete transfer method (DTM) [5]. The DTM method has some advantage due to its computational efficiency, easy application to complex geometries and implementation into the CFD codes. The method can be very accurate, but it has a shortcoming, as it is not easy to assess the accuracy of the net radiative heat flux on the individual surface in the complex enclosure. Even if the total energy balance in the enclosure is conserved, it is not self-evident that the radiation heat flows on individual surfaces are calculated correctly. To assess and improve the accuracy of the CFD simulation by the DTM method, a validation against the simple example with the exact analytical solution that resembles the real case is highly recommended.

The following sections give a description of analytical and numerical methods. Analytical method is based on the view factor calculation, reciprocity and summation relations in an enclosure. Description of numerical method considers DTM method as implemented in the CFD code ANSYS CFX [6]. The accuracy of numerical results is validated on the two analytical examples, cylinder and closed ring. The geometry and dimension of the latter approximately resemble the realistic case of the heat radiation inside the vacuum vessel of fusion reactor, the results of which are presented in the last section. The chapter ends with the summary of main conclusions.

2. Analytical solution method for thermal radiation in an enclosure

The theory of radiation exchange between the surfaces described in this section is based on two assumptions. First, the surfaces form an enclosure and second, they are separated by a medium that does not participate in thermal radiation. Radiatively non-participating media has no effect on the radiation transfer between the surfaces. There is no scattering, emission or absorption in the medium. Such conditions occur in vacuum, and also in monatomic and most diatomic gases at low and moderate temperatures, before the ionization and dissipation occurs. In fact, in many engineering applications medium does not affect the radiation heat transfer [2].

The radiation heat exchange between the two differently oriented black surfaces with final dimensions can be expressed as [2, 7]:

$$P_{ij} = F_{i\to j}A_i\,\sigma T_i^{\,4} - F_{j\to i}A_j\,\sigma T_j^{\,4} = A_iF_{ij}\sigma\left(T_i^{\,4} - T_j^{\,4}\right), \tag{1}$$

where $\sigma = 5.67{\cdot}10^{-8}\ Wm^{-2}\ K^{-4}$ is the Stefan-Boltzman constant and F_{ij} is the view factor. The net radiation heat exchange between the surface A_i at absolute temperature T_i and an enclosure of N black surfaces at absolute temperature T_j may be described by the energy balance on the opaque surface A_i (**Figure 1**):

$$j_i = j_{i,\,emission} - j_{i,\,absorption}. \tag{2}$$

Absorption depends on irradiation, which depends on emission from other surfaces including those far away from the observed surface. To calculate the total radiative energy balance the entire enclosure has to be considered. Thus, all radiation contributions are accounted for. An open enclosure is in practice closed by introducing artificial surfaces. For example, opening can be considered as a surface with zero reflectivity or as a radiation source when presenting environmental radiation. Enclosure is usually composed of complex geometries which may complicate the calculations. To deal with the complex geometry, the enclosure can be approximated by several simple surfaces that are assumed to be isothermal, as indicated in **Figure 1**.

Radiation exchange between the surfaces in addition to their radiative properties and temperatures strongly depends on the surface geometries, orientations and distances between them.

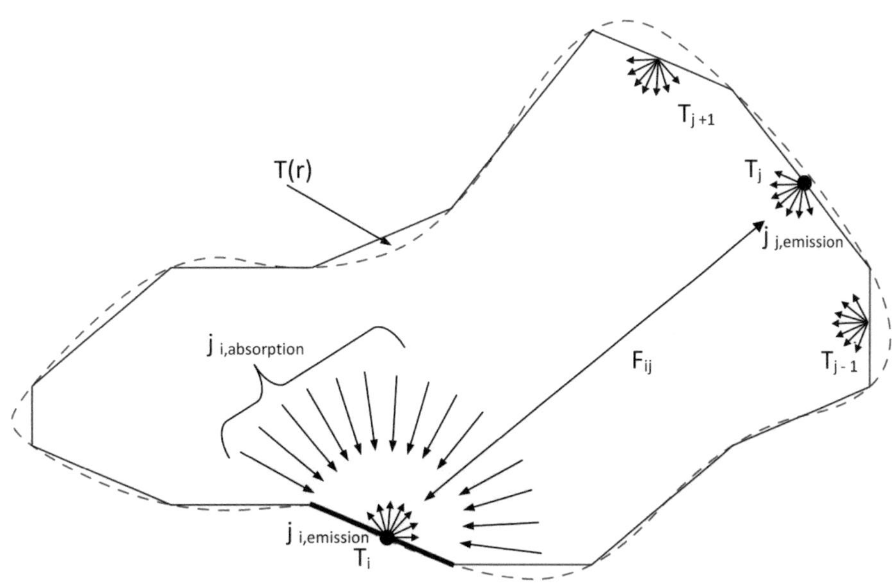

Figure 1. Real and idealized enclosure.

This leads to the development of a geometric function known as view factor. The view factor F_{ij} is defined as the fraction of the radiation leaving the surface A_i that is intercepted by the surface A_j and is calculated a general expression [3]:

$$F_{ij} = \frac{1}{A_i} \int\limits_{A_i} \int\limits_{A_j} \frac{\cos\theta_i \cos\theta_j}{\pi\, S^2} dA_i dA_j, \tag{3}$$

where S is the distance between the surfaces A_i and A_j, θ_i and θ_j are the angles between the surface normal vectors n_i, n_j and S. (**Figure 2**).

Two very useful view factor relations are valid for the enclosure. The first one is the reciprocity relation ($A_i F_{ij} = A_j F_{ji}$) and the second is the summation rule $\left(\sum_{j=1}^{N} F_{ij} = 1\right)$ [7]. The latter relation follows the conservation requirement that all radiation leaving the surface A_i must be intercepted by some other surface A_j. As shown later, these two relations are not useful only for the analytical calculations but are also important for the assessment of the accuracy of numerical methods. Reciprocity relation can be used to check the accuracy of individual view factor and the summation rule can be used for validation of the energy conservation. The view factor F_{ii} deserves a special consideration. If the surface is convex, no radiation leaving A_i will strike itself so $F_{ii} = 0$. If the surface is concave, it sees itself and part of radiation leaving A_i will be intercepted by itself so $F_{ii} \neq 0$.

To calculate the radiation exchange in the enclosure of N surfaces a total of N^2 view factors are required. The view factors on N surfaces can be written in a matrix form:

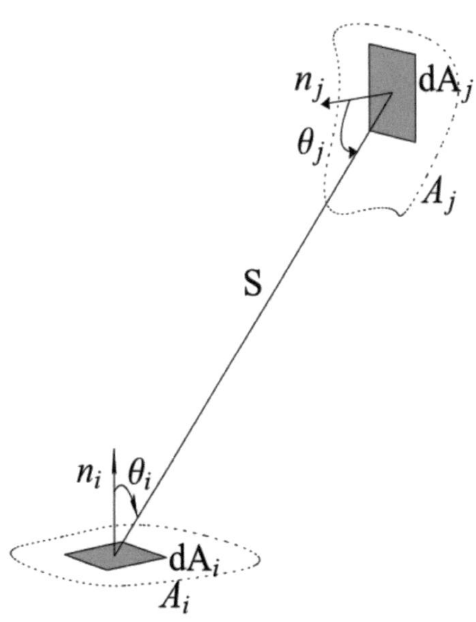

Figure 2. View factor-Radiation exchange between two surfaces.

$$\begin{bmatrix} F_{11} & F_{12} & \cdots & F_{1N} \\ F_{21} & F_{22} & & \\ \vdots & & \ddots & \\ F_{N1} & & & F_{NN} \end{bmatrix} \tag{4}$$

However, all view factors do not have to be calculated directly. Most of them can be derived from the view factor relations. N view factors can be obtained from the N equations of the summation relations and $\frac{N(N-1)}{2}$ view factors may be obtained from the reciprocity relations. When the enclosure in the model includes M surfaces that cannot see itself ($F_{ii} = 0$), the total number of the view factors that need to be calculated directly amounts:

$$N^2 - N - \frac{N(N-1)}{2} - M = \frac{N(N-1)}{2} - M \tag{5}$$

In some cases, the use of symmetry can additionally reduce the number of directly calculated view factors.

Even for simple geometries the analytical calculation of view factors, Eq. (3) is not easy. The calculation of view factor between the two finite surfaces requires solving of the double area integral, or fourth-order integration. Such integrals are difficult to evaluate analytically except for very simple geometries. For practical use, the view factors can be generated from already known solutions for the frequently used geometries that are collected in the form of tables and charts. The most complete set of solutions is given in a catalog of Howell [3]. For more complicated geometries, the view factors need to be calculated by numerical integration that can be computationally expensive, depending on the complexity of the geometry.

The energy balance on the selected surface A_i in an enclosure of N black surfaces is calculated using Eq. (3), where the emission $j_{i,emission}$ and absorption are expressed as:

$$j_{i,emission} = \sigma T_i^{\,4}, \tag{6}$$

$$j_{i,absorption} = \sum_{j=1}^{N} j_{j,emission} F_{ij}. \tag{7}$$

The net radiation heat from the surface A_i at absolute temperature T_i to the enclosure of N black surfaces at absolute temperature T_j can be expressed as [2]:

$$P_i = \sigma \sum_{j=1}^{N} A_i F_{ij} \left(T_i^{\,4} - T_j^{\,4}\right), \tag{8}$$

where F_{ij} is the view factor between the surface A_i and one of the enclosing surfaces A_j. Radiation properties such as emissivity ($\varepsilon_i = 1$ for the black body) and temperatures are set, surface areas are easy to calculate, whereas the calculation of the view factors F_{ij} can be a very difficult task for complex geometries.

2.1. The cylinder example

To demonstrate the use of analytical methods, the radiation heat transfer in a simple model of an open cylinder is calculated. Diameter and height of the cylinder are 50 and 150 mm, respectively. The cylinder is open at the top surface to a large surroundings at $T_{sur} = 27^{\circ}C$. The bottom and the side surfaces of the cylinder are approximated as black surfaces and are maintained at $T_F = 1500^{\circ}C$ and the opening at the top is approximated as blackbody at the temperature of the surroundings. The goal of the exercise is to calculate the heat exchanged between the surfaces assuming that the thermal radiation is the only heat transfer mode and the outer backs of surfaces are adiabatic. The sketch of the cylinder with the bottom surface A_1, the side surface A_2 and the top surface A_3 is shown in **Figure 3**.

To calculate the radiation exchange between the three surfaces, in general, nine view factors are needed. According to Eq. (5), only $\frac{N(N-1)}{2} - M$ view factors need to be calculated directly. Taking into account the two flat (A_1 and A_3 on **Figure 3**) which cannot see itself ($M = 2$) and one concave surface (A_2) only one view factor has to be calculated directly, the others can be derived using the view factor relations as follows:

$$\begin{bmatrix} F_{11} & F_{12} & F_{13} \\ F_{21} & F_{22} & F_{23} \\ F_{31} & F_{32} & F_{33} \end{bmatrix} \rightarrow \begin{bmatrix} 0 & F_{12} & F_{13} \\ \frac{A_1}{A_2} F_{12} & F_{22} & F_{23} \\ \frac{A_1}{A_3} F_{13} & \frac{A_2}{A_3} F_{23} & 0 \end{bmatrix} \tag{9}$$

where,

$$F_{12} = 1 - F_{13}$$

$$F_{23} = \left(1 - \frac{A_1}{A_3} F_{13}\right) \Big/ \frac{A_2}{A_3}$$

$$F_{22} = 1 - \frac{A_1}{A_2} F_{12} - F_{23}.$$

The unknown view factor F_{13} can be calculated analytically by solving the integral in Eq. (3). The inside-sphere method [2] was found to be very convenient for the considered geometry, detailed derivation of the view factor F_{13} is provided in [8]. The easiest way to obtain the view factor for the sought geometry is to generate it from the available database of already calculated view factors for similar geometry configurations. In the catalog of Howell [3], the view factor for the two parallel coaxial disks of unequal radius can be found (the case C-41). In our case the disks (surfaces A_1 and A_3) have equal radius, the view factor F_{13} then yields:

http://dx.doi.org/10.5772/intechopen.74420

$$F_{13} = 1 + 2H^2 - 2H\sqrt{1 + H^2}, \tag{10}$$

where H equals to $h/2r$. The remaining view factors obtained using the view factor relations (Eq. (9)) are given in **Table 1**.

Using the Eq. (8), we can calculate the net radiative heat transfer from each surface. The net radiation loss to the surroundings through the open surface A_3 amounts −1521.6 W. Analytical results for the net radiation heat from the surfaces are collected in **Table 2**. The heat flow balance in the enclosed cylinder is the sum of all radiation heat flows and should be equal to zero.

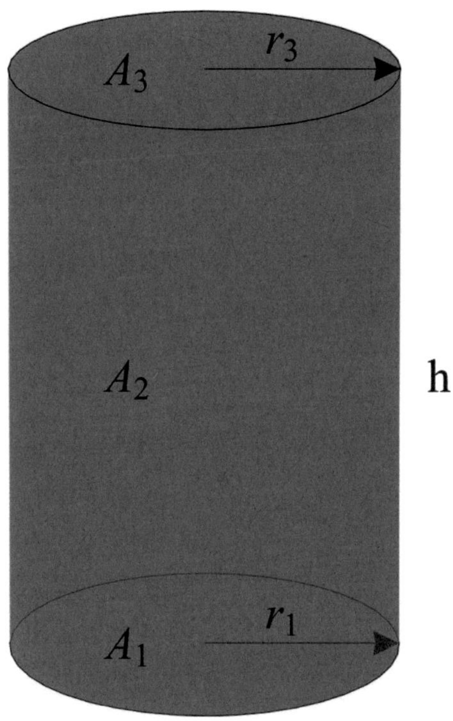

Figure 3. Cylinder case-Radiating surfaces.

The view factor
$F_{11} = F_{33} = 0$
$F_{12} = F_{32} = 2H\left(\sqrt{1 + H^2} - H\right)$
$F_{13} = F_{31} = 1 + 2H^2 - 2H\sqrt{1 + H^2}$
$F_{21} = F_{23} = \frac{\sqrt{1+H^2}-H}{2}$
$F_{22} = 1 + H - \sqrt{1 + H^2}$

Table 1. Calculated view factors for the cylinder case.

Surface	Net radiation transfer from the surface A_i [W]
A_1	40.07
A_2	1481.53
A_3	−1521.60
Heat flux balance	0

Table 2. Net radiation heat transfer from the surfaces for the cylinder case.

3. Numerical simulation methods

Thermal radiation exchange in complex geometry configurations is calculated using the numerical simulation methods. The accuracy of numerical simulations depends on the numerical mesh and on the type of numerical methods used. For accurate results it is important to understand the main characteristics of the method to efficiently reduce the accompanying numerical errors. To validate the accuracy of the selected method a comparison of the results on a simpler geometry that has an exact analytical solution is of great importance.

Different numerical methods are available for solving the complex thermal radiation problems. In general, the radiation transfer through a medium is affected by absorption, emission and scattering and can be described by the generic radiation transport equation:

$$\frac{dI\left(\vec{r},\vec{s}\right)}{ds} = K_a I_b(T) - K_a I\left(\vec{r},\vec{s}\right) - K_S I\left(\vec{r},\vec{s}\right) + \frac{K_S}{4\pi}\int_{4\pi} I\left(\vec{r},\vec{s}\,'\right)\phi(\hat{s}\cdot\hat{s}\,')d\Omega', \tag{11}$$

where I is radiation intensity which depends on position ($\vec{r}$) and direction ($\vec{s}$), I_b is blackbody emission intensity, K_a and K_S are the absorption and scattering coefficients of the medium, ϕ is the scattering function and s is the path length. Analytical solutions for Eq. (11) exist only for very simple cases. There are several numerical methods used to predict the thermal radiation, based on Eq. (11) [1, 9].

One of the most efficient methods for solving the thermal radiation between the surfaces is the discrete transfer method (DTM) developed by Lockwood and Shah [5]. The DTM solves the simplified form of the radiation transport equation (Eq. (11)) along rays (see **Figure 4**). To determine the direction of the rays, the unitary hemisphere over the element face is discretized using spherical coordinates. The span is divided into angles by the ray number, and rays directions are computed to pass through the center of the angles. In total, the square of ray number is traced from an element of the surface.

The rays are fired in prescribed directions from discrete point P_i, which is located at the center of a cell face on the boundary surface. Each ray is traced until it hits another boundary and Q_j is the impingement point. In general Q_j is not the central point of a boundary cell, but it is assumed that radiation intensity at Q_j and at the central point P_j are equal. The boundary conditions or initial values of radiation intensity at point Q_j for gray, diffuse surface depends on the incident radiation intensity at point P_j, which further depends on the radiation intensity of all other rays that reach point P_j (see **Figure 4**). Simplified radiation transport equation (Eq. (11)) is discretized in 3D finite volumes along the rays. The path along the ray is discretized using the sections formed from breaking the path at volume boundaries. The physical quantities in each volume are assumed to be uniform.

Due to the fixed sampling and ray discretization the physical quantities can be found at fixed points. The accuracy of the DTM is controlled by the number of rays and by the mesh density. For accurate results the control volumes (mesh) must be chosen in a way that the irradiation field is reasonably homogeneous inside them. The major problem of the DTM is the lack of error information. Large errors can be produced when the ray sample misses the sensitive area or object. This error is known as the ray effect.

Unless the surfaces are black ($\varepsilon_i = 1$), the radiation intensity of selected surfaces depends on the radiation intensity of all other surfaces (see **Figure 4**) and solution requires iterative calculation procedure. The detailed description of the numerical solution procedure can be found in [9, 10].

In this study the DTM method implemented within the ANSYS CFX [6] computational code is tested. The option that includes "non-participating media" is the most appropriate for the

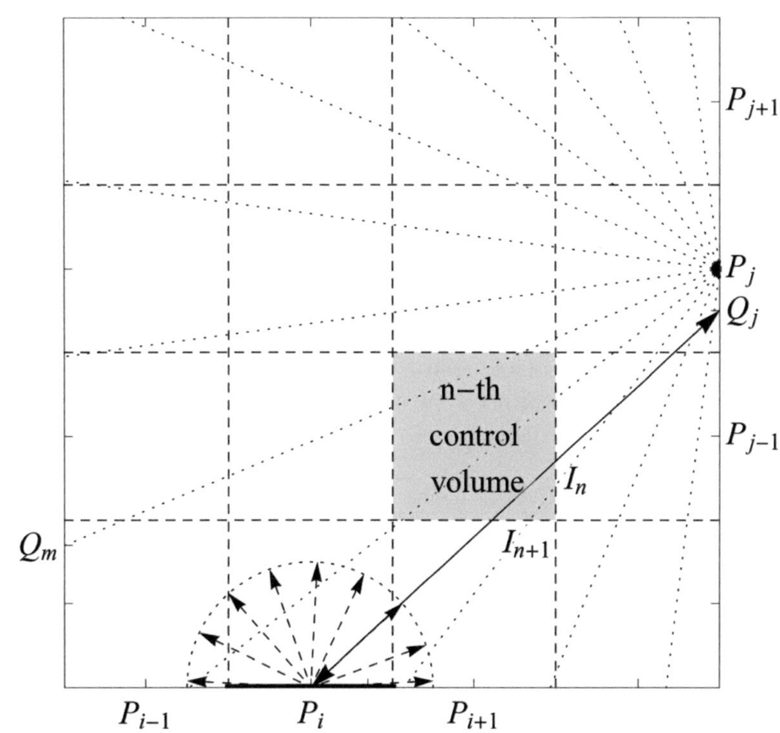

Figure 4. Discrete transfer method-Ray definition and discretization in 2D.

calculation of the thermal radiation inside the vacuum vessel of the fusion reactor. This option is known as surface to surface model (S2S). Here the volumetric emission, absorption and scattering are ignored. So all coefficients in Eq. (11) are zero and only the boundary conditions on the radiating surfaces are required.

3.1. Validation on the cylinder example

Analytical solutions for the net radiation heat transfer in an open cylinder are derived in Section 2.1 and the results are presented in **Table 3**. The numerical results of the same case are obtained by the ANSYS CFX code using the S2S thermal radiation model for the non-participating media. For the purpose of sensitivity analysis two different ray numbers on the single numerical mesh with 72,000 hexahedral mesh elements.

Numerical results presented in the third column of **Table 3** are calculated using the ray number 48 (actual number of rays is $48^2 = 2304$) on the mesh with 72,000 mesh elements and 7200 mesh faces. This means that 2304 rays are fired from the center of each face. Hence, the whole number of rays involved in the calculations is $7200 \times 2304 = 16\ 588\ 800$. The results in the **Table 3** show that the ANSYS CFX prediction of the radiative heat transfer from the surface A_1 is the least accurate. Better accuracy can be achieved by increasing the number of rays, which is evident from the results for the ray number 128 in the fourth column of the **Table 3**. The overall heat flow balance in the enclosure is very well preserved for both numerical simulations and as such cannot provide the information about the numerical accuracy for the individual heat exchanging surface. Only the comparison with the analytical solution can give this type of information.

3.2. Validation of the closed ring example

As a second validation case a closed ring geometry with a rectangular cross-section was selected (see **Figure 5**). The ring geometry tends to resemble the realistic model of the fusion reactor torus as far as possible but is still simple enough to allow the analytical solution that enables proper validation of numerical results. The inner radius (6 m), outer radius (12 m) and height (11 m) of the ring ensure that heat radiation surfaces of the ring approximately match the inner surfaces of the DEMO in-vessel components. In the **Figure 9** the ring dimensions (represented by the yellow rectangle) are compared with the dimensions of the DEMO tokamak. The ring geometry contains a small surface (red rectangular surface in **Figure 5**) representing the upper surface of the

Surface	Analytical	Numerical results			
		Ray number 48		Ray number 128	
	Net heat flow [W]	Net heat flow [W]	Relative error [%]	Net heat flow [W]	Relative error [%]
A_1	28.95	26.78	−7.5	28.42	−1.8
A_2	1070.36	1071.20	0.08	1069.30	−0.1
A_3	−1099.31	−1097.98	−0.1	−1097.78	−0.1
Heat flow balance	0 [W]	−0.0002 [W]		−0.06 [W]	

Table 3. Comparison of analytical and numerical results for the net radiation heat transfer for the cylinder example.

detached divertor cassette that is kept at a lower temperature (323 K) than the remaining ring surfaces (373 K). The imposed surface temperatures are the same as the temperatures in the DEMO reactor model (see the next Section 4). All surfaces of the ring enclosure are approximated as blackbodies. Analytical results for the net radiation heat transfer on different ring surfaces (as marked in **Figure 5**) are presented in **Table 4** and are calculated using the Eq. (8). Analytical values of view factors were obtained from the catalog of known view factor solutions [3] or directly calculated from Eq. (3).

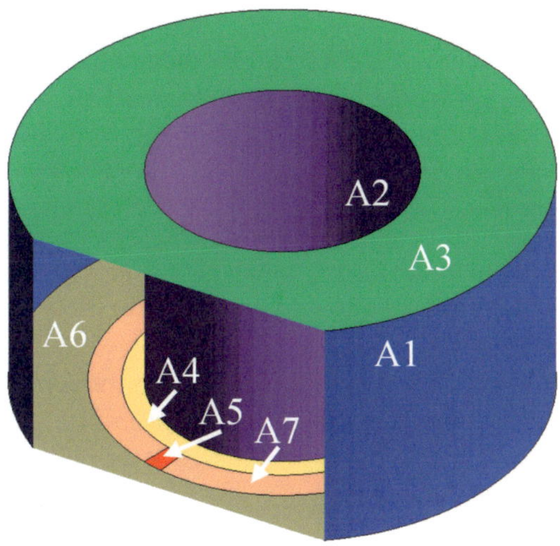

Figure 5. Geometry and surfaces of the closed ring example.

	Anal. results	Numerical results (ray number: 48)					
Mesh		**Coarse**		**Medium**		**Fine**	
t [s]		**127**		**711.5**		**4013**	
Surfaces	**Net heat flow [W]**	**Net heat flow [W]**	**Relative error [%]**	**Net heat flow [W]**	**Relative error [%]**	**Net heat flow [W]**	**Relative error [%]**
A1	321.338	320.100	0.190	321.133	0.032	321.671	−0.051
A2	176.113	175.733	0.058	175.850	0.040	175.988	0.019
A3	153.166	150.963	0.339	152.954	0.033	152.811	0.055
A4	0	0.093	−0.014	0.059	−0.009	−0.004	0.001
A5	−650.616	−647.569	−0.468	−650.421	−0.030	−650.440	−0.027
A6	0	0.516	−0.079	0.324	−0.050	−0.020	0.003
A7	0	0.163	−0.025	0.103	−0.016	−0.006	0.001
Heat flow balance	0.000	0.000		0.001		0.000	
Total rel. error (r_{tot})			0.2%		0.03%		0.02%

Table 4. Analytical and numerical results for the closed ring.

For validation purposes, the numerical results are computed on the three different hexahedral meshes using the ANSYS CFX code. The meshes are presented in **Figure 6**. The Coarse mesh in **Figure 6 (a)** has 12, 22 and 162 cells in r, z and φ directions, respectively. The other two meshes are refined by doubling the number of cells in each direction. The Medium mesh (b) consists of $24 \times 44 \times 324$ mesh cells and the Fine mesh (C) consists of $48 \times 88 \times 648$ mesh cells.

Analytical results of the net heat transfer on the ring surfaces are compared with the numerical results on **Table 4**. Numerical simulations are performed by ANSYS CFX using the discrete transfer model (DTM) for non-participating media (S2S). The negative net heat transfer value on the surface A5, which represents the detached cassette at 323 K, means that it receives the thermal radiation from other components that are at higher temperatures of 373 K. The other surfaces with positive net heat flow are net emitters. The simulation results in **Table 4** are obtained for three different mesh densities using the ray number 48 (actual number of rays per surface element is 2304).

The total sum of net heat flows (heat flow balance) is also shown in **Table 4**. Ideally (e.g. analytical solution) the net emitted heat flow in an enclosure is equal to the net absorbed heat flow leading to the zero heat flow balance. As shown the total heat flow balance is very accurately predicted for all mesh densities. This means that the energy flow inside the domain is well preserved in all cases, but does not provide any information about the accuracy of thermal radiation calculation. The total relative error r_{tot} of the net thermal radiation transfer in a model can be estimated using the equation:

$$r_{tot} = \frac{1}{N}\sum_{i=1}^{N} |r_i|, \quad r_i = \frac{P_{i,an} - P_{i,num}}{\sum_{i=1}^{N} |P_{i,an}|/2}. \tag{12}$$

where P_i is the net heat flow on the *i-th* surface. The relative error r_i on the *i-th* surface is defined as the ratio between the difference of analytical $P_{i,an}$ and numerical $P_{i,num}$ net heat flow

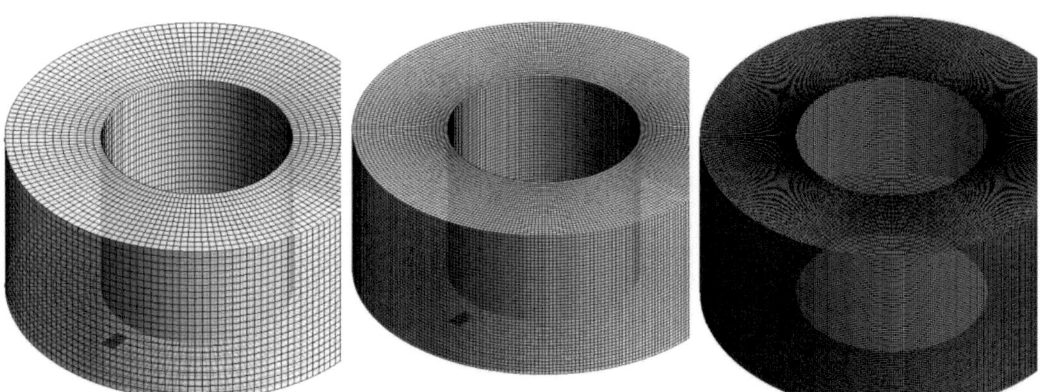

(a) *(b)* *(c)*

Figure 6. Numerical meshes of the closed ring case: Coarse (a), Medium (b) and Fine mesh (c).

and the average value of all exchanged heat in the enclosure. The averaged exchanged heat in the enclosure is represented by the sum of absolute values of the net emitted and net absorbed heat flows divided by 2. The results in **Table 4** show that the total relative error r_{tot} of the numerical simulation reduces with the mesh refinement. The computational time increases with denser mesh (2nd row in **Table 4**). The Medium mesh provides the best compromise between the accuracy of results and the computational time.

The results for three different meshes and three different ray numbers are shown in **Table 5**. The computational time expectedly increases with the number of rays on each mesh. In the case of a Coarse and Fine mesh the increased ray number improves the accuracy of the results, whereas this does not seem to be the case for the Medium mesh.

The dependence of numerical solution on the number of rays, computational time and mesh resolution is presented in **Figure 7**. The ray number – square root of the real number of rays is shown in the abscissa of **Figure 7**. Corse, Medium and Fine mesh are marked by blue, green and red color, respectively. Medium mesh is tested in detail by using denser sampling of ray numbers, also including the odd number of rays.

Model	**S2S**								
Mesh	**Coarse**			**Medium**			**Fine**		
N. elements	43,768			342,144			2,737,152		
ray number	48	64	128	48	64	128	48	64	128
t [s]	127	222	893	711	1063	4384	4013	5858	21,410
Tot. rel. error [%]	0.2	0.1	0.1	0.03	0.06	0.04	0.02	0.01	0.01

Table 5. Closed ring: Mesh and ray number comparison.

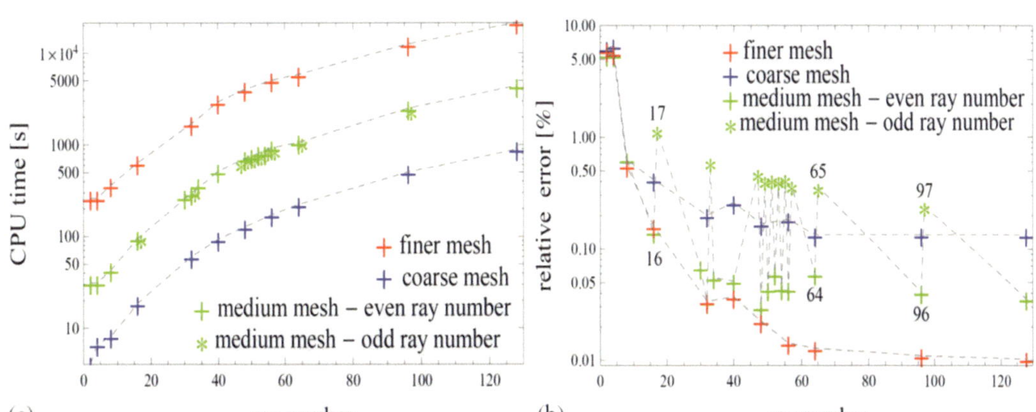

Figure 7. (a) Required computing time and (b) relative errors per number of rays.

From the **Figure 7a**, it can be seen that the computing time increases with the number of rays and with the mesh density. In general the curve of the computing time dependence has the same shape for all mesh densities only the curves are shifted upwards with the higher number of mesh cells. Finer mesh requires high computational resources already at the low ray number.

The accuracy of the DTM method is controlled by the number of rays. **Figure 7b** it shows that the relative error decreases with increasing number of rays and that the refinement of the mesh reduces the numerical error. Further it can be seen that the relative error decreases very slowly and then remains at a certain level when the number of rays increases over the specific number of rays (e.g. 30 for the medium mesh – green curve). In the case of Medium mesh it can be seen that the difference in the accuracy of the even (e.g. 64) an odd (e.g. 65) ray number is relatively high. Obviously the odd number of rays decreases the numerical accuracy.

The number of rays has to be determined in a way that the accuracy is acceptable and that the simulation is not computationally too demanding. The number of rays used in the following simulations is obtained by comparing the **Figure 7a** and **7b**. Fine mesh requires more computational time than the Medium mesh for the same relative error. Fine mesh has less than two times better accuracy than the Medium mesh at the same number of rays (ray number 48) but requires almost six times more computing time. With the ray number 48 it is feasible to achieve a good accuracy with the reasonable computing time. The total relative error for the numerical solution is equal to 0.03% and is approximately evenly distributed over all surfaces.

4. Application on the vacuum vessel of the fusion reactor

The results of the validation cases can provide a useful information on the mesh density and the ray number for the DTM method. Based on the closed ring case, a similar mesh density and ray number are applied in the ANSYS CFX input model for the real DEMO tokamak geometry.

The presented practical example focuses on the heat loading of the divertor cassette immediately after the shutdown of the DEMO fusion reactor. The demonstration fusion reactor DEMO is planned to be the last major step before the commercial fusion power plant (**Figure 8a**). During the reactor operation the divertor (**Figure 8b**) is subjected to a high incident heat flux of removed plasma particles with the values above 10 MW/m^2. Such heat loads may eventually cause severe damaging and consequently the need for regular replacement of the divertor cassettes that is envisaged at a 2-year cycle.

The component of interest is one of the 54 divertor cassettes that is being replaced during the regular maintenance. The cassette under replacement (**Figure 8c**) is unplugged from the cooling pipes, while the remaining cassettes and the blanket remain actively cooled. Because of the lack of internal cooling the detached cassette is heated up due to the decay heat in activated structure materials. Thermal radiation from the colder surfaces of the surrounding divertor cassettes and blankets represents the only cooling mode of the detached cassette.

The assumed conditions during the maintenance (see **Figure 9**) are defined as follows [11]:

- Blankets are actively cooled at $T = 373\ K$.
- All the remaining divertor cassettes are actively cooled at $T = 373\ K$.
- Vacuum vessel is set at room temperature $T = 300\ K$.

Boundary conditions on the detached divertor cassette (see **Figure 10**) are set for the purpose of this study:

- At the plasma facing surface the temperature is either fixed at 323 K or the passively cooled surface is assumed.
- Temperature of the side surfaces facing actively cooled components is set to 373 K.
- At surface facing the vacuum vessel the temperature is set to 300 K.
- The decay heat just after the shutdown is adopted as 25 kW/m^3 [11].

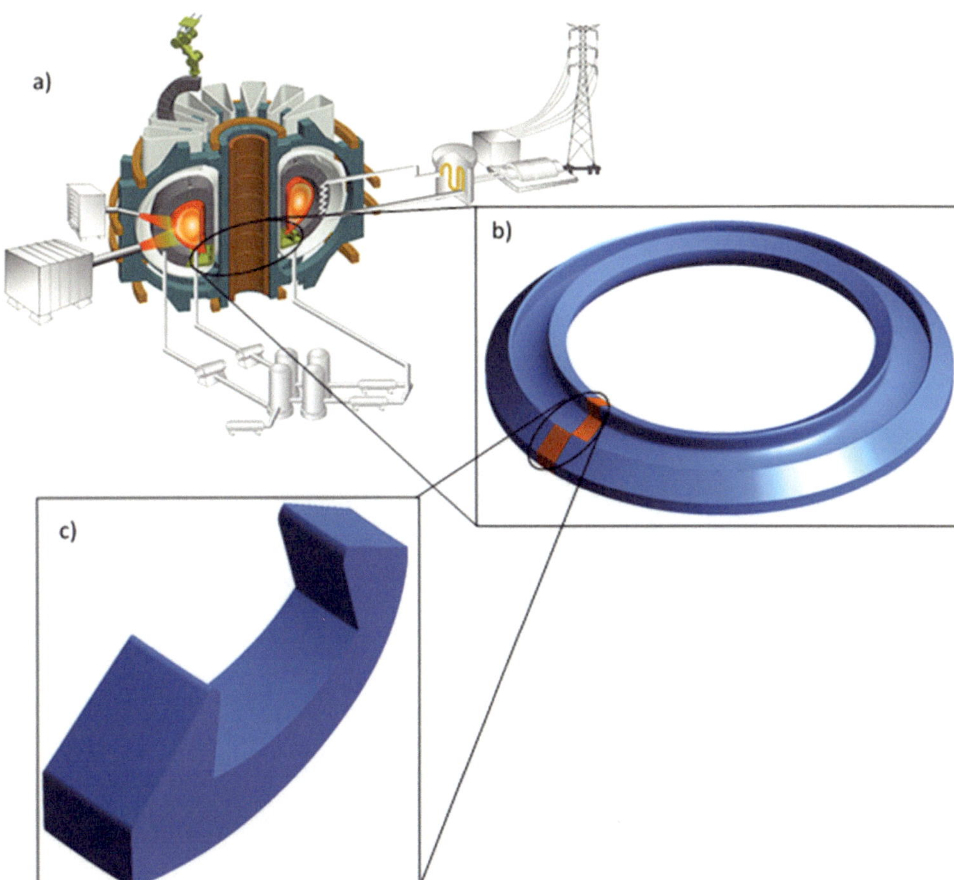

Figure 8. (a) Demonstration fusion power plant DEMO, (b) divertor, (c) divertor cassette.

Taking into account the detached divertor cassette the toroidal symmetry of the tokamak is not preserved, therefore the full 360° geometry of the vessel has to be modeled. Since the heat load on the detached divertor cassette needs to be evaluated, it is sufficient to consider only the in-vessel components (blanket and divertor) and vacuum vessel without the manway ports. The ANSYS CFX model of DEMO vacuum vessel with one detached divertor cassette (out of 54) is shown in **Figure 9**. Components included in the model are additionally simplified: all gaps between the actively cooled divertor cassettes, blanket and vacuum vessel are neglected as they have a negligible effect on the solution. The gap around the detached divertor cassette has been considered with the appropriate boundary conditions. The simulation model is set up as a closed cavity radiation problem with surfaces approximated as gray bodies ($\varepsilon_i = 0.25$) [11].

In addition to the thermal radiation between the system surfaces also the heat conduction in the detached cassette and internal heat generation due to the decay heat in activated structure materials has to be considered.

Three cases have been simulated by applying different boundary conditions on the detached cassette. The simulation results are presented in **Table 6**. The most realistic case is the Case 1, where the internal heat generation is considered inside the cassette body and its top surface is modeled as a passive boundary condition. The second case (Case 2) is the most similar to the conditions of the closed ring validation case, therefore it does not include the internal heat generation and the top surface temperature of the detached cassette is set to 323 K (50°C). In the Case 3, besides the internal heat generation, an additional surface cooling of the detached

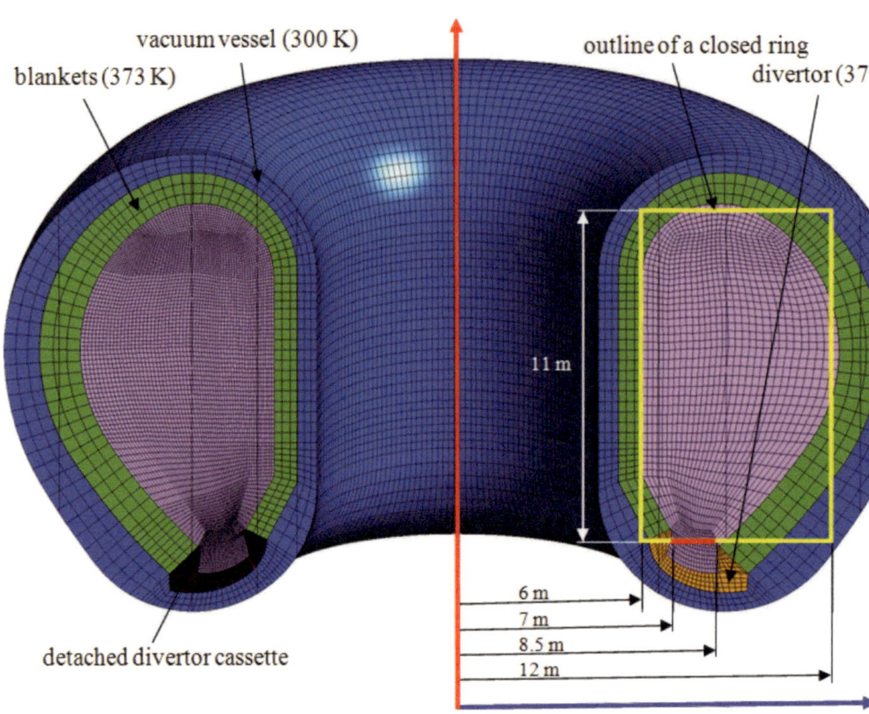

Figure 9. Half-cut of the DEMO tokamak model used in simulations.

Cases	Case 1	Case 2	Case 3
Max. T (body) [K]	423.0	373	395.1
Max. T (surface) [K]	420.6	323 (373)	323 (373)
Net heat flow [W]	Simulation results	Simulation results	Simulation results
	Error estimation	Error estimation	Error estimation
S1: detached cassette	154.531	−341.501	−341.501
	154.53 ±0.05	−341.5±0.1	−341.±0.1
S2: blankets above the detached cassette	−7.636	17.866	17.866
	−7.636±0.002	17.866 ±0.005	17.866 ±0.005
S3: blankets	−112.115	249.773	249.773
	−112.12 ±0.03	249.77±0.08	249.77±0.08
S4: divertor	−34.780	73.862	73.862
	−34.78 ± 0.01	73.86 ±0.02	73.86 ±0.02
Total sum of heat flows	0.000	0.000	0.000

Table 6. Numerical results and estimated heat flows for three different cases.

cassette is modeled by imposing its top surface temperature to the value 323 K. Comparing to the Case 1, the Case 3 aims to evaluate the effect of external surface cooling on the reduction of maximum temperature inside the cassette body. The results presented in **Table 6** are calculated by the ANSYS-CFX using the surface-to-surface thermal radiation model with the ray number 48. The mesh shown on **Figure 9** has 698,898 mesh cells.

The **Table 6** includes the heat flow balance inside the tokamak enclosure that shows perfect energy conservation in the tokamak (the total sum of emitted and received heat flows is zero in all cases). The overall error represents only 10^{-6}% or less of all exchanged radiation heat inside the tokamak. Based on the results obtained by comparing the numerical and analytical solutions of the closed ring example, we are aware that the accurately calculated heat flow balance does not mean that the heat flows on an individual surface are equally precisely predicted. Taking into account the results of the closed ring example, we may assume that the estimated error for an individual surface is of the same order as the error in the benchmark model, considering that the sufficiently dense mesh is used. Also, it is assumed that the emissivity does not affect the error, since the emissivity is not a geometry dependent parameter. Relative error for the individual net heat flow is thus estimated at 0.03%.

The calculated heat flows in the Cases 2 and 3 are exactly the same, as the cases have the same radiative boundary conditions. The negative heat flow on the detached divertor cassette in cases 2 and 3 means that the detached cassette (323 K) receives the radiation heat from other components at a higher temperature (373 K), which is also in accordance with the results of the closed ring example. In the Case 1, the heat flow on the detached cassette is positive, which means its top surface is hotter than the surrounding surfaces. In this case, the detached cassette is passively cooled by the thermal radiation. However, the temperature at the surface is still rather high (420 K).

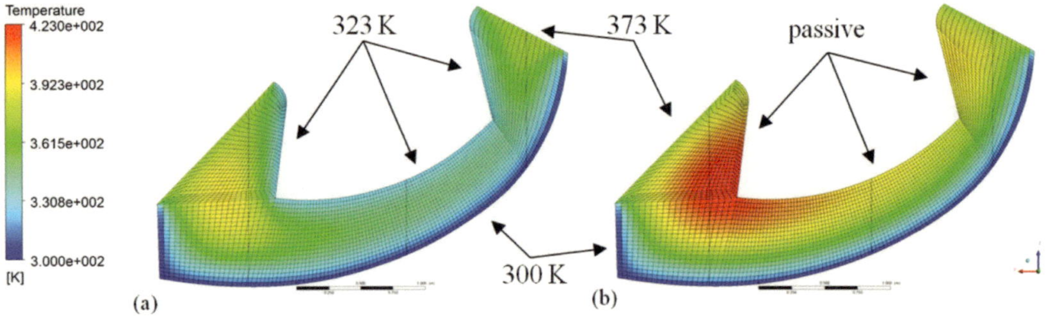

Figure 10. Temperature distribution in the detached cassette body. (a) Cassette with the imposed surface temperature at 323 K. (b) Cassette with the passive top surface.

In addition to the net heat flows, **Table 6** also shows the maximum temperature inside the detached cassette body and on its top surface. In the Cases 1 and 3, it can be seen that the maximum temperature increases due to the internal heat generation and amounts to 423 and 395 K for the Case 1 and Case 3, respectively.

The temperature distribution inside the detached cassette for the Cases 1 and 3 is presented in **Figure 10**. Due to the external cooling (imposed temperature of 323 K on the top surface), the region of maximum temperature in the Case 3 is lower and displaced toward the inside of the cassette body (**Figure 10a**). The temperature peak for the Case 1 is significantly higher (423 K) and located closer to the upper surface (see **Figure 10b**).

5. Conclusions

To justify the use CFD codes for the applications with dominant radiation heat transfer, the implemented models need to be first appropriately validated on simple analytical examples.

The practical application under consideration is the heat load on the detached divertor cassette inside the DEMO vacuum vessel after the reactor shutdown. The detached cassette is subjected to the internal heating and is cooled solely by thermal radiation from the surrounding in-vessel components. The thermal analysis is performed with the CFD code using the discrete transfer method (DTM) for the thermal radiation modeling.

Two representative analytical examples (cylinder and closed ring) were used to evaluate the accuracy of the real case simulations performed by DTM numerical method. The analytical approach uses the view factor calculation, reciprocity and summation relations in an enclosure to calculate the exact radiation heat transfer on each radiating surface. The closed ring example resembled the geometry and dimensions of the DEMO vacuum vessel with the detached cassette. It was shown that the conservation of energy in the enclosure does not guarantee the accuracy of individual surface heat flows. The accuracy of separate surface heat flows can be assessed only by comparison of analytical and numerical results. With proper selection of mesh and number of rays the DTM numerical method achieved good agreement with analytical solution at a reasonably low computational time. It has been demonstrated, that the DTM method applied within the ANSYS CFX code gives accurate predictions of the thermal radiation in the complex geometry configurations.

Acknowledgements

The authors acknowledge the financial support from the Slovenian Research Agency (research core funding No. P2-0026 "Reactor engineering").

Author details

Boštjan Končar* and Luka Klobučar

*Address all correspondence to: bostjan.koncar@ijs.si

Jožef Stefan Institute, Ljubljana, Slovenia

References

[1] Viskanta R, Mengug M. Radiation heat transfer in combustion systems. Progress in Energy and Combustion Science. 1987;**13**:97-160

[2] Modest MF. Radiative Heat Transfer. New York: McGraw-Hill; 1993

[3] Howell JR. A Catalogue of Radiation Heat Transfer Configuration Factors. Available from: http://www.thermalradiation.net/indexCat.html [Accessed 14 April, 2017]

[4] Howell JR. The Monte Carlo method in radiative heat transfer. ASME Journal of Heat Transfer. 1998;**120**:547-560

[5] Lockwood FC, Shah NG. A new radiation solution method for incorporation in general combustion prediction procedures. In: Proc. of 18th Symp. on Combustion; 1981. 1405

[6] ANYSY CFX documentation, 17.1

[7] Siegel AR, Howell JR. Thermal Radiation Heat Transfer. Washington: Taylor & Francis; 1992

[8] Klobučar L, Končar B, Thermal Radiation Heat Transfer between Surfaces [Seminar Work]. Faculty of Mathematics and Physics, University of Ljubljana; 2016

[9] Coelho PJ, Carvalho MG. A conservative formulation of discrete transfer method. Journal of Heat Transfer. 1997;**119**:118

[10] Klobučar L. Simulation of Thermal Loading on Divertor Cassette in Fusion Reactor [Msc Thesis]. Faculty of mathematics and physics, University of Ljubljana; 2016

[11] Končar B, Draksler M, Costa Garrido O, Vavtar I. Thermal Analysis of DEMO Tokamak 2015 [Report]. Jozef Stefan Institute; 2016

Chapter 3

Probabilistic Heat Transfer Problems in Thermal Protection Systems

Kun Zhang, Jianyao Yao, Jianqiang Xin and Ning Hu

Additional information is available at the end of the chapter

http://dx.doi.org/10.5772/intechopen.74240

Abstract

Thermal protecti.on system (TPS) is one of the most important subsystems of hypersonic vehicles which are subjected to severe aerodynamic heating. The reliability and structural integrity of TPS are crucial to the structural safety and integrity of hypersonic aircrafts. During the design and service stages of the TPS, there are numerous inevitable uncertainties in aerothermal environment, material properties, manufacture and assembly errors, analysis modeling errors, etc., which have great impact on the reliability assessment of the TPS. In this chapter, the probabilistic heat transfer for TPS is presented to achieve light-weight TPS with high reliability. The uncertainties and their modeling and sampling methods are introduced at first. Then the finite element model and the precise integration method for transient probabilistic heat transfer of TPS are given. Based on the theoretical and numerical methods, the probabilistic design and reliability assessment procedure for TPS are finally formed. A typical multilayer ceramic composite material TPS is used as example to show the effectiveness of the proposed method.

Keywords: thermal protection system, uncertainty, probability method, reliability assessment, precise time integration

1. Introduction

When the vehicle flies in the atmosphere with hypersonic speed, the air is strongly compressed and rubbed, which converts kinetic energy into heat, resulting in a sharp increase of the ambient air temperature. The air temperature in stagnation point of the vehicle can reach 400°C when Ma ≥ 3.As the flight Mach number further increases, a strong bow shock will be formed in front of the head. The gas temperature in the shock layer is very high, causing the dissociation and ionization of air molecules, generating thermal radiation effects. Elevated temperatures can easily lead to damage to crew members and instruments in the cabin and adversely affect the structural integrity and reliability of the projectile. Meanwhile, the heat radiation of the shock wave layer and the optical window is also the infrared interference source of the optical system. Therefore, thermal protection is always a major issue for the design of hypersonic vehicles.

There are different types of TPSs according to the aerothermal conditions. NASA has classified the TPS into three categories, namely the passive, semi-passive, and active thermal protection system. The structure and technology of the active TPS are more complicated, which is the key and a difficult point for the research of the reusable hypersonic vehicle in the future. Currently, ablative and non-ablative passive (semi-passive) TPSs are being used extensively in hypersonic vehicles, both of which are typical multilayer TPSs. The non-ablative passive or reusable TPSs mainly include the rigid ceramic tile TPS, the metallic TPS and the ceramic matrix composite (CMC) shingle TPS.

The ablative thermal protection structure uses carbon, silicon, phenolic, and other heat-resisting ablative composites to cover the surface of vehicles. The material undergoes complex physical and chemical changes such as decomposition, melting, evaporation, sublimation, and erosion. The heat is taken away by the consumption of the surface material [1], and the working mechanism is shown in **Figure 1**. Under the action of heat flow, the resin matrix and other inside materials are pyrolyzed after absorbing heat. The generated gas flocks to the boundary layer to prevent heat from transmitting into the inside material, causing a heat blocking effect. At the same time, the resin and other carbonized products caused by heating are deposited on the surface to form a "carbonization layer," which acts as an outward radiation.

Rigid ceramic tile TPS consists of ceramic tiles (high temperature and emissivity coating), nomex flexible strain isolation pad (SIP), room temperature vulcanizing (RTV) silicone, and

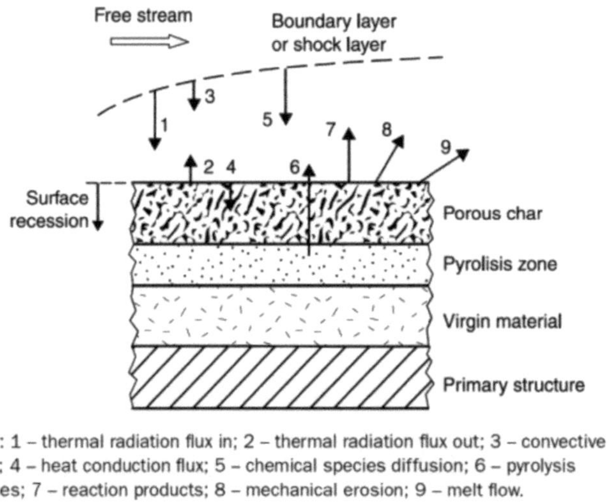

Figure 1. Typical ablative thermal protection system profile [1].

gap filler, and the multilayer structure is shown in **Figure 2**. The latest generation of this kind of TPS is the toughened uni-piece fibrous reinforced oxidation-resistant composite (TUFROC), which is used in the X-37B spacecraft. This new ceramic composite structure not only withstands the heat generated during reentry but also solves the bottleneck problems of ceramic tiles in the high-temperature environment, such as thermal cracking and oxidation. TUFROC comprises a treated carbonaceous cap composed of refractory oxidation-resistant ceramic carbon insulation (ROCCI), which provides dimensional stability to the outer mold line, while the fibrous base material provides maximum thermal insulation for the vehicle structure. This structure can be reused, and its structure is shown in **Figure 3**.

The research on metallic TPS started in the late 1950s to early 1960s. Metallic TPS has the advantages of light quality, durability, strong maneuverability, and low cost. At present, the main structure of metallic TPS is superalloy honeycomb TPS panel, which can be used for windward or leeward surfaces of hypersonic vehicles with 600–1200°C. The structure of an improved metallic TPS design, adaptable robust metallic operable reusable (ARMOR) insulation, is shown in **Figure 4**. In this structure, the honeycomb sandwich panels are simultaneously subjected to aerodynamic heating, pneumatic pressure, and sound pressure. In order to solve the connection difficulties, the outer honeycomb sandwich panel is connected to support structure at each corner of the panel box by the metal bracket. The pressure load is transmitted through the four flexible supports and the beam structure of the panel box, not only avoiding thermal interference but also to allow free thermal expansion of the outer surface [4, 5].

The concept of CMC shingle TPS was originally preferred by HERMES. Compared with the traditional rigid ceramic tile TPS, it has the advantages of strong impact resistance, easy

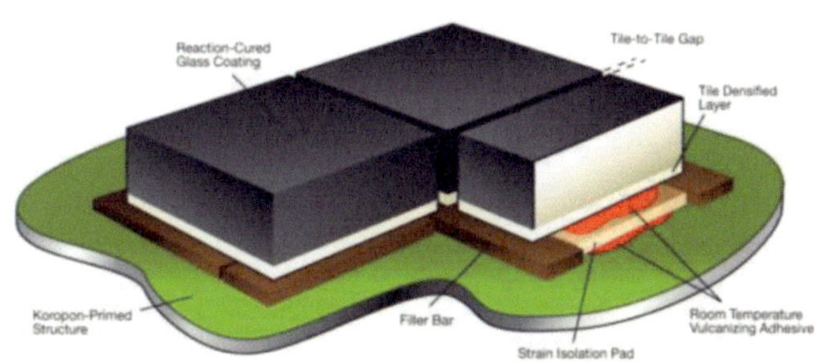

Figure 2. Schematic of rigid ceramic tile TPS [2].

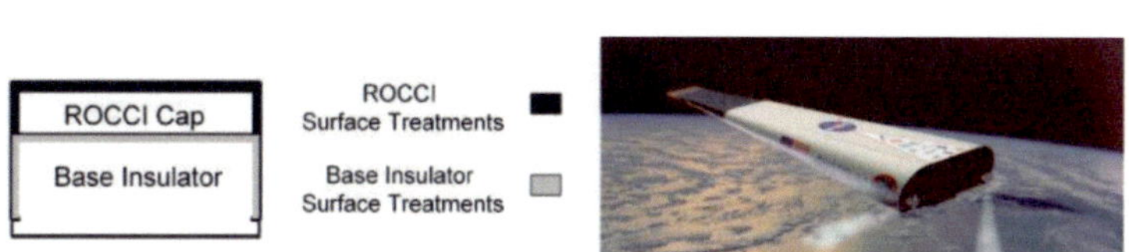

Figure 3. Schematic and application of TUFROC [3].

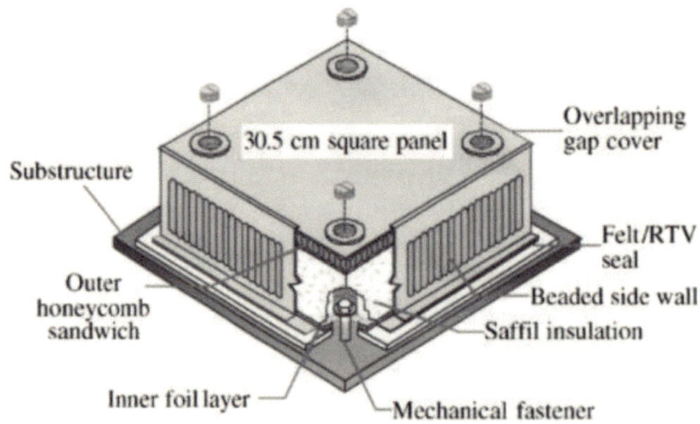

Figure 4. Schematic of ARMOR metallic TPS [5].

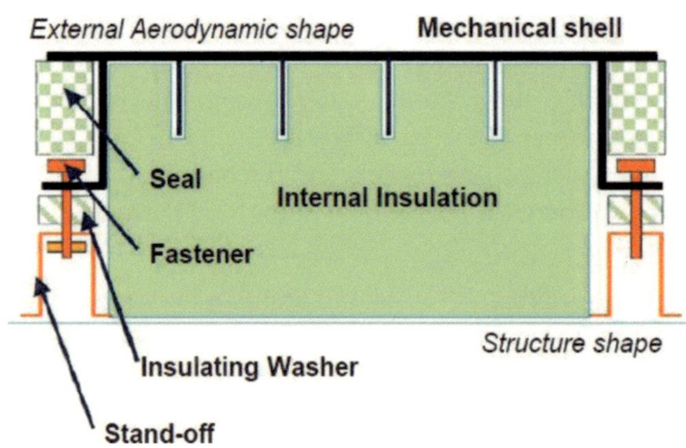

Figure 5. Schematic of CMC shingle TPS [6].

loading and unloading, easy to monitor, and so on. Compared with metallic TPS, it has the advantages of high carrying capacity, high temperature resistance, producing larger size components, and lower maintenance costs. As shown in **Figure 5**, this kind of TPS consists of CMC shingle, sealing materials, internal insulation materials, and connectors. The structural elements of CMC shingle TPS can be divided into two categories, namely, the element to withstand mechanical load (mechanical enclosure, fasteners, etc.) and element to withstand heat load (internal insulation, insulation felt, etc.). In the design stage, the mechanical and thermal protection functions can be treated separately so that the best material can be selected in each field.

For a variety of TPSs, design and maintenance is a comprehensive system issue that includes the acquisition of aerodynamic thermal, determination of materials and sizing, numerical and experimental verification, flight tests, health monitoring and management, maintenance, and other aspects. Various uncertainties exist in all stages, such as ballistic deviation, deviation of material property, model errors, data measurement errors, system assembly deviations, external loading conditions, and other unknown uncertainties. How to fully consider the influence of these uncertainties in the process of design and use and to set system margin and safety factor reasonably is the key issue for the reliability and structural integrity of TPS. Therefore, in the following sections, the treatment of various uncertainties and the probabilistic design method for TPS will be systematically introduced.

2. Uncertainties in TPS and their analysis methods

2.1. Classification and acquisition of uncertainties

There are two kinds of uncertainties in the design and service stage of TPS. One is based on the nature of the uncertainty and the other is based on the source of the uncertainty. According to the nature of uncertainties, they can be divided into the following three categories [7]:

1. Stochastic variability: It means that uncertainty can be clearly recognized but cannot be reduced by means of experiment or analysis, such as natural fluctuations in the atmospheric environment and so on. If the quantitative change law of this kind of uncertainty can be obtained or it can be parameterized, it can be analyzed by Monte Carlo simulation.
2. Structural uncertainty: It mainly refers to the errors and uncertainties caused by mathematical models and numerical discretization, such as the simplification of physical processes and boundary conditions, temporal and spatial discretization errors, and so on. These kinds of uncertainties require reliable validation and verification of the model to control their effect on the design and application of TPS.
3. Parametric uncertainty: It mainly refers to the uncertainties of the various input parameters required in the analysis and design, such as the catalytic and emissivity of the wall, the rate of chemical reaction, transition criteria, and so on.

According to the sources of the uncertainties, they can be roughly divided into the following categories:

1. Trajectory uncertainties: It is mostly due to navigation uncertainty and orbital maneuver uncertainty, which can be modeled by the overshoot and undershoot of trajectory design.
2. Aerodynamic heating uncertainties: In hypersonic flight, complex ionization and chemical reaction occur in the boundary layer. Besides, uncertainties in flow conditions and boundary conditions are also factors causing aerodynamic uncertainties. The uncertainty of aerodynamic heating is one of the key factors in the size design of TPS.
3. Material uncertainties: For non-ablative TPS, the deviation of parameters, such as thermal conductivity, heat capacity, surface emission coefficients, and other thermal physical parameters, is mainly considered. For ablative TPS, the abovementioned uncertainties must be considered both before and after ablation. In general, material uncertainty is obtained by experiment. Material uncertainty is also one of the major factors in the size design of TPS.

4. Uncertainties of processing and assembly error: The error of processing and assembly refers to the geometric uncertainty, such as thickness error, geometric error, gap in assembly process and so on.

5. Analysis model uncertainties: This kind of uncertainty refers to the error introduced in the description, modeling, and numerical calculation process of TPS. The uncertainty of the numerical calculation part can be controlled and estimated by the validation and verification of the model.

6. Uncertainties of ground test: This type of uncertainty mainly refers to the ground test equipment that cannot accurately reproduce the true working process of aerodynamic thermal environment. In general, the impact of this type of uncertainty can be eliminated by reasonably setting the thickness margin of the TPS.

7. Other unknown uncertainties: Other uncertainties not included above.

As mentioned above, there are many uncertainties in the design, analysis, and use of TPS. Due to its complexity, it is almost impossible to accurately obtain the degree and distribution of uncertainty through a large amount of test data. Therefore, in probability analysis, it is often necessary to combine various methods to estimate the uncertainty of the input parameters. In general, there are six ways to get the uncertainty of the system parameters [8]. These ways are:

1. A combination of a variety of data sources to estimate the mean and standard deviation of the parameters. This is the only approach that is statistically rigorous and should be used wherever possible.

2. Performing an independent assessment of the data source. This method can be used when only a single data source is available.

3. Using recommended data in relevant reviews or reports.

4. Using the uncertain data given or cited in the original data.

5. Adopting the similarity rules.

6. Relying on the expert judgments.

2.2. Sampling methods and surrogate models

2.2.1. Sampling methods

In the probability and reliability analysis of TPS, the sampling procedure is always inevitable. Therefore, how to obtain the statistical characteristics of the system accurately with a few samples is the key to improve the overall analysis efficiency. We introduce three commonly used sampling methods in probability analysis, namely the Monte Carlo sampling method, the Latin hypercube sampling method, and the adaptive importance sampling method.

(1) Monte Carlo sampling method.

Monte Carlo sampling can be directly used to simulate the real process of various projects. When the input uncertainty can be quantified, this method can simulate any process of the TPS in practice. A significant disadvantage of this method is that there is no memory feature for the sampling process, which may lead to a concentration or agglomeration of sampling points around a local region of the sample space. If the sample data are distributed in the entire input parameter space unevenly, those concentrated data points in the simulation cycle are equivalent to double counting, which does not provide effective reference value any more.

(2) Latin hypercube sampling method.

Latin hypercube sampling method has the feature of sampling memory, which can effectively overcome the problem of low efficiency caused by data concentration in Monte Carlo sampling method. Besides, it forces that the sampling points must be discretely distributed throughout the sampling space. Therefore, in general, for the same precision results of same problem, Latin hypercube sampling method can reduce the simulation cycle number by 20–40% than direct Monte Carlo sampling method. It needs to be pointed out is that the number of simulation cycles is determined by the problem itself to achieve a desired accuracy or confidence level.

In Latin hypercube sampling method, the range of each random variable is divided into several intervals. A representative value is randomly selected from each interval of a random variable. These representative values form a sample of this random variable, which ensures that all possible values are extracted during the sampling process, called the memory feature in the sampling process.

(3) Adaptive importance sampling method.

In the analysis of system probability properties, the distribution of output parameters in the whole sample space is concerned. Therefore, we should avoid sample concentration in a certain area when sampling. But in the analysis of system failure probability or reliability, more attention should be paid to the samples near the failure critical state. However, the failure probability of the TPS is very low. If the direct Monte Carlo or the Latin hypercube method is used, the samples near the failure state are sparse and thus the final failure probability estimation is not accurate enough. In this case, important sampling methods can be used to change the sampling center of gravity in order to increase the number of samples in the failure state.

2.2.2. Surrogate models

In the analysis of probability and reliability of TPS, the direct calculation using the sampling method introduced in the previous article consumes a large amount of computational resources. However, it is an effective way to improve computational efficiency by replacing the complex relationship between random input and stochastic output using the existing data to construct the surrogate model. At present, the surrogate models commonly used in the engineering field include response surface method (RSM), Kriging model, and so on.

(1) Response surface method.

The basic idea of RSM is to use a proper function to approximate a function that cannot be explicitly expressed. The main purpose is to reduce computation as much as possible with acceptable accuracy. To achieve this goal, the probabilistic features of the original state surface must be described as accurately as possible, especially near the design point. RSM assumes that the influence from random input variables can be expressed using mathematical functions. In general, the function is a polynomial of one or two orders, such as:

$$\hat{y} = g(x) = \beta_0 + \sum_{i=1}^{k} \beta_i x_i \tag{1}$$

$$\hat{y} = g(x) = \beta_0 + \sum_{i=1}^{k} \beta_i x_i + \sum_{i=1}^{k} \beta_{ii} x_i^2 + \sum_{i} \sum_{j>i} \beta_{ij} x_i x_j \tag{2}$$

where β_0, β_i, β_{ii}, and β_{ij} represent polynomial coefficients obtained by fitting the sample points, usually determined by least square method.

For TPS, although it is unrealistic to accurately replace the real function by the response surface function, sensitivity analysis based on the response surface can be easily performed to provide guidance for subsequent parameter designs. In practice, the RSM is often used to solve the probability and sensitivity near the checking point where the performance function is easier to fit. Meanwhile, the simplified form of the response surface also helps to reduce the computational efforts of system analysis when analyzing the sensitivity of a TPS with multiple random variables.

Although RSM requires fewer samples than the Monte Carlo method, it is not as versatile as Monte Carlo simulation techniques. In order to use RSM, the function relationship among random variables must be smooth and continuous. If this case is not satisfied, this method cannot be used. Besides, the number of required cycles depends on the number of random input variables. If there are many random variables, fitting the response surface requires a large number of samples, which lead to the loss of efficiency. As an unbiased estimate, the response surface function value calculated at each sample point has an error, even at the design points.

(2) Kriging model.

The Kriging model method is a statistical prediction method based on the stochastic process. It can find the optimal, linear, and unbiased estimates of regional variables with the smoothing effect and the least statistical variance of estimation. The Kriging model method has the characteristics of local estimation, which makes it easier to obtain the ideal fitting effect in solving the problem of higher nonlinearity.

The Kriging model is composed of the global model and the local deviation, which can be expressed as

$$Y(x) = f(x) + Z(x) \tag{3}$$

where $Y(x)$ is unknown approximate model, $f(x)$ usually is a fitted polynomial function, and $Z(x)$ is a random process whose mean value is zero, variance is σ^2, and covariance is not zero. The covariance matrix can be expressed as:

$$\mathrm{COV}\left[Z(x^i), Z(x^j)\right] = \sigma^2 R\left[R(x^i, x^j)\right] \tag{4}$$

where R is correlation matrix and $R(x^i, , x^j)$ represents the correlation function of sample point x^i, x^j. Here is an example of Gaussian correlation function, and its expression is

$$R(x^i, x^j) = \prod_{k=1}^{n_{dv}} \exp\left(-\theta_k \left|x_k^i - x_k^j\right|^2\right) \tag{5}$$

where n_{dv} represents the number of design variables, and θ_k is unknown related parameters. If the related functions are determined, the response estimation of arbitrary test point x is:

$$\tilde{y} = \widehat{\beta} + r^T(x)R^{-1}\left(y - f\widehat{\beta}\right) \tag{6}$$

where y is a column vector of length n_s (sampling number). When $f(x)$ is constant, f is unit column vector of length n_s. $r^T(x)$ is the relevance vectors of length n_s between test point x and sampling point $\{x^1, , x^2, \cdots, , x^{n_s}\}$, which is expressed as:

$$r^T(x) = \left[R\left(x, x^1\right), R\left(x, x^2\right), \cdots, R(x, x^{n_s})\right]^T \tag{7}$$

$\widehat{\beta}$ in Eq. (6) can be estimated by:

$$\widehat{\beta} = \left(f^T R^{-1} f\right)^{-1} f^T R^{-1} y \tag{8}$$

The variance of the global model is estimated as:

$$\widehat{\sigma}^2 = \frac{\left(y - f\widehat{\beta}\right)^T R^{-1}\left(y - f\widehat{\beta}\right)}{n_s} \tag{9}$$

The relevant parameters θ_k are determined by the maximum likelihood estimation by solving the following nonlinear unconstrained optimization problem:

$$\max_{\theta_k > 0}\left(\frac{\left[n_s \ln\left(\widehat{\sigma}^2\right) + \ln|R|\right]}{2}\right) \tag{10}$$

When θ_k is obtained, $r^T(x)$ can be obtained by Eq. (7). After obtaining the response value by Eq. (6), Kriging model could be established. Although the construction process of Kriging model is more complex than classical polynomial RSM, it can better approximate complex functions with strong nonlinear features.

3. Structural response analysis considering uncertainties

When designing and analyzing a TPS, the impact of uncertainties on the structural responses should be taken into account due to divergence of material property parameters and geometry uncertainties from manufacture errors. These uncertainties must be properly introduced into the probabilistic finite element model.

3.1. Characterization of uncertainties and probabilistic finite element model

Parametric uncertainties of TPS structures, once acquired, should be used for building up a finite element model using commercial software packages like ANSYS and ABAQUS. Taking a CMC shingle TPS, for example, it is a typical multi-plate structure, where the top part is a C/C composite cover plate with a radiation layer, followed by an insulation felt, and a high-temperature and a low-temperature insulating layer. And there is a strain isolation pad between the low-temperature insulating layer and the vehicle surface. The finite element model is shown in **Figure 6**. Here, the uncertainties in material properties and geometries are considered.

Firstly, we consider the uncertainties of thermal conductivity and specific thermal capacity. These parameters are nonlinear dependent on the working temperature. Here, fitted polynomial functions are applied to relate the parameter and the temperature, where only the constant component varies to represent the uncertainties. For example, a 4-order polynomial is used to fit 10 experimental data of thermal conductivity shown in **Figure 7 (a)** and **(b)** demonstrates its upper and lower bounds. Such characterization can be generalized for other nonlinear parameters that vary with temperature.

3.2. Precise time integration for transient thermal conduction

The temperature distribution can be obtained through transient analysis once the finite element model is built. The analysis efficiency is one of the most crucial factors since massive samples and calculations have to be carried out during the probabilistic analysis of a system. Usually, the direct integration is used for the transient heat conduction analysis, which shows good generality but high dependency of precision and stability on time step length. To improve the precision, efficiency, and stability of transient thermal analysis, the precise time integration (PTI), which is based on the exponential integrator, is applied here.

The discretization equation of thermal conduction is

$$C\dot{T} + KT = P \tag{11}$$

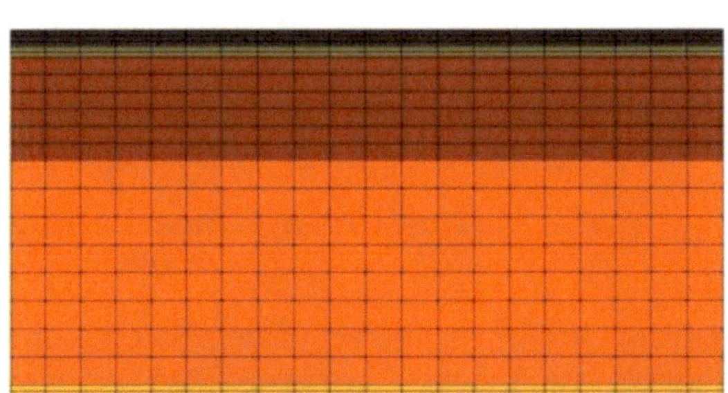

Figure 6. The finite element model of typical multilayer TPS.

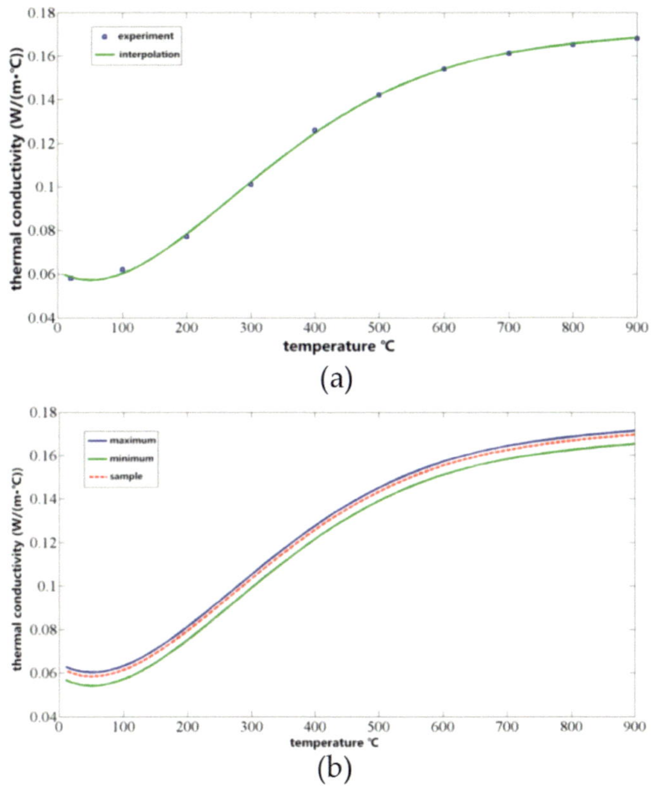

Figure 7. Characterizing the uncertainties of thermal conductivity. (a) Fitting thermal conductivity and (b) samples of thermal conductivity.

where C is heat capacity matrix, K is heat conduction coefficient matrix, P is heat load vector, T is node temperature vector, and $\dot{T}$ is the derivative, respectively. This equation can be rewritten as

$$\dot{T} = HT + r \tag{12}$$

where

$$H = -C^{-1}K, \quad r = C^{-1}P \tag{13}$$

The analytical transient solution can be written as

$$T(t) = \Phi(t, t_0)T_0 + \int_{t_0}^{t} \Phi(t, t_1)f(t_1)\mathrm{d}t_1 \tag{14}$$

where T_0 represents the initial temperature, and

$$\Phi(t, t_1) = \exp\left[H \cdot (t - t_1)\right] \tag{15}$$

Generally, we only need to carry out those calculations at the exact moments and do not need to start from t_0. Instead, it is acceptable to calculate t_{k+1} based on t_k [9, 10], which is

$$T_{k+1} = \exp(\tau H)T_k + \int_0^\tau \exp[(\tau - s)H] \cdot r(t_k + s)\mathrm{d}s \tag{16}$$

where $\tau = t_{k+1} - t_k$ represents the time interval for the integration. Assuming a linear relation between the inhomogeneous terms and time during such an interval, we have

$$r(t) = r_0 + r_1(t - t_k) \tag{17}$$

$$T_{k+1} = \exp(\tau H)\left[T_k + H^{-1}\left(r_0 + H^{-1}r_1\right)\right] - H^{-1}\left[r_0 + H^{-1}r_1 + r_1\tau\right] \tag{18}$$

Noting that C and K may be time-varying matrices under nonlinear circumstances, it can be assumed that they are time invariant during every interval so that Eq. (18) still works for transient thermal conduction. Besides, due to the existence of inhomogeneous terms, inversion of H, which is strongly recommended to be avoided in solving large-scale problems, is needed in Eq. (18). Therefore, dimensional expanding method can be applied to remove the matrix inversion by adding the inhomogeneous terms as a part of the state variable vector. Let

$$\phi = \begin{Bmatrix} T \\ P \end{Bmatrix} \tag{19}$$

and Eq. (11) can be rewritten as

$$\dot{\phi} = H^* \cdot \phi, \qquad H^* = \begin{bmatrix} H & C_1 \\ 0 & D_1 \end{bmatrix} \tag{20}$$

where $C_1 = C^{-1}$ and D_1 rely on the relations between the inhomogeneous terms and their temporal derivatives. If the inhomogeneous terms remain constant during the integration interval, D_1 can be written as

$$D_1 = [0]_{n\times n} \tag{21}$$

If the inhomogeneous terms P vary linearly during the integration interval, the augmented vector can be written as

$$\phi = \begin{Bmatrix} T \\ P \\ \dot{P} \end{Bmatrix} \tag{22}$$

and

$$\dot{P}_k = \frac{P(t_{k+1}) - P(t_k)}{\tau}, \qquad D_1 = \begin{bmatrix} 0 & I_{n\times n} \\ 0 & 0 \end{bmatrix} \tag{23}$$

Now we get an equivalent homogeneous format as

$$\dot{T} = H^* \cdot T, \qquad H^* = \begin{bmatrix} H & C_1 & 0 \\ 0 & 0 & I_{n\times n} \\ 0 & 0 & 0 \end{bmatrix} \tag{24}$$

And the solutions can be calculated from

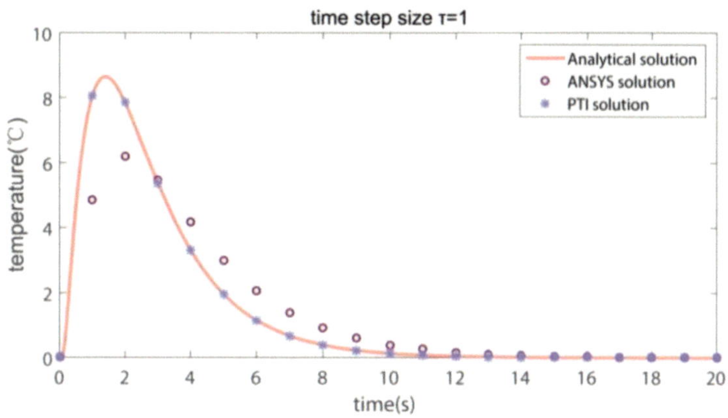

Figure 8. A comparison between precise time integration and time difference methods.

$$T_{k+1} = \exp(\tau H^*) T_k \tag{25}$$

The equality above expands the dimensions, but avoids matrices inversions, which makes the algorithm more efficient. The key to this algorithm is how we acquire the matrix exponential $\exp(\tau H^*)$ or the product of matrix and vector $\exp(\tau H^*)T_k$.

The precise time integration for transient thermal conduction has some impressive advantages, including the insensitivity of time step length, unconditional stability, and high efficiency. **Figure 8** shows a comparison between PTI algorithm and the conventional time difference method for a transient heat transfer problem with analytical solutions. It can be easily observed that the precise integration method is far more accurate than the conventional time difference method used in ANSYS .

4. Probabilistic design and reliability assessment method for thermal protection system

The main failure mode of TPS for thermal reliability assessment is the overtemperature of bond-line. Based on the theories and methods introduced in the previous sections, the procedures for probabilistic design and reliability evaluation are shown in **Figure 9**. The three main modules are introduced in detail as the following.

(1) Deterministic design.

Deterministic design, also known as nominal design, is often used to determine the initial geometry of TPS. First, thermal load at different locations needs be predicted numerically or experimentally according to the aerothermal environment using the initial configuration. Secondly, the proper material types and structural forms of TPS according to estimated thermal load need to be determined. We can finally get an initial thickness h through transient thermal analysis.

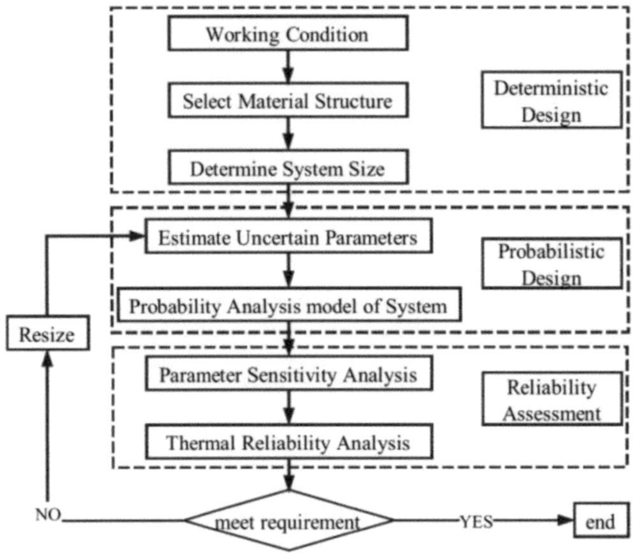

Figure 9. Probabilistic design and reliability assessment process of TPS.

$$h(x, y, z) = f(Q(t), MAT) \tag{26}$$

where $Q(t)$ represents aerodynamic flux and MAT represents the TPS material properties, respectively.

(2) Probabilistic design.

Firstly, the source, range, and distribution of the system uncertainties are determined, and the parameters are reasonably set using the methods in Section 2. Then, the probability structural thermal response is analyzed by the methods described in Section 2 and 3. Finally, the probability characteristics of the structure temperature field are obtained. The highest value of bondline overtemperature is usually taken as a random output variable, which is:

$$T_{\max} = \mathbb{T}(s_1, s_2, \cdots) \tag{27}$$

where $\mathbb{T}$ is the function of random input and output, and $s_1, s_2, \cdots$ are the random input variables.

(3) Reliability assessment.

The thermal reliability of a TPS can be defined as:

$$R = 1 - P_F = 1 - \int_{T_{\lim} \le T_{\max}} f(s) \mathrm{d}s \tag{28}$$

where P_F is failure probability, $T_{\lim}$ is the limiting temperature, $T_{\max}$ is maximum temperature of system, and $f(s)$ is probability density function of random input variables, respectively. For discrete samples, the reliability can be obtained directly from the statistical output:

$$R = \left(1 - \frac{N(T_{\lim} \le T_{\max})}{N}\right) \times 100\% = \frac{N(T_{\lim} > T_{\max})}{N} \times 100\% \tag{29}$$

In the probabilistic design, it is necessary to set a reasonable range of reliability. If the system meets the reliability requirements, the design process ends, or else the system parameters need to be adjusted to carry out the reliability evaluation again.

5. Application of probabilistic design in thermal protection system

The probabilistic design of a typical CMC shingle TPS is used as an example to show the effectiveness of the design procedure in the previous section. Here, the uncertainty of the material properties and thickness is considered.

Variable	Physical meaning	Mean	Std. dev.	Lower bound	Upper bound
H1	Thickness of CMC cover	4.0E-3	1.33E-4	3.8E-3	4.2E-3
H2	Thickness of fiber cloth	3.0E-3	1.0E-4	2.85E-3	3.15E-3
H3	Thickness of high-temp insulation layer	2.0E-2	6.67E-4	1.90E-2	2.10E-2
H4	Thickness of low-temp insulation layer	4.5E-2	1.5E-3	4.275E-2	4.725E-2
H5	Thickness of strain isolation pad	3.0E-3	1.0E-4	2.85E-3	3.15E-3
H6	Thickness of skin	4.0E-3	1.33E-4	3.8E-3	4.20E-3
A4	Sp.ht. of CMC cover	815.4	27.18	774.63	856.17
B4	Sp.ht. of fiber cloth	1000.0	33.33	950.0	1050.0
D5	Sp.ht. of high-temp insulation layer	900.30	30.10	855.28	945.32
E2	Sp.ht. of low-temp insulation layer	990.0	33.00	940.50	1039.5
F4	Sp.ht. of strain isolation pad	1000.0	33.33	950.0	1050.0
G2	Sp.ht. of skin	840.0	28.00	798.0	882.0
L5	Heat conductivity of CMC cover	47.4	1.59	45.32	50.09
M3	Heat conductivity of fiber cloth	0.23	7.67E-3	0.2185	0.2415
O7	Heat conductivity of high-temp insulation layer	6.125E-2	2.04E-3	5.82E-2	6..43E-2
P6	Heat conductivity of low-temp insulation layer	5.195E-2	1.73E-3	4.94E-2	5.45E-2
Q4	Heat conductivity of strain isolation pad	6.10E-2	2.03E-3	5.80E-2	6.41E-2
S2	Heat conductivity of skin	0.65	2.16E-2	0.6175	0.6825

Table 1. Random input parameters.

5.1. Model uncertainty

The typical composite TPS investigated here is shown in **Figure 6** of Section 3.1, assuming that 18 random input variables follow the truncated normal distribution. The standard deviation is 3% of the mean value, and the range of the value is mean $\pm$ 5%. The random input parameters are listed in **Table 1**. Taking H1 (thickness of CMC cover) as an example, the probability density and cumulative probability distribution curve are shown in **Figure 10**.

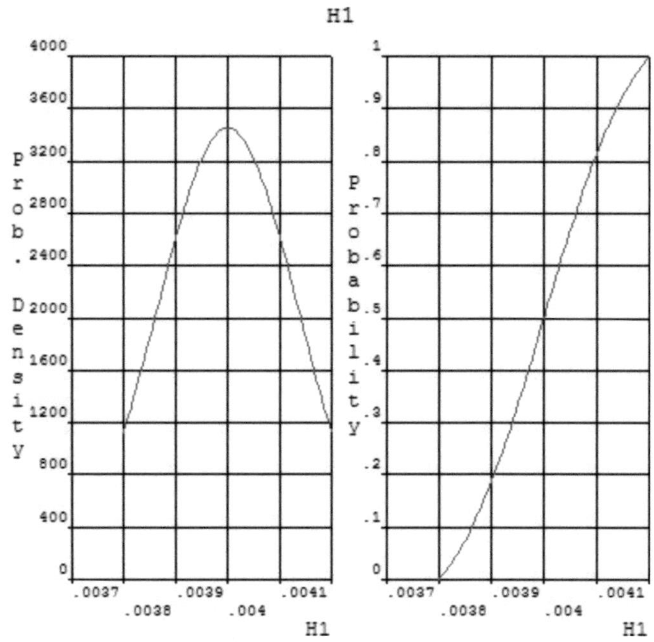

Figure 10. Probability density and cumulative probability distribution curve of H1.

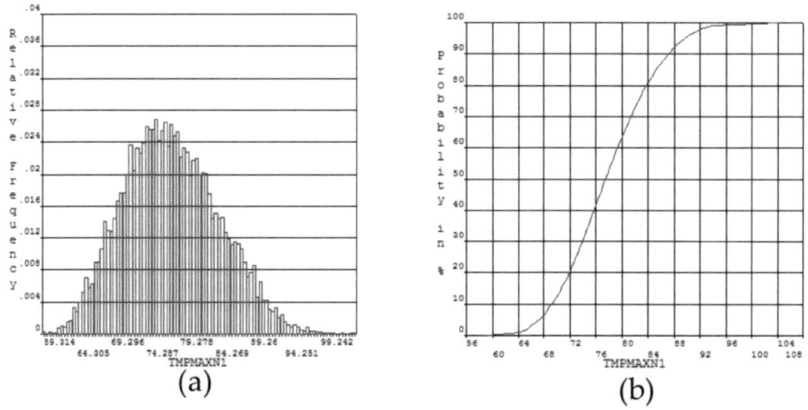

Figure 11. Probability characteristics of the maximum bondline temperature. (a) Probability histogram and (b) cumulative probability.

http://dx.doi.org/10.5772/intechopen.74240

5.2. Probability characteristics and sensitivity analysis

For a given heat flux, a nonlinear transient thermal analysis is conducted considering uncertainties listed in **Table 1**. The highest bondline temperature during the computational interval is selected as the random output parameter. In order to reduce the computational load, Latin hypercube sampling method is used. Then, the response surface is fitted with the calculated samples. Next, the probability characteristic of random output is obtained by sampling again on the response surface. In order to ensure the accuracy of the fitting, the number of samplings for fitting the response surface needs to be at least twice the number of samples, that is, $((18 + 1) \times (18 + 2)/2) \times 2 = 380$ times.

Sampling 10,000 times on response surface, the histogram and the cumulative probability distribution are shown in **Figure 11**. The results show that the maximum bondline temperature is basically in accordance with the normal distribution with a mean of 77.83°C, a standard deviation of 6.765, a maximum of 104.23°C, and a minimum of 59.31°C. Based on the results of the probabilistic analysis, the reliability can be evaluated using the method described in Section 4.

The sensitivity of the random output to the random input parameters can be obtained by probability analysis, as shown in **Figure 12**. The top five input variables that have great impact on the output variables of the TPS are H4, E2, H3, P6, and H6, respectively, namely the thickness of low-temperature insulation layer, the specific heat capacity of the low-temperature insulation layer, thickness of high-temperature insulation layer, heat conductivity of low-temperature layer, and the thickness of the vehicle structure. Among them, the thickness of the low-temperature insulation layer has the greatest impact on the maximum bondline temperature with negative correlation. In the first five parameters, only P6 is positively correlated.

According to the sensitivity analysis result, in order to improve the reliability of the investigated TPS, manufacturing errors and the deviation of specific heat of low-temperature insulation layer should be strictly controlled.

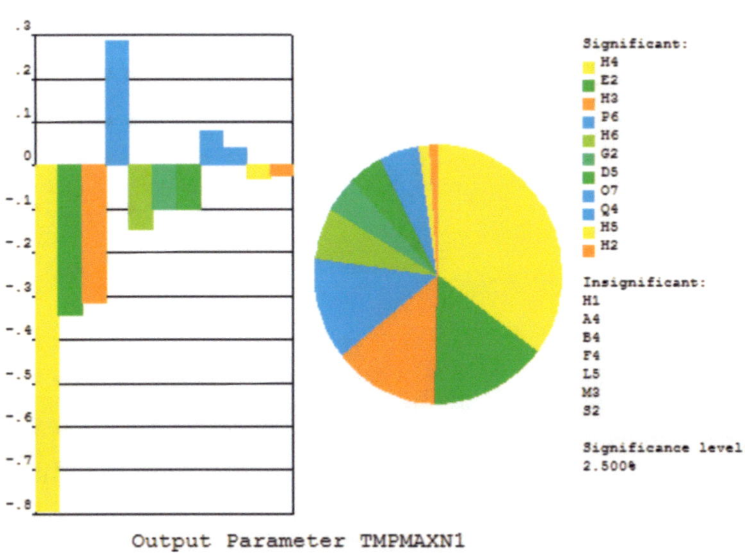

Figure 12. Sensitivity of maximum bondline temperature to random input parameters.

6. Conclusion

With the rapid improvement of experimental and computational technologies, the uncertainties of thermal protection system are more easily acquired and modeled, which makes the probabilistic design of highly reliable TPS possible. In this chapter, the types and sources of uncertainties during the design and service stages of the thermal protection system are summarized. The mathematical treatment and finite element modeling methods of various uncertainties are introduced. The dimensional expanding precise time integration method is used to improve the numerical accuracy and efficiency of transient heat conduction analysis. Based on the theoretical and numerical methods, the probabilistic design and analysis framework is finally formed and is validated via an example of a typical CMC shingle TPS.

The following is recommended for the future research: (1) theoretical and experimental methods for uncertainty acquisition of TPS; (2) probabilistic modeling and fast analysis method of structural thermal response considering uncertainty; (3) the influence of structural cracks and defects on the reliability of TPS; and (4) reliability and risk assessment method of TPS based on probability analysis.

Acknowledgements

This work is supported by the National Natural Science Foundation of China (Grant No. 11502037).

Author details

Kun Zhang[1], Jianyao Yao[1]*, Jianqiang Xin[2] and Ning Hu[1]

*Address all correspondence to: yaojianyao@cqu.edu.cn

1 College of Aerospace Engineering, Chongqing University, Chongqing, China

2 Research and Development Center, China Academy of Launch Vehicle Technology, Beijing, China

References

[1] Meseguer J, Pérez-Grande I, Sanz-Andrés A. Thermal protection systems. In: Meseguer J, Pérez-Grande I, Sanz-Andrés A, editors. Spacecraft Thermal Control. 1st ed. Holand: Academic Press, Elsevier; 2012. pp. 305-325. DOI: 10.1533/9780857096081.305

[2] Rodriguez AC, Snapp CG. Orbiter thermal protection system lessons learned. In: AIAA SPACE 2011 Conference & Exposition; 27–29 September 2011; Long Beach, California. United States: AIAA; 2011

[3] Stewart DA, Leiser DB. Lightweight TUFROC TPS for hypersonic vehicles. In: 14th AIAA/AHI Space Planes and Hypersonic Systems and Technologies Conference; 6–9 November 2006; Canberra, Australia. America: AIAA; 2006

[4] Blosser ML. Fundamental modeling and thermal performance issues for metallic thermal protection system concept. Journal of Spacecraft and Rockets. 2004;**41**(2):195-206. DOI: 10.2514/1.9182

[5] Poteet CC, Blosser ML. Improving metallic thermal protection system hypervelocity impact resistance through numerical simulations. Journal of spacecraft and rockets. 2004;**41**(2):221-231. DOI: 10.2514/1.9193

[6] Pichon T, Barreteau R, Soyris P, et al. CMC thermal protection system for future reusable launch vehicles: Generic shingle technological maturation and tests. Acta Astronautica. 2009;**65**(1):165-176. DOI: 10.1016/j.actaastro.2009.01.035

[7] Glass DE. Ceramic matrix composite (CMC) thermal protection systems (TPS) and hot structures for hypersonic vehicles. In: 15th AIAA Space Planes and Hypersonic Systems and Technologies Conference; 28 April-1 May 2008, Dayton, Ohio. United States: AIAA; 2005

[8] Wright MJ, Bose D, Chen YK. Probabilistic modeling of aerothermal and thermal protection material response uncertainties. AIAA Journal. 2007;**45**(2):399-410. DOI: 10.2514/1.26018

[9] Zhong WX, Williams FW. A precise time step integration method. Proceedings of the Institution of Mechanical Engineers, Part C: Journal of Mechanical Engineering Science. 1994;**208**(6):427-430. DOI: 10.1243/PIME_PROC_1994_208_148_02

[10] Lin J, Shen W, Williams FW. A high precision direct integration scheme for structures subjected to transient dynamic loading. Computers & structures. 1995;**56**(1):113-120. DOI: 10.1016/0045-7949(94)00537-D

Section 2

Heat Transfer in Complex Systems

Chapter 4

Gas-Liquid Stratified Flow in Pipeline with Phase Change

Guoxi He, Yansong Li, Baoying Wang,
Mohan Lin and Yongtu Liang

Additional information is available at the end of the chapter

http://dx.doi.org/10.5772/intechopen.74102

Abstract

When the natural gas with vapor is flowing in production pipeline, condensation occurs and leads to serious problems such as condensed liquid accumulation, pressure and flow rate fluctuations, and pipeline blockage. This chapter aims at studying phase change of vapor and liquid-level change during the condensing process of water-bearing natural gas characterized by coupled hydrothermal transition and phase change process. A hydrothermal mass transfer coupling model is established. The bipolar coordinate system is utilized to obtain a rectangular calculation domain. An adaptive meshing method is developed to automatically refine the grid near the gas-liquid interface. During phase change process, the temperature drop along the pipe leads to the reduction of gas mass flow rate and the rise of liquid level, which results in further pressure drop. Latent heat is released during the vapor condensing process which slows down the temperature drop. Larger temperature drop results in bigger liquid holdup while larger pressure drop causes smaller liquid holdup. The value of velocity with phase change is smaller than that without phase change while the temperature with phase change is bigger. The highest temperature locates in gas phase. But near the pipe wall the temperature of liquid region is higher than gas region.

Keywords: hydrocarbons pipeline, vapor/condensation-stratified flow, heat transfer, phase change, multi-component

1. Introduction

Condensation occurs when the natural gas with vapor is flowing in production pipeline and leads to serious problems such as condensed liquid accumulation, pressure and flow rate fluctuations, and pipeline blockage or corrosion. The pipe flow together with phase change is commonly encountered in various heat and mass transfer processes over the past four decades, for instance, in petroleum and chemical processing industry, steam-generating equipment, nuclear reactors, geothermal fields, heat exchangers, cooling systems, and solar energy system [1–4]. In petroleum transportation, two-phase flow characterization is a very common and economic technique, where vapor-liquid two-phase stratified flow is often observed in horizontal or slightly inclined systems [5, 6].

There exist several problems in the pipeline network system that the saturated vapor in gas would condense due to pressure and temperature drop [7, 8]. The condensate would attach to the pipe wall as a form of film or droplet [9, 10]. The condensation will decrease the effective cross-sectional area and cause the increase of pressure drop which may lead to system shutdown [11, 16]. Generally, the condensed water accumulates at the lower parts of the pipeline due to the hilly pipeline route topography, which results in a continuous change of liquid holdup along the pipeline [12–14]. The changing liquid holdup and flow area are bounded to affect the flow patterns which inevitably influence the operating pressure and temperature inversely. Thus, the flow of condensed water and water-bearing gas in production pipelines is a complex process with coupling of hydraulic, thermal, and phase change phenomena [14–17]. Researchers have investigated the gas-liquid two-phase pipe flow system by experiments or hydrodynamic and thermodynamic models.

It has been observed experimentally that when phase change occurs during the saturated vapor pipeline transportation process, the thermal gradients are created in the wall of the pipeline that lead to severe liquid condensation and stratified vapor-liquid two-phase flow [17, 18]. The fundamental engineering parameters are the pressure drop, liquid holdup, phase fraction, phase flow rate, temperature, thermo-physical properties of the fluids, and pipeline geometry [19].

Not limited to hydraulic parameters, more recent attention has turned to non-isothermal flow in a pipe or plane channel where some numerical studies are also found [20–27]. The detailed characteristic of heat transfer is taken into consideration in these mentioned models instead of an average value being represented for the temperature profiles in a pipe [28, 29]. According to their studies, the wall temperature distribution is different from the assumption of fully developed isothermal state [30]. The energy transfer model has been taken into the flow progress for the optimization of transportation, estimation of corrosion, or prediction of wax deposition [31, 32]. Concretely, the two-dimensional (2D) momentum and three-dimensional (3D) energy equations for both phases have been established for dynamic and thermal numerical simulation [8, 27, 30, 33]. The smooth or wavy interface between phases was obtained in a different range of flow rates [17]. For such a two-phase non-isothermal stratified flow, analytical and numerical heat transfer solutions limited to laminar flow and without phase change have been obtained for fully developed stratified flow under different thermal boundary conditions [27]. Then, solutions which are more applicable to fully developed turbulent gas-liquid smooth stratified flow have been obtained through the use of high Reynolds model [30]. Recently, the steady-state axial momentum and energy equations coupled with a low Reynolds model were established and solved [34]. The pressure drop, liquid height, and temperature field which are included in the solutions could match well with the experimental data.

However, although equation of state (EOS) was utilized in previous one-dimensional (1D) models to calculate the phase fraction [10, 12, 16, 22, 34, 41, 44], the flow rate, temperature, and pressure were not coupled with the varying liquid level.

According to the description of the physical process of gas-liquid two-phase pipe flow, the model can be divided into isothermal model, gas-liquid two-phase pipe flow model coupled with heat and gas-liquid two-phase pipe flow model coupled with heat and mass transfer. In the gas-liquid two-phase pipe flow model under the isothermal condition, it is assumed that all phases are in thermodynamic equilibrium state without considering the heat transfer process between pipe flow and environment, and the physical parameters of gas-liquid two-phase are just the single value function of pressure. In the gas-liquid two-phase flow model coupled with heat, the heat transfer of gas-liquid and surrounding environment is considered. In the gas-liquid two-phase flow model coupled with heat and mass transfer, the coupling effect of flow, heat transfer, and mass transfer are considered simultaneously.

The two-dimensional (2D) or three-dimensional (3D) stratified gas-liquid two-phase model including mass conversation equation, momentum conversation equation, turbulence model equation, boundary conditions, and related auxiliary equations for model closure were applied to describe the flow in pipelines. Differences among them mainly existed in two aspects. On the one hand, different turbulence models were built including Spalart-Allmaras Model (SAM), $k-\varepsilon$series model, Reynolds Stress Model (RSM), Direct Numerical Simulation (DNS), and so on. On the other hand, different gas-liquid interface configuration models were built including volume of fluid (VOF) method lattice Boltzmann method, level set method, smooth interface model, wavy interface model, and curve interface model. The relationship between pressure drop, liquid holdup, velocity distribution, turbulent viscosity, and shear stress has been determined through these formula or methods. The flow pattern and flow regime and even secondary flow have also been discussed under different boundary conditions. However, the energy conservation equation and phase change of fluid components were not taken into consideration in those studies.

Recently, attempts have been made to introduce energy equation into the improved model and the detailed solutions about temperature distribution have also been worked out by considering potential energy, kinetic energy, heat transfer, and Joule-Thomson effect [17, 35–38]. The phase change was ignored in these models. Although equation of state (EOS) was utilized in previous studies to calculate gas condensation of gas-condensate flow in pipelines [12–15, 17–20, 39], the flow rate, temperature, pressure were not coupled with liquid level. Turbulent flow is not considered, which would lead to different numerical results in their one-dimensional (1D) model. Moreover, 1D model could not present the detailed distribution of hydraulic and thermal parameters at pipe cross-section.

This chapter mainly introduces the different turbulence models, the interface shape model, and the phase transition model in the process of gas-liquid two-phase stratified flow in the horizontal pipeline. The turbulence model mainly includes the $k-\varepsilon$ model, the large eddy simulation (LES) model, and the turbulence model under the low Reynolds number condition at the near wall surface. The interface shape mainly includes the flat interface model, the wavy interface model, the curve interface model, and the wavy-curve interface model. As a result of the phase transition, a varying wavy curve interface will be produced. If the phase transition is large, the varying wavy curve interface may gradually become varying wavy flat interface, and finally lead to the change of flow pattern, no longer the stratified gas-liquid two-phase flow. The phase transformation model mainly includes the empirical formula of water vapor phase transformation, the Peng-Robinson (PR) equation of hydrocarbon mixture, the SRK equation, and the PT flash model of water hydrocarbon impurities. Also, this chapter compares the different research model, enriches the contents of the book, demonstrates the development process of different mathematical models in different research points, and finally introduces the development direction of each point in the future and the relationship between the complex model and research points.

2. Numerical modeling of stratified gas-liquid wavy pipe flow with phase change

Since pressure and temperature drop along the pipeline, the saturated vapor of hydrocarbons would condenses gradually, as shown in **Figure 1**. The condensed liquid would attach to the pipe wall as liquid film or accumulate at the lower part of the pipeline, which could result in continuous change of the liquid level [17, 29, 30].

Thus, some models of stratified vapor-liquid two-phase flow coupled with phase change were proposed. Assumptions could be made as follows: (1) precipitation of condensed liquid is a flash evaporation equilibrium process which occurs in a short moment; (2) regardless of the attachment to inner wall of the pipe, all the condensed liquid accumulates at the bottom of the pipe; (3) two-phase flow in vapor-liquid pipeline has a stratified flow pattern as well as a stable developing flow area in every calculated pipeline segment; (4) the smooth vapor-liquid interface model is adopted to describe the interface shape; (5) the heavy components of hydrocarbon are simplified as pseudo-componentC_7^+; and (6) without considering the effect of gravity on the P–T flash process.

Several kinds of flow patterns are likely to form for vapor/condensation flowing in pipeline: stratified flow, slug flow, annular flow, and stratified-dispersed flow. It is hard to calculate the liquid level, the exact position of liquid film, and the migration patterns of condensed liquid (as shown in **Figure 1**) via one-dimensional model which could not present the detailed

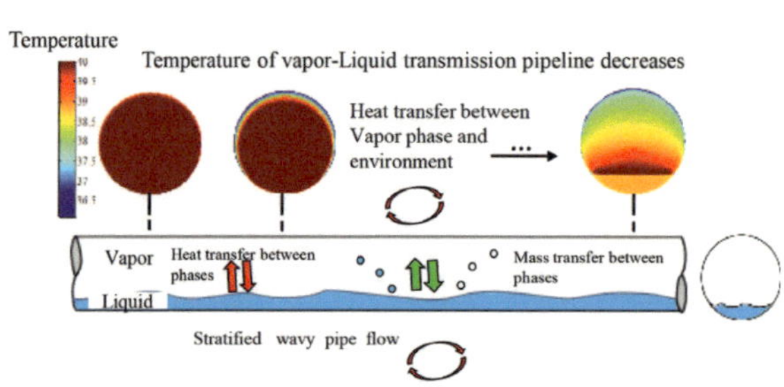

Figure 1. Vapor-liquid two-phase flow coupled with heat and mass transfer.

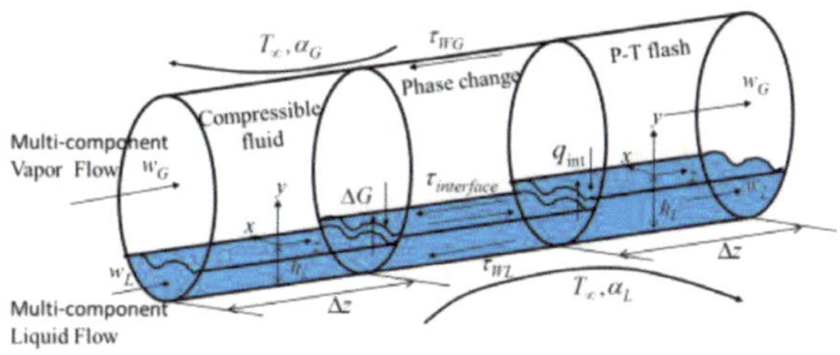

Figure 2. Schematic illustration of stratified vapor–liquid flow with phase change.

distribution of hydraulic and thermal parameters at pipe cross section. Moreover, the 1D model does not consider the turbulent flow which would lead to different numerical results. Hence, the control volume in three-dimensional coordinates is adopted to discretize the calculating area, as shown in **Figure 2**.

In this section, the governing equations based on physical conservation are chosen and established in three-dimensional coordinates, which include mass conservation equation, momentum conservation equation, energy conservation equation, turbulent flow model, and phase change model.

2.1. Mass conservation equation

The total mass flow rate G is the sum of vapor and liquid mass flow rates and keeps constant during the flowing process:

$$G_G + G_L \equiv G \tag{1}$$

The mass transfer rate of phase-change within pipe segment has been taken into consideration due to the vapor phase gradually condensing or the liquid evaporating during the flow process. The change of vapor mass flow rate ΔG_G is equal to opposite number of that of liquid$-\Delta G_L$:

$$\Delta G_G + \Delta G_L = \Delta G \equiv 0 \tag{2}$$

The liquid volume fraction H_L, vapor volume fraction $H_G = 1 - H_L$, pressure, vapor, and liquid velocities as well as temperature are the principal unknowns in the cross-sectional area along the pipeline. The derivation of mass conservation equation is presented here based on two-fluid approach. The mass conservation equations for each phase within the control volume are given as:

$$H_L \rho_L \frac{\partial w_L}{\partial z} + \rho_L w_L \frac{\partial H_L}{\partial z} + H_L w_L \left.\frac{\partial \rho_L}{\partial P}\right|_T \frac{\partial P}{\partial z} + H_L w_L \left.\frac{\partial \rho_L}{\partial T_L}\right|_P \frac{\partial T_L}{\partial z} = \frac{\Delta G_L}{A_{pipe} \Delta z} \tag{3}$$

$$H_G \rho_G \frac{\partial w_G}{\partial z} + \rho_G w_G \frac{\partial H_G}{\partial z} + H_G w_G \left.\frac{\partial \rho_G}{\partial P}\right|_T \frac{\partial P}{\partial z} + H_G w_G \left.\frac{\partial \rho_G}{\partial T_G}\right|_P \frac{\partial T_G}{\partial z} = \frac{\Delta G_G}{A_{pipe} \Delta z} \tag{4}$$

2.2. Momentum conservation equation

The law of momentum conservation is a universal principle for any flow system that the varying rate of momentum is equal to the sum of the forces imposed on the control volume. Considering the compressibility of vapor and liquid, the equation of momentum is as follows:

$$\frac{\partial(\rho_{G,L}w_{G,L})}{\partial t} + \text{div}\left(\rho_{G,L}w_{G,L}\ \overrightarrow{u_{G,L}}\right) = -\frac{\partial p}{\partial z} + \frac{\partial \tau_{xz,G,L}}{\partial x} + \frac{\partial \tau_{yz,G,L}}{\partial y} + \frac{\partial \tau_{zz,G,L}}{\partial z} + F_{G,L} \quad (5)$$

here dp/dz is the pressure gradient in the axial direction, Pa/m. In the condition of steady flow state, the axial pressure gradient dp/dz in liquid is balanced by the shear stress at wall τ_{WL} and interface τ_{int}, respectively, in vapor by τ_{WG} and τ_{int}.

$\tau_{xz,G,L} = \mu_{m,G,L}\left(\frac{\partial u_{G,L}}{\partial z} + \frac{\partial w_{G,L}}{\partial x}\right)$ and $\tau_{yz,G,L} = \mu_{m,G,L}\left(\frac{\partial v_{G,L}}{\partial z} + \frac{\partial w_{G,L}}{\partial y}\right)$ represents viscous stress introduced by molecular viscosity; μ_m denotes molecular dynamic viscosity; $F_{G,L} = \rho_{G,L}\, g \sin\theta$.

According to the theory of CFD, the N-S equation is applicable to any kind of flow. A direct calculation of N-S equation requires a high computer capacity which is not practical in engineering. Hence, an item of *Reynolds* is introduced, and the final momentum equation for vapor and liquid stratified flow is as follows:

$$\frac{\partial}{\partial x}\left(\Gamma_{w,G,L}\left(\frac{\partial w_{G,L}}{\partial x}\right)\right) + \frac{\partial}{\partial y}\left(\Gamma_{w,G,L}\left(\frac{\partial w_{G,L}}{\partial y}\right)\right) = \frac{dp}{dz} - \rho_{G,L} g \sin\theta + \frac{\partial \rho_{G,L} w_{G,L} w_{G,L}}{\partial z} \quad (6)$$

where $\Gamma_{w,G,L}$ means effective diffusion coefficient, which is the effective viscosity given as the sum of the molecular and eddy viscosity, $\Gamma_{w,G,L} = \mu_{m,G,L} + \mu_{t,G,L}$.

2.3. Turbulent flow model

Following a similar approach as Xiao et al. [36] and Reboux et al. [37, 38] for two-phase turbulent flow, the eddy viscosity is modeled using the large eddy simulation (LES) turbulence model based on the assumption of non-isotropic turbulence. Meanwhile, changes are made to take account of the progressive attenuation of turbulence close to the wall. LES model is applicable to the flow with different *Reynolds* number. The subgrid scale viscosity is written as:

$$\mu_{t,G,L} = \rho_{G,L} f_{\text{int},W}(C_S D_S \Delta)^2 |S|_{G,L} \quad (7)$$

The *Smagorinsky* constant is $C_S = 0.1 - 0.2$, while the resolved rate-of-strain tensor is:

$$|S|_{G,L} = \left(2S_{ij}S_{ij}\right)^{1/2}_{G,L} = \left[\left(\frac{\partial w_{G,L}}{\partial x}\right)^2 + \left(\frac{\partial w_{G,L}}{\partial y}\right)^2\right]^{1/2} \quad (8)$$

The filter width Δ is defined as $\Delta = (Vol)^{1/3}$, where *Vol* is the volume of the computational cell.

Damping functions have been introduced into the expressions of turbulent viscosity. Near a wall, a wall-damping function is required to perish the eddy viscosity on the wall.

$$D_S = 1 - \exp\left(\frac{d^+}{25}\right); d^+ = \frac{\left(\rho\tau_{\text{int},W}\right)^{0.5} d}{\mu_m}; d = \min\left(\frac{D}{2} - \sqrt{x^2 + \left(y - \frac{D}{2} + h_L\right)^2}, |d|\right) \tag{9}$$

$$f_{\text{int},W} = 1 - \exp\left[-1.3 \times 10^{-4} d^+ - 3.6 \times 10^{-4}\left(d^+\right)^2 - 1.08 \times 10^{-5}\left(d^+\right)^3\right] \tag{10}$$

And d^+ is the dimensionless distance to pipe wall or vapor-liquid interface. Fulgosi et al. [39] provided an exponential dependence of f_{int} on the dimensionless distance to the interface or pipe wall. Parameter d means the distance to pipe wall or vapor-liquid interface.

The choice of turbulence model is crucial for this sort of study, due to the presence of these secondary flows. Nallasamy and Rodi explained that the well-known $k - \varepsilon$ model assumes an isotropic eddy viscosity, which makes it unsuitable for problems where the anisotropic nature of the turbulent viscosity is important in the calculation of the flow field. This may become significant in stratified gas-liquid flow as the height of the interface approaches the center of the pipe, as stressed by Newton and Behnia. FLUENT offers different options when using this model. Based on the current flow conditions, the linear pressure-strain model and standard wall functions were selected. Turbulence model used by different researchers is shown in **Table 1**.

2.4. Heat transfer model

It is of great significance to study turbulent flow and heat-transfer mechanism due to the frequent occurrence in many industrial applications, such as heat exchangers, vapor turbines, cooling systems, and nuclear reactors [27]. In this study, a quasi-steady state of temperature profile is calculated where axial thermal conduction is neglected. Since the operating temperature is influenced by ambient temperature, fluid properties, and hydraulic parameters such as flow rate and liquid level; a heat-transfer model for vapor-liquid flow is established.

Energy equation means the increase of energy, which is equal to the result of heat flux entering the representative-element volume deducting the work from internal force. Ignoring axial heat conduction and viscous dispersive item, the equation of energy conservation is as follows:

$$\frac{\partial}{\partial x}\left(\Gamma_{T,G,L}\left(\frac{\partial T_{G,L}}{\partial x}\right)\right) + \frac{\partial}{\partial y}\left(\Gamma_{T,G,L}\left(\frac{\partial T_{G,L}}{\partial y}\right)\right) = \frac{\partial \rho_{G,L} c_{P,G,L} w_{G,L} T_{G,L}}{\partial z} \tag{11}$$

$\Gamma_{T,G,L}$ means effective diffusion coefficient which is the effective heat transfer coefficient and defined as, $\Gamma_{T,G,L} = \frac{\mu_{m,G,L} c_{P,G,L}}{\Pr_{m,G,L}} + \frac{\mu_{t,G,L} c_{P,G,L}}{\Pr_{t,G,L}} . \frac{\mu_{m,G,L} c_{P,G,L}}{\Pr_{m,G,L}}$ and $\frac{\mu_{t,G,L} c_{P,G,L}}{\Pr_{t,G,L}}$ are respectively molecular thermal diffusion and eddy diffusion of heat transfer. $\Pr_m = c_P \frac{\mu_m}{\lambda}$ is molecular *Prandtl* number. $\Pr_t$ is turbulent *Prandtl* number, which is assumed by Jones and Launder as 0.9 [41]. However, Hishida et al. obtained the conclusion that the value of $\Pr_t$ increases with the distance increase to pipe wall [42]. Only if $d^+ > 30$, $\Pr_t$ is 0.9. When $d^+ = 10$, the value of $\Pr_t$ rises up to 1.6

Formula name	expression
Duan et al. [5]	Low Reynolds number model: $\mu_t = f_\mu C_\mu k^2/\varepsilon$ $\frac{\partial}{\partial x}\left(\left(\mu_m + \frac{\mu_t}{\sigma_k}\right)\frac{\partial k}{\partial x}\right) + \frac{\partial}{\partial y}\left(\left(\mu_m + \frac{\mu_t}{\sigma_k}\right)\frac{\partial k}{\partial y}\right) + G_k - \rho\varepsilon = 0$ $\frac{\partial}{\partial x}\left(\left(\mu_m + \frac{\mu_t}{\sigma_\varepsilon}\right)\frac{\partial \varepsilon}{\partial x}\right) + \frac{\partial}{\partial y}\left(\left(\mu_m + \frac{\mu_t}{\sigma_\varepsilon}\right)\frac{\partial \varepsilon}{\partial y}\right) + C_1 f_1 G_k - C_2 f_2 \frac{\varepsilon}{k} = 0$ $C_1 = 1.92\ C_2 = 1.3\ C_\mu = 0.09 \sigma_k = 1.0\ \sigma_\varepsilon = 1.3$ $f_\mu = \left[1 - \exp\left(-0.0165\mathrm{Re}_y\right)\right]^2 \times \left(1 + \frac{20.5}{\mathrm{Re}_t}\right)$ $f_1 = 1 + \left(\frac{0.05}{f_\mu}\right)^3 f_2 = 1 - \exp\left(-\mathrm{Re}_t^2\right)$
Jiang et al. [40]	LES model: $\mu_{t,G,L} = \rho_{G,L} f_{\mathrm{int},W} (C_S D_S \Delta)^2 \lvert S \rvert_{G,L}$

Table 1. Different turbulent models.

according to the measurement of circular tube of turbulent heat transfer experiments [33]. Another calculation of turbulent *Prandtl* number is given by Kays and Mansoori based on the direct numerical simulation [27]:

$$\mathrm{Pr}_{t,G,L} = \left[0.5882 + 0.228\left(\frac{\mu_{t,G,L}}{\mu_{m,G,L}}\right) - 0.0441\left(\frac{\mu_{t,G,L}}{\mu_{m,G,L}}\right)^2\left(1 - \exp\left(-5.165\right)\left(\frac{\mu_{m,G,L}}{\mu_{t,G,L}}\right)\right)\right]^{-1} \tag{12}$$

2.5. Phase change model

The pressure and temperature drop along pipelines, which incurs the change of ratio of gas mass volume and liquid mass volume. It is prone to evoke P–T flash evaporation in each component. The mass percentage would change accordingly, influencing the parameters like molar mass, density, viscosity, heat capacity and thermal conductivity. The vapor enthalpy would change during the process of phase change and dissipate into the vapor or liquid in the form of phase change heat which results in variation of fluid temperature [22, 32, 45].

Cong Guo et al. considered a modification of the original model, in which the rate of conductive heat through the tube wall due to temperature difference can be calculated [43]. Peneloux et al. proposed the concept of volume translation. They argue that the volume obtained using SRKEOS is a "pseudo-volume" and proposed a method of calculating the "pseudo volume" [44]. Sadegh et al. proposed an equilibrium criterion for the Peng-Robinson equation of state (PREOS) based on the volume translated Peng-Robinson equation of state (VTPREOS) and a translated functional relationship is used based on the theory of Peneloux et al. to discuss the volume translation technique [45]. The study of Li Zhang et al. shows that the condensation heat transfer coefficient reduces with the increase of wall subcooling from around 2 to 14°C. With the rise in the wall subcooling, the heat flux increases, resulting in an increasing rate of steam condensation, which brings forth a thicker condensate film on the tube surface. The thicker condensate film around the tube offers a higher thermal resistance to steam condensation and in turn reduces the condensation heat transfer coefficient [22]. Considering the volume addition of LSI phase due to the coalescence, Kai Yan et al. used the additional velocity and considered all the conditions when some portion of SSI phase can come into LSI phase [46]. Bonizzi proposed a model for calculating the atomization flux and the bulk concentration based on the recommendation by Williams et al. and Pan and Hanratty [47–51]. Vinesh et al. showed the physical model, considered for phase change which corresponds to hydrodynamically as well as thermally developing vapor-liquid stratified flow in a plane channel, with heating from the top and cooling from the bottom wall [52] (**Table 2**).

The equation of state in P–T Flash involves Peng-Robinson (PR) equation, which was proposed by Peng-Robinson in 1976. It can predict the molar volume more accurately than SRK equation and can be applied for polar compounds. Apart from that, the equation is applicable to vapor and liquid at the same time and widely used in the calculation of phase equilibrium.

During the P–T flash calculation process, the required parameters are the component of light hydrocarbon, gasification rate, density, molar mass, and enthalpy. The relation between input and output is explained in **Figures 3** and **4**.

The fluid in the pipe includes *Nc* sorts of hydrocarbon, $i = 1, 2, \ldots, Nc$. The total energy keeps constant during the phase change process. When vaporization occurs, excess heat of vapor

Formula name	Expression
Cong Guo et al. [43]	The rate of conductive heat through the tube wall: $Q_2 = \Delta m \left[C_{p,2-2}\left(T_{v,2-2} - T_{f,2-2}\right) + h_{fg}\right] - C_{p,2-2} m_{1-1}\left(T_{f,2-2} - T_{f,1-1}\right)$ The temperature of the condensate around circumferential wall: $T_f = T_{w,in} + 0.31(T_v - T_{w,in})$
Peneloux et al. [44]	The "pseudo-volume" obtained using SRKEOS: $v_{actual} = v_{SRKEOS} - c$ The definition of "pseudo volume": $\tilde{V} = V + \sum_{i=1}^{n} c_i z_i$
Sadegh et al. [45]	The "pseudo partial volume" can be defined as: $\tilde{v}_i = \left(\partial \frac{\tilde{V}}{\partial z_i}\right)_{T,P,z_j} = v_i + c_i \quad j \neq i; i = 1, n$ With this definition "pseudo fugacity coefficients" can be defined as: $\ln \tilde{\phi}_i = \int_0^P \left(\frac{\tilde{v}_i}{RT} - \frac{1}{P}\right) dP \quad i = 1, n$
Kai Yan et al. [46]	The expression of the additional velocity $\vec{\mathbf{U}}_{Add}$: $\vec{\mathbf{U}}_{Add,s} = \begin{cases} -\alpha_3 \max\left[\left(\vec{\mathbf{U}}_3 - \vec{\mathbf{U}}_m\right)_s, 0\right], & \text{if } \frac{\partial \alpha_2}{\partial s} > 0 \\ -\alpha_3 \min\left[\left(\vec{\mathbf{U}}_3 - \vec{\mathbf{U}}_m\right)_s, 0\right], & \text{if } \frac{\partial \alpha_2}{\partial s} < 0 \end{cases}$
Williams et al. & Pan and Hanratty [47, 48]	The atomization flux: $R_A = \frac{k_A u_G^2 (\rho_G \rho_L)^{1/2}}{\sigma_{GL}} (\Gamma_Z - \Gamma_{Z,C})$ In which $k_A \approx 2.0 \times 10^{-6}$ [8, 9].
Bonizzi et al. [51]	The flux of droplet deposition is usually expressed as: $\langle R_D \rangle = k_D C_B$ $C_B = \frac{EW_L}{Q_G S} = \rho_L \frac{\alpha_D}{\alpha_G}$

Table 2. Different phase change models.

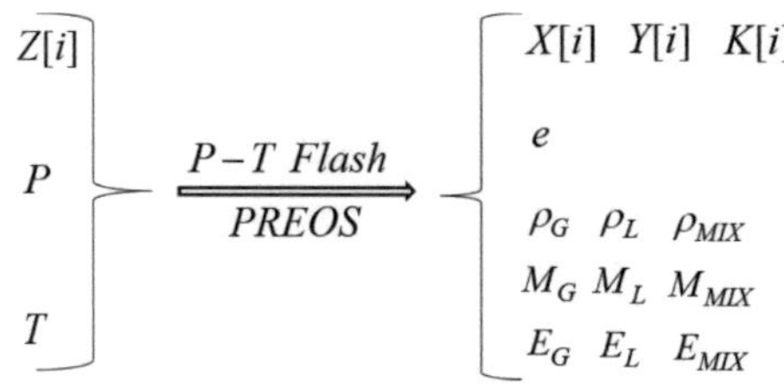

Figure 3. Input and output of P–T flash calculation.

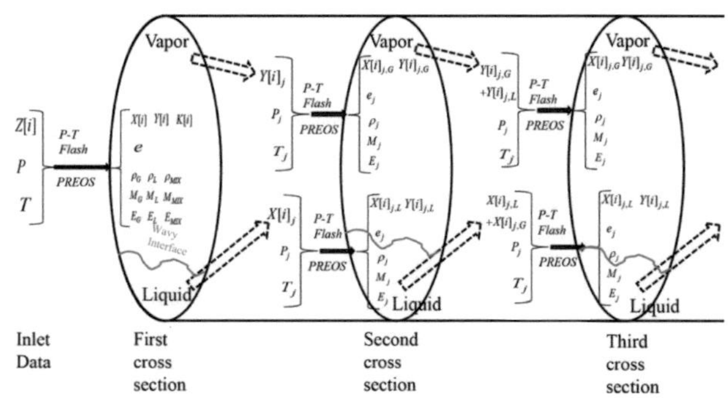

Figure 4. The illustration of the way to obtain the mole fraction by P–T flash in each grid cell.

phase is absorbed from the liquid phase. When condensation occurs, excess heat is released from the vapor phase. L_H represents the latent heat of vaporization or condensation, which is calculated by [32]:

$$L_H = \begin{cases} E_G - E_L \ , & \text{when condensation occurs} \\ E_L - E_G \ , & \text{when vaporization occurs} \end{cases} \tag{13}$$

Therefore, the latent heat may result in the temperature change of the vapor-liquid system. The enthalpy difference in the process of vaporization or condensation can be obtained by virtue of P–T flash calculation. The latent heat helps to retard the temperature change. The process can be calculated as follows:

$$\int_{A_{G,L}} \rho_{G,L} w_{G,L} L_H l_\eta l_\xi d\eta d\xi = c_{P,G} G_G \Delta T_G + c_{P,L} G_L \Delta T_L \tag{14}$$

The parameters of flow state and related dimensionless parameters are given by:

$$G_G = \int_{A_G} \rho_G w_G l_\eta l_\xi d\eta d\xi, \ \ G_L = \int_{A_L} \rho_L w_L l_\eta l_\xi d\eta d\xi \tag{15}$$

The properties of fluids are calculated by:

$$(e, X, Y, \rho_G, \rho_L, \rho_{MIX}, M_G, M_L, M_{MIX}, E_G, E_L, E_{MIX}) = PREOS(Z, P, T) \tag{16}$$

$$\mu_{m,G} = \mu_{m,G}(T_G, \rho_G), \lambda_G = \lambda_G(T_G), c_{P,G} = c_{P,G}(P_G, T_G, M_G) \tag{17}$$

$$\mu_{m,L} = \mu_{m,L}(P_L, T_L, M_L, \rho_L, X), \lambda_L = \lambda_L(\rho_L, T_L), c_{P,L} = c_{P,L}(\rho_L, T_L) \tag{18}$$

Both gas and liquid phases are not ideal fluids. Vapor phase is a mixture of multi-component light hydrocarbons. Thus, the property of vapor phase is a combination of their quality weighting. The PREOS equation is currently acknowledged as the most accurate formula to calculate the density of mixed vapor. If the pressure, temperature, and relative density of light hydrocarbon are known, the viscosity can be calculated using experimental formulas. Both the thermal conductivity and the heat capacity at constant pressure for vapor mixture are related to temperature and pressure, which also can be calculated using the experimental formulas. For the liquid phase, the density is calculated by PR EOS. The other physical properties including viscosity, thermal conductivity, and specific heat capacity are respectively a function of temperature, density, and other critical properties of components.

The relationship between liquid level $h_L$and liquid circulation areaA_L is as follows:

$$A_{pipe}H_L = A_L = \frac{1}{4}D^2\arccos\left(1 - \frac{2h_L}{D}\right) - \frac{1}{4}D^2\left(1 - \frac{2h_L}{D}\right)\sqrt{1 - \left(1 - \frac{2h_L}{D}\right)^2} \tag{19}$$

The value of$H_L$as well as the calculating method differs in each cross section, as depicted in **Figure 6**.

As shown in **Figure 6**, the evaporation fraction, density, and molar weight of vapor and liquid phase can be obtained with the known mole fraction of each component since the temperature and pressure are the same as inlet data in the first cross section. The liquid hold-up can be calculated as follows:

$$H_L = \frac{1}{1 + \frac{\rho_L M_G}{\rho_G M_L} \frac{e}{(1-e)}} \tag{20}$$

The temperature, pressure, and mole fraction vary in the second and other subsequent cross section due to the influence of thermal conduction, pressure drop, and phase change but can be derived from the former section. Hence, it requires the calculation of the evaporation fraction, density, and molar weight in each grid cell. The liquid holdup can be calculated as follows:

$$H_L = \frac{\sum_j \frac{\rho_{MIX,j} M_{L,j}}{M_{MIX,j} \rho_{L,j}} A_j(1 - e_j)}{\sum_j \frac{\rho_{MIX,j}}{M_{MIX,j}} \left[\frac{M_{L,j}}{\rho_{L,j}} A_j(1 - e_j) + \frac{M_{G,j}}{\rho_{G,j}} A_j e_j\right]} \tag{21}$$

Then, the liquid level h_L can be obtained by Eq. (19).

2.6. Wavy gas-liquid interface model

Δy is effective level shift of the gas-liquid interface. It increases the length scale of turbulence, which further increases the shear stress near the rough pipe wall. The shift is caused by the increase of flow resistance by the rough interface, which varies the velocity distribution at the interface.

Dimensionless shift at wavy gas-liquid interface is defined as:

$$\Delta y^+_{\text{int},G,L} = \rho_{G,L}\Delta y_{\text{int},G,L}w_{\tau,\text{int},G,L}/\mu_{G,L} \tag{22}$$

It can be calculated by (Cebeci and Smith, 1974):

$$\Delta y^+_{\text{int},G,L} = 0.9\left(\sqrt{\varepsilon^+_{\text{int},G,L}} - \varepsilon^+_{\text{int},G,L}e^{-\varepsilon^+_{\text{int},G,L}/6}\right), \quad 4.54 < \varepsilon^+_{\text{int},G,L} < 2000 \tag{23}$$

The abovementioned correlation is based on the data fit, the lower limit corresponds to a smooth surface, where $\Delta y^+_{\text{int},G,L} \approx 0$. $\varepsilon^+_{\text{int},G,L}$ is the dimensionless equivalent roughness, expressed by:

$$\varepsilon^+_{\text{int},G,L} = \rho_{G,L}\varepsilon_{\text{int}}w_{\tau,\text{int},G,L}/\mu_{G,L} \tag{24}$$

The ε_{int} and $u_{\tau,\text{int}}$ are calculated by:

$$-3.6\log_{10}\left[\frac{6.9}{\text{Re}_G} + \left(\frac{\varepsilon_{\text{int}}}{3.7D_G}\right)^{1.11}\right] = \sqrt{\frac{1}{f_{\text{int}}}} \tag{25}$$

$$w_{\tau,\text{int},G,L} = \left(\overline{\tau_{\text{int}}}/\rho_{G,L}\right)^{1/2} \tag{26}$$

where $\text{Re}_G = \rho_G u_G D_G/\mu_G, D_G = 4A_G/(S_G + S_{\text{int}}).S_G$ is the gas phase wetted perimeter length; m.f_{int} and $\overline{\tau_{\text{int}}}$ are calculated as:

$$f_{\text{int}} = \frac{2\overline{\tau_{\text{int}}}}{\rho_G(\overline{w_G} - \overline{w_L})^2} \tag{27}$$

$$\overline{\tau_{\text{int}}} = \overline{\tau_{\text{int},G,L}} = \frac{1}{a}\int_0^a \Gamma_{w,\text{int},G,L}\frac{l_\eta}{l_\xi}\frac{\partial w_{G,L}}{\partial \xi}\bigg|_{\xi=\pi} d\eta \tag{28}$$

The flow geometry of stratified flow in circular receiver is very complex. In order to simplify the model, many researchers use different models in their researches. The following are some models.

2.7. Boundary conditions

The boundary conditions included in vapor-liquid two-phase stratified pipe flow with heat transfer and phase change involve vapor-liquid interface condition, wall boundary condition, symmetrical boundary condition and inlet boundary condition. As for the vapor-liquid interface, in order to illustrate the mutual influences among flow, heat transfer and phase change,

<table>
<tr><th>Formula name</th><th>expression</th></tr>
<tr><td>Duan et al. [5]
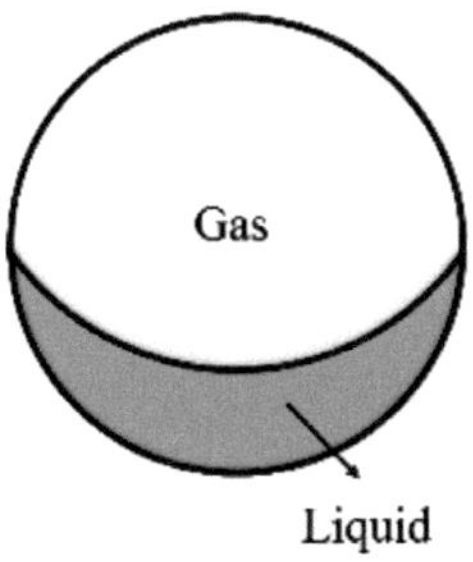
</td><td>$$\frac{\Delta E}{L}=\frac{1}{L}\Delta(E_P+E_S)=R^3\rho_{Oil}g\left(1-\frac{\rho_{Gas}}{\rho_{Oil}}\right)$$ $$\left(\begin{array}{l}\left(\frac{\sin^3\theta_0}{\sin^2\theta^*}(\arctan\theta^*-\arctan\theta_0)(\pi-\theta^*+0.5\sin(2\theta^*))+\frac{2}{3}\sin^3\theta_0^P\right)\\ +\frac{2}{Bo}\left(\sin\theta_0\frac{\pi-\theta^*}{\sin\theta^*}-\sin\theta_0^P+\cos\alpha\left(\theta_0^P-\theta_0\right)\right)\end{array}\right);$$ $$B_o=\frac{\Delta\rho g R^2}{\sigma};\theta_0^P=\cos^{-1}\left(1-\frac{2h_L}{R}\right);H_L=\frac{\theta_0^P-\sin\theta_0^P}{2\pi};\theta_0\text{-}\sin\theta_0\cos\theta_0=H_L\pi$$</td></tr>
<tr><td>Newton and Behnia [20]
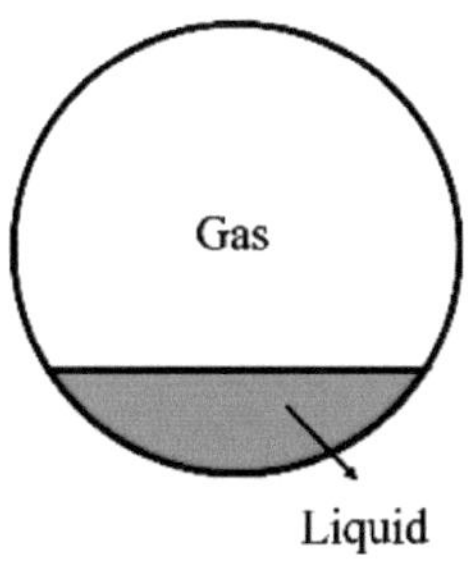
</td><td>The noncircular liquid and gas domains in stratified pipe flow are conveniently modeled with the bipolar coordinate system described analytically by:
$x=\frac{c\cdot\sinh(r)}{\cosh(r)-\cos(\theta)},\ y=\frac{c\cdot\sinh(\theta)}{\cosh(r)-\cos(\theta)};\ \begin{array}{l}\gamma<\theta<\pi,\ -\infty<r<+\infty\\ \pi<\theta<\pi+\gamma,\ -\infty<r<+\infty\end{array};$
where γ is equal to half the angle subtended by the center of the pipe and the gas-liquid interface and is given by:
$\gamma=\cos^{-1}\left(1\text{-}\frac{2h_L}{D}\right)$</td></tr>
<tr><td>Badie et al. [53]
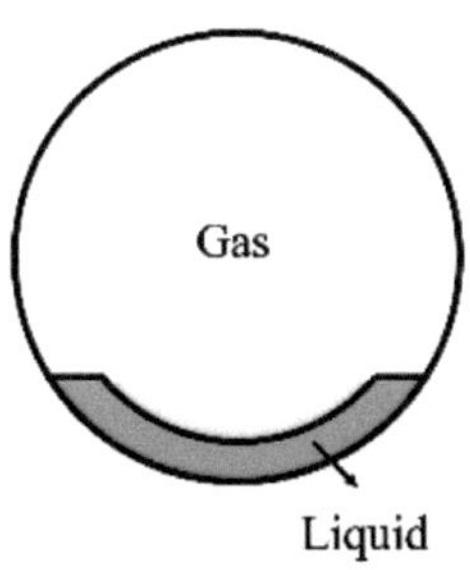
</td><td>The apparent rough surface (ARS) model:
$\frac{\varepsilon_L}{1\text{-}\varepsilon_L}=\frac{u_{L,S}}{u_{G,S}}\left(1+\left(\frac{f_L\rho_L}{f_i\rho_G}\right)^{\frac{1}{2}}\right)$
$\varepsilon_L\leq 0.06$
$\frac{f_L}{f_i}=108\mathrm{Re}_{SL}^{-0.726}$</td></tr>
<tr><td>Badie et al. [53]
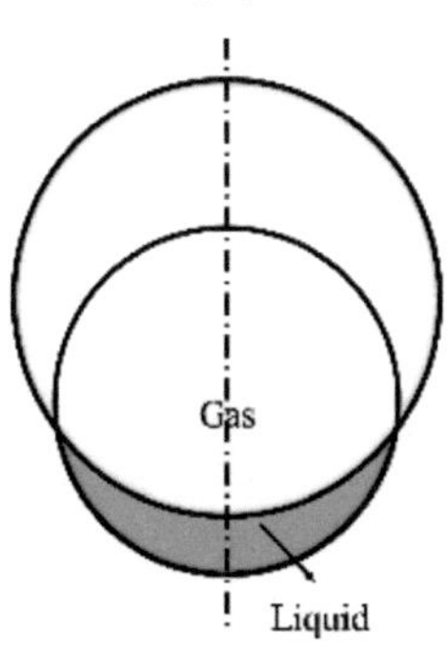
</td><td>The double-circle model:
$P_G=(\pi-\theta)D;\ P_L=\theta D;\ P_i=\theta_i D_i;$
$S_G=(1-\varepsilon_L)\frac{\pi D^2}{4};\ S_L=\varepsilon_L\frac{\pi D^2}{4};$
$D_i=D\frac{\sin\theta}{\sin\theta_i};\ \theta_i=\left(\frac{\sin\theta}{\sin\theta_i}\right)^2\left(\theta+\frac{\sin^2\theta}{\tan\theta_i}-\frac{\sin 2\theta}{2}-\pi\varepsilon_L\right)$</td></tr>
</table>

Table 3. Different gas-liquid interface models.

the equal interfacial shear stress has been prescribed as the vapor-liquid interface condition which is related to the fluid properties and velocity distribution and can be calculated as:

$$\tau_{\text{int},G} = (\mu_m + \mu_t)_G \frac{\partial w_G}{\partial n}\bigg|_{\text{int}} = (\mu_m + \mu_t)_L \frac{\partial w_L}{\partial n}\bigg|_{\text{int}} = \tau_{\text{int},L} \tag{29}$$

The temperature and heat flux of two phases are respectively equal at vapor liquid interface (**Table 3**).

$$q_{\text{int},G} = q_{\text{int},L}, T_{\text{int},G} = T_{\text{int},L} \tag{30}$$

At the pipe wall, the non-slip condition is applied for velocity in both two phases.

$$w_{W,G,L} = 0 \tag{31}$$

The temperature boundary condition at pipe wall of vapor and liquid phases are convective heat transfer and its coefficient of pipeline outer wall remains constant.

$$\alpha_{G,L}(T_W - T_{G,L}) = \lambda_{G,L} \frac{dT_{G,L}}{dn}\bigg|_W \tag{32}$$

The gradients of velocity and temperature are zero at the symmetrical boundary.

$$\frac{\partial w_{G,L}}{\partial n}\bigg|_{symmetrical} = 0, \quad \frac{\partial T_{G,L}}{\partial n}\bigg|_{symmetrical} = 0 \tag{33}$$

At the inlet of pipeline, the velocity and temperature filed of two phases are respectively equal to the pipe inlet values.

$$w_{G,L} = w_{inlet,G,L}, T_{G,L} = T_{inlet} \tag{34}$$

3. Results and discussion

Based on the theory of flow and heat transfer, turbulent flow and phase equilibrium, the model is solved by multi-physical field coupling numerical simulation. The non-circular liquid and vapor domains in stratified pipe flow can be simply modeled with the bipolar coordinate system, which is helpful in solving the problem caused by the inhomogeneity of boundaries. Bipolar cylindrical coordinate is composed of two orthogonal circles in rectangular coordinate. As the flow field in both phases is bounded by a circular pipe wall and a plane interface, the calculation domain has been converted to rectangle form from the anomalous physical domain by adopting the bipolar coordinate system.

With the increase of axial distance, the liquid level in pipeline changes constantly and leads to the change of flow area in both phases. The grid size changes adaptively along with the flow area, where the flow area is determined by the height of gas-liquid interface.

Compound	CH_4	C_2H_6	C_3H_8	i-C_4H_{10}	n-C_4H_{10}	i-C_5H_{12}	n-C_5H_{12}	n-C_6H_{14}	C_7^+
Mole percent (%)	78.03	4.73	5.98	3.05	3.54	2.85	0.54	0.69	0.59

Table 4. Chemical composition of the light hydrocarbons used in current study.

Variable-size grid has advantages in calculating the changing interface. The grid number remains unchangeable. The location of the gas-liquid interface is obtained by the secant method, and the convergence condition is that the conservation of the mass flow rate of the gas-liquid phase and the total mass flow rate equal to the inlet mass flow rate. In this way, the interface is detected.

Vapor-liquid two phase flow and heat transfer coupled with phase change have been simulated in this section. The simulated pipeline is with inner diameter of 100 mm and total length of 6000 m. The superficial velocities of vapor and liquid are respectively $U_{SG} = 6.223$ m/s and $U_{SL} = 0.016$ m/s. The pressure gradient is about $dp/dz = -21$ Pa/m and the liquid holdup at pipe inlet is $H_L = 0.0343$. The mass flow rate of the fluid at pipe inlet is 5.892 kg/s. The fluid of vapor-liquid mixture at pipe inlet also has a pressure of 10.343 MPa and a temperature of 48°C. The convective heat transfer coefficient is about 10 $W/(m^2 \cdot K)$ and the ambient temperature is 15°C. The simulated pipeline is divided into 600 segments where the cross-sectional area of each segment has a mesh grid scale of 84 × 122 for numerical calculation. The fluid is a mixture of multi-component hydrocarbons, which are shown in **Table 4**.

3.1. Mole fraction, density distribution, and liquid level along the pipeline

The mole fractions of each component, density distribution, temperature distribution, and liquid level along the pipeline are obtained in the condition of condensation production.

Mole fractions of each component at pipe inlet in both phases are shown in **Table 5**. In vapor phase, the mole fraction of methane is larger than all the other light hydrocarbons (C_2^+). In liquid phase, the mole fractions of the other light hydrocarbons are bigger than that in vapor phase, but the methane is still the main component.

Mole fractions of each component in both vapor and liquid phase are shown in **Figure 5**. The content of methane in vapor phase increases when flowing in the pipe while the content of the other light hydrocarbons become less and less in vapor phase. During the condensing process which is dominated by temperature drop, the methane keeps evaporating; while during the evaporating process which is dominated by pressure drop, the other light hydrocarbons keep condensing. Meanwhile, the bigger the molar mass is, the faster the condensing rate is.

Compound	CH_4	C_2H_6	C_3H_8	i-C_4H_{10}	n-C_4H_{10}	i-C_5H_{12}	n-C_5H_{12}	n-C_6H_{14}	C_7^+	Total
Mole fraction in vapor phase (%)	80.54	4.69	5.61	2.71	3.05	2.25	0.41	0.44	0.30	100
Mole fraction in liquid phase (%)	43.31	5.35	11.05	7.82	10.30	11.21	2.32	4.10	4.54	100

Table 5. Mole fraction of each component at pipe inlet in both phases.

Six pipe cross sections located at every 1000 m along the pipeline are selected to illustrate the change of density and temperature distribution, as shown in **Figures 6** and **7**. It is exactly because the pressure on the cross section is the same, so the uneven distribution of the temperature leads to the uneven distribution of the fluid density. In the two phases, the density distribution is opposite to the temperature distribution. The high temperature means low density. The temperature value distributed at every single cross section can be ranked in

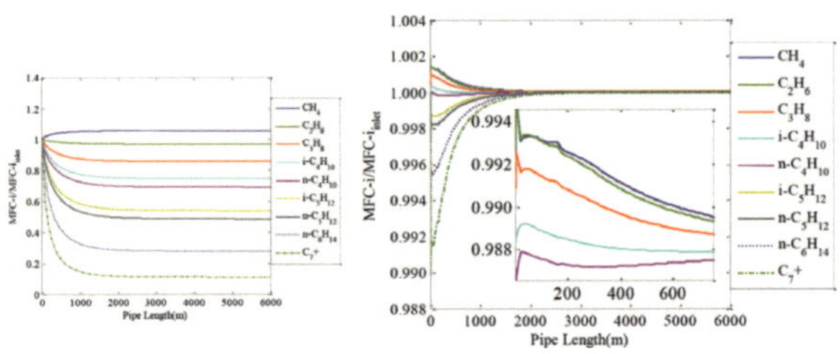

Figure 5. Mole fraction of each component changes along pipeline. (left) vapor phase; (right) liquid phase.

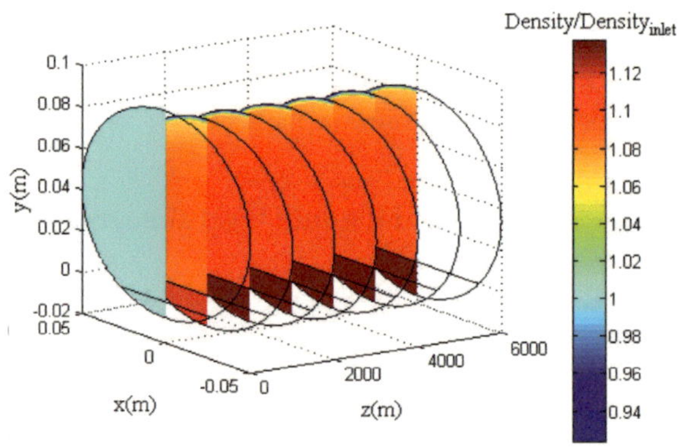

Figure 6. Density (dimensionless) distribution depicted in 6 pipe cross sections along the pipeline (every 1000 m).

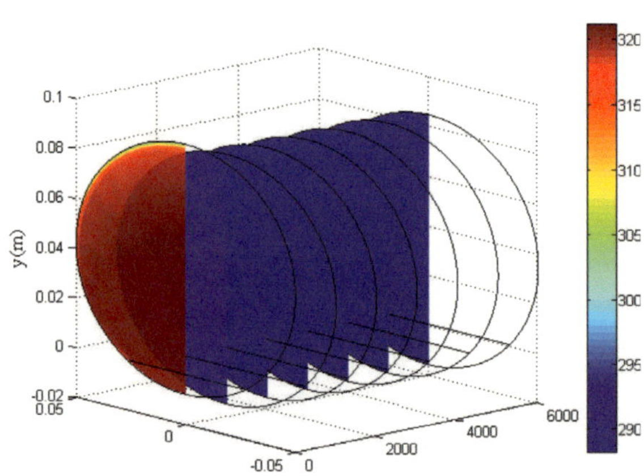

Figure 7. Temperature (K) depicted in 6 pipe cross sections along the pipeline (every 1000 m).

descending order: the interior of liquid phase, most parts in vapor phase, vapor phase near the top wall. However, the descending rank of density is liquid phase, vapor phase near the top wall, and the interior of vapor phase. Along the pipeline, the temperature in the area referenced above decreases gradually while the density increases gradually. Therefore, the temperature of the sequential cross sections tends to be the same and the density distribution within the two phases gradually becomes uniform.

Figure 8 illustrates the selected 12 pipe cross sections located at every 500 m along the pipeline, where the varying trend of liquid level are presented in three-dimensional coordinate system. The minimum liquid level is 7.82 mm at pipeline inlet. The liquid level at outlet is about 16.18 mm and keeps declining trend, which can also be found in **Figure 9(b)**.

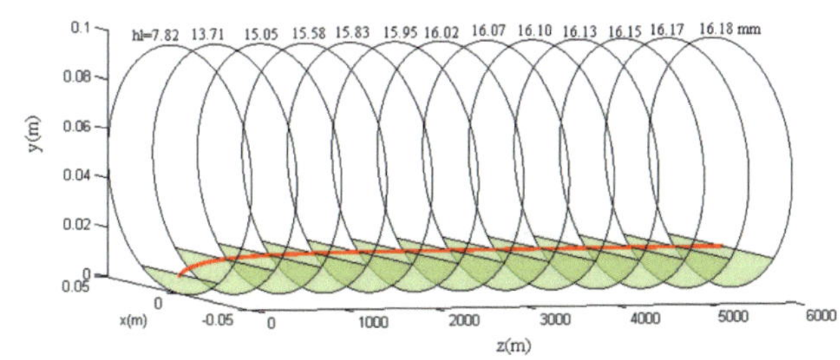

Figure 8. Liquid level depicted in 12 pipe cross sections along the pipeline (every 500 m).

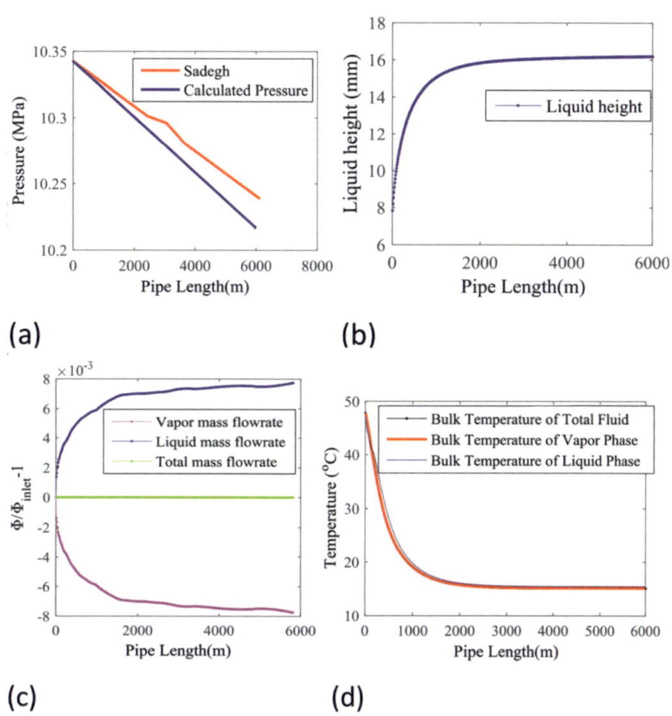

Figure 9. Pressure gradient, liquid level, fluid mass flow rate, and temperature along the pipeline. (a) pressure gradient, (b) liquid level, (c) mass flow rate, and (d) bulk temperature.

3.2. Pressure gradient, liquid level, fluid mass flow rate, and temperature along the pipeline

The pressure gradient, liquid level, fluid mass flow rate, and temperature along the pipeline obtained in the condition of condensation production have been compared with that found in literature.

When the phase change behavior is considered along pipe flow, the vapor in gas phase starts to condense to liquid which begins from the pipe inlet due to the significant temperature drop at pipe wall. The pressure drop fit well with the simulated results presented by Sadegh, as shown in **Figure 9(a)** [16].

During the condensing process, vapor mass flow rate gradually reduces, as shown in **Figure 9(c)**, and the liquid mass flow rate increases due to the constant total mass flow rate. The increase of liquid mass flow rate leads to further rise of liquid level, as shown in **Figure 9(b)**.

The liquid holdup firstly increases until it reaches a maximum value and then gradually decreases. The reason behind this is as follows: The increase of liquid holdup results from the liquid precipitation caused by dominant temperature drop. Due to the large difference between fluid temperature and ambient temperature, the amount of liquid precipitation is greater than liquid evaporation. On the contrary, the decrease of liquid holdup is leaded by liquid evaporation due to dominant pressure drop. Being same to liquid holdup, the liquid mass flow rate maintains the same trend, that is, gradually increasing to reach a maximum value and then gradually reduced, which is depicted in **Figure 9(c)**. As the total mass flow is constant, the mass flow rate of the vapor phase decreases first and then increases. When compared with the process of evaporation, the precipitation process caused by temperature drop is transient and intense, which is related to the temperature difference between the inside and outside of the pipeline and to the convective heat transfer coefficient.

The tendency of temperature drop is similar to that in the existing research [16]. But there also exists difference between vapor bulk temperature, liquid bulk temperature, and the total bulk temperature, which cannot be revealed by one-dimensional model. The liquid bulk temperature is always higher than the vapor bulk temperature while the vapor bulk temperature is almost always equal to the total bulk temperature. Latent heat is revealed during the vapor condensing process which slows down the temperature drop, as shown in **Figure 9(d)**.

3.3. Velocity and temperature distribution at pipe length of 3000 m

Through solving the model, with phase change happening, it can be obtained that the pressure gradient is 21.29 Pa/m and the liquid level is 16.02 mm when axial distance reaches 3000 m.

Figure 10(a) shows that the velocity of vapor phase slows down while approaching either the pipe wall or the vapor-liquid interface because of the hindering effects and fluid viscosity. The velocity of liquid phase keeps increasing from pipe wall to the interface. The maximum velocity at the pipe cross section occurs within the vapor phase.

In **Figure 10(b)** shows that the temperatures of both vapor and liquid phase drop while approaching the pipe wall because of the lowest ambient temperature and the convective heat transfer effects. The temperature of liquid phase keeps increasing from pipe wall to the interface. The maximum temperature at the pipe cross section exists within the liquid phase near the interface.

Figure 10(c) shows that the temperature at pipe wall of vapor phase is lower than that of liquid phase when convective heat transfer exists due to the smaller heat carried by the vapor phase than liquid phase. Thus, lower specific heat capacity results in bigger temperature drop at pipe wall of vapor phase. The thermal conductivity of the liquid phase is greater than vapor phase, hence, the temperature gradient in liquid phase is smaller than that in vapor phase, and the bulk average temperature of the liquid phase is higher than the vapor phase. Heat is transferring from

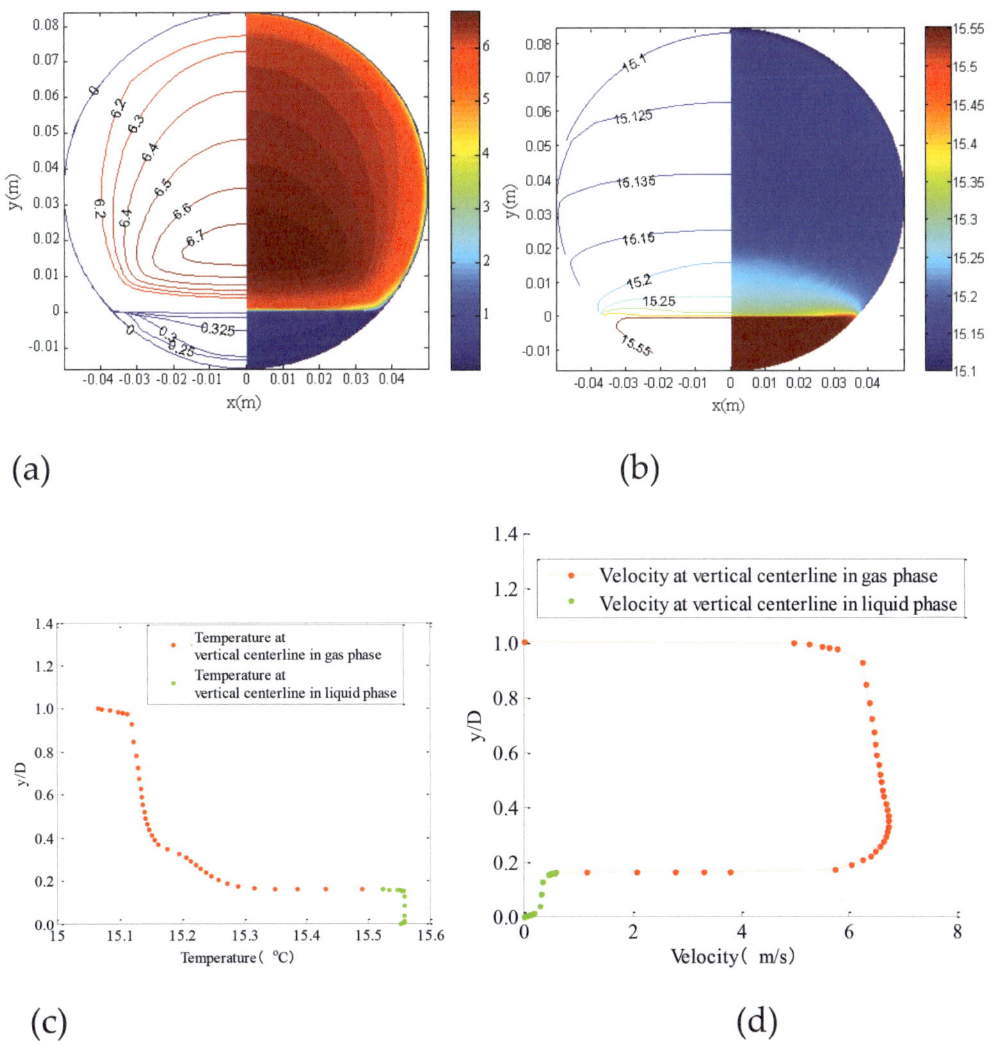

Figure 10. Velocity and temperature distribution at pipe length of 3000 m. (a) Velocity profile at pipe cross section; (b) temperature profile at pipe cross section; (c) velocity profile at vertical centerline; and (d) temperature profile at vertical centerline.

liquid phase to vapor phase through the interface, which makes the temperature drop of the liquid phase and reduces the temperature difference between the two phases.

Figure 10(d) reveals the velocity distribution at the centerline of the pipe. By the dragging force of the interface, the velocity of liquid phase reaches the maximum value at the interface while the velocity of vapor phase reaches the maximum value at the location between the interface and its bulk center. The liquid phase slows down while approaching the pipe wall because of the hindering of the pipe wall and its high viscosity.

4. Conclusion

The vapor-liquid two-phase pipe flow and heat transfer are studied by virtue of numerical simulation in light hydrocarbon transportation pipeline coupled with hydraulics, thermodynamics, and phase change. A three-dimensional non-isothermal vapor-liquid stratified flow model including phase change model in bipolar coordinate system has been established, where LES turbulence model is utilized to simulate the turbulence flow and the wall attenuation function is used to describe the inadequacy performance of vapor-liquid interface. The vapor phase and the liquid phase are both considered to be compressible and the PR equation of state is chosen for the vapor-liquid equilibrium calculation where the multi-component hydrocarbon flash calculation is used to evaluate the physical properties, gasification rate, and enthalpy departure of the phases. The P–T flash calculation has been applied to predict the varying liquid level and the multi-component mass fraction in each phase during the process of vapor/liquid stratified pipe flow. The axial pressure gradient, liquid holdup, velocity, and temperature fields have been presented. The fluid mass flow rate, mole fraction, density distribution, and liquid level along the pipeline are also given out.

The simulation results indicate that the influence of pressure and temperature on liquid holdup is different. During the light hydrocarbon transportation process in pipeline, the temperature drop leads to the reduction of vapor mass flow rate and the rise of liquid level as well as mass flow rate. Larger temperature drop results in bigger liquid holdup while larger pressure drop causes smaller liquid holdup due to the change of physical properties and phase equilibrium. After the increase of liquid holdup caused by dominant temperature drop reaching the maximum value, then the decrease of liquid holdup maintains its trend till the pipe outlet, which results from liquid evaporation due to dominant pressure drop.

The highest velocity locates in vapor phase while the highest temperature locates in liquid phase. The liquid bulk temperature is always higher than the vapor bulk temperature. The vapor bulk temperature is almost always equal to the total bulk temperature, which cannot be revealed by one-dimensional model. Latent heat is revealed during the vapor condensing process which slows down the temperature drop. The average velocity of liquid is lower than that of vapor, but the temperature of liquid is higher than vapor.

When the fluid flows in the pipeline, the content of methane in vapor phase increases all the time while the content of the other light hydrocarbons (C_2^+) become less and less.

The condensing rates of the other light hydrocarbons have a positive correlation with their molar mass. The temperature value distributed at every single cross section can be ranked in descending order: the interior of liquid phase, most parts in vapor phase, vapor phase near the top wall. However, the descending rank of density is liquid phase, vapor phase near the top wall, and the interior of vapor phase. Along the pipeline, the temperature in the area referenced above decreases gradually while the density increases gradually. Therefore, the temperature of the sequential cross-sections tends to be the same and the density distribution within the two phases gradually becomes uniform. As for the varying trend of liquid level presented in three-dimensional coordinate system, it has the same trend of the liquid holdup which firstly increases until it reaches a maximum value and then gradually decreases.

Thus, models in this chapter can be utilized to accurately predict pressure gradient, velocity, temperature field, liquid holdup, fluid physical properties, and mole fraction, which are essential to the determination of pipe size, design of downstream equipment, and guarantee of flow assurance.

Acknowledgements

This work was supported by the National Natural Science Foundation of China [Grant number 51474228]; and the Beijing Scientific Research and Graduate Joint Training Program [Grant number ZX20150440].

Author details

Guoxi He*, Yansong Li, Baoying Wang, Mohan Lin and Yongtu Liang

*Address all correspondence to: heguoxicup@163.com

Beijing Key Laboratory of Urban Oil and Gas Distribution Technology, China University of Petroleum-Beijing, Beijing, P.R. China

References

[1] Akansu SO. Heat transfers and pressure drops for porous-ring turbulators in a circular pipe. Applied Energy. 2006;**83**:280-298

[2] Siavashi M, Bahrami HRT, Saffari H. Numerical investigation of flow characteristics, heat transfer and entropy generation of nanofluid flow inside an annular pipe partially or completely filled with porous media using two-phase mixture model. Energy. 2015;**93**: 2451-2466

[3] Revellin R, Lips S, Khandekar S. Local entropy generation for saturated two-phase flow. Energy. 2009;**34**(9):1113-1121

[4] Ferreira RB, Falcão DS, Oliveira VB. Numerical simulations of two-phase flow in an anode gas channel of a proton exchange membrane fuel cell. Energy. 2015;**82**:619-628

[5] Duan J, Liu H, Gong J, Jiao G: Heat transfer for fully developed stratified wavy gas–liquid two-phase flow in a circular cross-section receiver. Solar Energy. 2015;**118**: 338-349

[6] Lun I, Calay RK, Holdo AE. Modeling two-phase flows using CFD. Applied Energy. 1996; **53**:299-314

[7] Oliemans RVA. Modeling of gas-condensate flow in horizontal and inclined pipes. In: Proc. of the ASME Pipeline Engineering Symposium-ETCE Dallas; 1987

[8] Adewumi MA, Nor-Azlan N, Tian S. Design approach accounts for condensate in gas pipelines. SPE Eastern Regional Meeting, Society of Petroleum Engineers. 1993

[9] Zhou J, Adewumi MA. Transients in gas-condensate natural gas pipelines. Journal of Energy Resource Technology. 1998;**120**:32-40

[10] Schouten JA, Janssen-van Rosmalen R, Michels JPJ. Condensation in gas transmission pipelines. International Journal of Hydrogen Energy. 2005;**30**:661-668

[11] Jin T. Network modeling and prediction of retrograde gas behavior in natural gas pipeline systems [thesis]. Doctoral dissertation. The Pennsylvania State University; 2013

[12] Deng D, Gong J. Prediction of transient behaviors of gas-condensate twophase flow in pipelines with low liquid loading. In: 2006 International Pipeline Conference. American Society of Mechanical Engineers; 2006

[13] Vincent PA, Adewumi MA. Engineering design of gas-condensate pipelines with a compositional hydrodynamic model. SPE Production Engineers. 1990:5381-5386

[14] Mucharam L, Adewumi MA, Watson RW. Study of gas condensation in transmission pipelines with a hydrodynamic model. SPE Production Engineers. 1990:5236-5242

[15] Sadegh AA, Adewumi MA. Temperature distribution in natural gas/condensate pipelines using a hydrodynamic model. SPE Eastern Regional Meeting; Society of Petroleum Engineers. 2005

[16] Abbaspour M, Chapman KS, Glasgow LA. Transient modeling of nonisothermal, dispersed two-phase flow in natural gas pipelines. Applied Mathematical Modelling. 2010; **34**:495-507

[17] Dukhovnaya Y, Adewumi MA. Simulation of non-isothermal transients in gas/condensate pipelines using TVD scheme. Powder Technology. 2000;**112**:163-171

[18] Almanza R, Lentz A, Jimeenez G. Receiver behavior in direct steam generation with parabolic troughs. Solar Energy. 1997;**61**:275-278

[19] Newton CH, Behnia M. A numerical model of stratified wavy gas–liquid pipe flow. Chemical Engineering Science. 2001;**56**:6851-6861

[20] Newton CH, Behnia M. Numerical calculation of turbulent stratified gas–liquid pipe flows. International Journal of Multiphase Flow. 2000;**26**:327-337

[21] Berthelsen PA, Ytrehus T. Numerical modeling of stratified turbulent two- and three-phase pipe flow with arbitrary shaped interfaces. In: The 5th International Conference on Multiphase Flow; 30 May–4 June, 2004; Yokohama, Japan

[22] Zhang L, Yang S, Xu H: Experimental study on condensation heat transfer characteristics of steam on horizontal twisted elliptical tubes. Applied Energy. 2012;**97**:881-887

[23] Gada VH, Datta D, Sharm A. Analytical and numerical study for two-phase stratified-flow in a plane channel subjected to different thermal boundary conditions. International Journal of Thermal Science. 2013;**71**:88-102

[24] Manabe R. A comprehensive mechanic heat transfer model for two-phase flow with high pressure flow pattern validation [thesis]. Department of Petroleum Engineering: University of Tulsa; 2001

[25] Fontoura VR, Matos EM, Nunhez JR. A three-dimensional two-phase flow model with phase change inside a tube of petrochemical pre-heaters. Fuel. 2013;**110**:196-203

[26] Mansoori Z, Yoosefabadi ZT, Saffar-Avval M. Two dimensional hydro dynamic and thermal modeling of a turbulent two phase stratified gas–liquid pipe flow. In: ASME 2009 Fluids Engineering Division Summer Meeting. American Society of Mechanical Engineers; 2009. p. 753-758

[27] Singh AK, Goerke UJ, Kolditz O. Numerical simulation of non-isothermal compositional gas flow: application to carbon dioxide injection into gas reservoirs. Energy. 2011;**36**:3446-3458

[28] Ebadian MA, Vafai K, Lavine A. Single and multiphase convective heat transfer. Applied Energy. 1992;**43**:291-292

[29] Gong G, Chen F, Su H, Zhou J. Thermodynamic simulation of condensation heat recovery characteristics of a single stage centrifugal chiller in a hotel. Applied Energy. 2012;**91**: 326-333

[30] Hu H, Zhang C. A modified k–e turbulence model for the simulation of two-phase flow and heat transfer in condensers. International Journal of Heat and Mass Transfer. 2007;**50**: 1641-1648

[31] Sarica C, Panacharoensawad E. Review of paraffin deposition research under multiphase flow conditions. Energy Fuel. 2012;**26**:3968-3978

[32] Duan J, Gong J, Yao H. Numerical modeling for stratified gas–liquid flow and heat transfer in pipeline. Applied Energy. 2014;**115**:83-94

[33] Ullmann A, Brauner N. Closure relations for two-fluid models for two-phase stratified smooth and stratified wavy flows. International Journal of Multiphase Flow. 2006;**32**(1): 82-105

[34] Vincent PA, Adewumi MA. Engineering design of gas-condensate pipelines with a compositional hydrodynamic model. SPE Production Engineers. 1990;**5**(04):381-386

[35] Haaland SE. Simple and explicit formulas for the friction factor in turbulent pipe flow. Fluids Engineering. 1983;**105**:89-90

[36] Helgans B, Richter DH. Turbulent latent and sensible heat flux in the presence of evaporative droplets. International Journal of Multiphase Flow. 2016;**78**:1-11

[37] Vargaftik NB, editor. Handbook of Physical Properties of Liquids and Gases-pure Substances and Mixtures. Hemisphere Pub; 1975

[38] DenHerder T. Design and simulation of PV super system using simulink [thesis]. San Luis Obispo: California Polytechnic State University; 2006

[39] Zaghloul JS. Multiphase analysis of three-phase (gas-condensate-water) flow in pipes (Doctoral dissertation [thesis]). Pennsylvania State University; 2006

[40] Jiang X, Siamas GA, Jagus K, et al. Physical modeling and advanced simulations of gas–liquid two-phase jet flows in atomization and sprays. Progress in Energy & Combustion Science. 2010;**36**(2):131-167

[41] Jones WP, Launder BE. The calculation of low-Reynolds-number phenomena with a two-equation model of turbulence. International Journal of Heat and Mass Transfer. 1973; **16**(6):1119-1130

[42] Hishida M, Nagano Y, Tagawa M. Transport processes of heat and momentum in the wall region of turbulent pipe flow. In: Proceedings of the 8th International Heat Transfer Conference. Hemisphere Publishing Corp; 1986;Washington, DC; 1986. p. 925-930

[43] Guo C, Wang T, Hu X, et al. Experimental investigation of the effects of heat transport pipeline configurations on the performance of a passive phase-change cooling system. Experimental Thermal and Fluid Science. 2014;**55**:21-28

[44] Péneloux A, Rauzy E, Fréze R. A consistent correction for Redlich-Kwong-Soave volumes. Fluid Phase Equilibria. 1982;**8**(1):7-23

[45] Sadegh A A, Adewumi M A. Temperature distribution in natural gas/condensate pipelines using a hydrodynamic model. In: SPE Eastern Regional Meeting. Society of Petroleum Engineers; January, 2005

[46] Yan K, Zhe D. A coupled model for simulation of the gas–liquid two-phase flow with complex flow patterns. International Journal of Multiphase Flow. 2010;**36**(4):333-348

[47] Williams LR, Dykhno LA, Hanratty TJ. Droplet flux distributions and entrainment in horizontal gas–liquid flows. International journal of multiphase flow. 1996;**22**(1):1-18

[48] Pan L, Hanratty TJ. Correlation of entrainment for annular flow in horizontal pipes. International Journal of Multiphase Flow. 2002;**28**(3):385-408

[49] Laurinat JE, Hanratty TJ, Jepson WP. Film thickness distribution for gas–liquid annular flow in a horizontal pipe. PhysicoChemical Hydrodynamics. 1985;**6**(1):79-195

[50] Pitton E, Ciandri P, Margarone M, Andreussi P: An experimental study of stratified–dispersed flow in horizontal pipes. International Journal of Multiphase Flow. 2014;**6**: 92-103

[51] Bonizzi M, Andreussi P. Prediction of the liquid film distribution in stratified-dispersed gas–liquid flow. Chemical Engineering Science. 2016;**142**:165-179

[52] Gada VH, Datta D, Sharma A. Analytical and numerical study for two-phase stratified-flow in a plane channel subjected to different thermal boundary conditions. International Journal of Thermal Sciences. 2013;**71**:88-102

[53] Badie S, Hale CP, Lawrence CJ, et al. Pressure gradient and holdup in horizontal two-phase gas–liquid flows with low liquid loading. International Journal of Multiphase Flow. 2000;**26**(9):1525-1543

Chapter 5

Boiling Heat Transfer: Convection Controlled by Nucleation

Irakli Shekriladze

Additional information is available at the end of the chapter

http://dx.doi.org/10.5772/intechopen.74418

Abstract

Due to the peculiar way of evolution of boiling heat transfer research, a model "theater of director" (MTD), pumping effect of growing bubble (PEGB) and MTD-based universal correlation (UC) remain beyond the attention of researchers for more than half a century. In parallel, there are periodic fundamental events, demonstrating the irrationality of such indifference. Since the 1980s, not having found a way to enhance boiling heat transfer, other than that uncovered by the MTD-UC, high-performance boiling surfaces are being developed by artificially increasing effective radius (ER) of nucleation centers (bypassing the reference to the relevant theoretical basis). In 2009, an independent review declares transient conduction and microconvection as the dominant boiling heat transfer mechanism, not knowing that this is just the PEGB. In 2014–2017, the real versatility and accuracy of the UC is confirmed by independent studies, which involve extensive databases on the pool and flow boiling (with some interpretation problems). Assessing the current status of the study, the chapter emphasizes the complete fiasco of traditionally adopted approaches, models and theories, led to the dominance of purely empirical relationships written in a dimensionless form. Heat transfer research community is invited to gain will and rid of the heavy burden of the past.

Keywords: boiling, nucleation, heat transfer, pumping effect, effective radius, correlation

1. Introduction

Importance of boiling heat transfer research is determined by the implementation of this process in the most energy-intense components of technical systems in nuclear and thermal power, space, aviation, cryogenics, refrigeration, chemical and other technologies. The complexity of the boiling phenomenon is due to the combination of turbulence and phase conversion with intricate feedbacks generating complex irregular dissipative structure with various thermo-hydrodynamic effects.

Eventually, all this translates into the exceptional peculiarities of this type of convective heat transfer. Because of this and some other reasons a great number of studies carried out for almost a century after the classic works of Max Jacob and his colleagues [1] have not yet led to generally recognized theory of boiling heat transfer, still remaining as the central unsolved problem of heat transfer theory.

By today's view, boiling heat transfer research has produced a huge agglomeration of experimental facts, analytical and numerical models, and countless helpless correlations, not unified by any single ideology. This agglomeration not only rendered meaningless numerous concrete real scientific achievements but also buried the real boiling fundamentals, the pumping effect of growing bubble (PEGB), the model theater of director (MTD), the universal correlation (UC) capable of serving as a robust framework for solving the problem as a whole.

Of course, understanding of revolutionary new model requires some time. However, half a century of delay with confirmation of the validity of the UC, which requires simple arithmetic calculations, cannot be explained in terms of conventional scientific practice. In this regard, one fact deserves attention. A very short abstract of the most accessible at that time international publication of the MTD-UC [2] claims to submit the UC of developed boiling heat transfer covering all groups of liquids including liquid metals without matching different constants and powers to different surface-liquid combinations.

Boiling heat transfer researchers simply had to check the above claim for a scientific breakthrough. Despite this, during the past almost 40 years, except for self-citations, the article has never been cited by any researcher. The author's systematic calls for a change in irrelevant approaches to the MTD-UC also prove futile. Obviously, such a situation is not an indicator of the overall focus on the effective solution of the boiling problem. Moreover, this clearly indicates a complete suppression of fair scientific debate and competition in the field.

Later, taking into account the above features of the current situation, we pay particular attention to comprehensive examination of the rare belated episodes of de facto validation of the boiling fundamentals by independent studies.

2. The MTD as an alternative to traditional approaches

The uniqueness of boiling heat transfer manifests itself in the independence of the intensity of heat transfer in developed boiling mode from the macro-hydrodynamic parameters of the two-phase medium. Drastic changes of these parameters with a change in the acceleration of gravity by several orders or significant supercooling of the bulk liquid practically do not affect the superheat of the heating surface relative to the saturation temperature.

These features of the boiling phenomenon, paradoxical from the standpoint of traditional concepts of convective heat transfer, were fully uncovered only in the 1960s. Before this, the classics of the theory of boiling heat transfer Jakob [1], Kruzhilin [3], Rohsenow [4], and other researchers have had to develop boiling heat transfer theory based on a traditional approach of convective heat transfer theory connecting heat transfer to the intensity of certain cooling mechanism or combination of certain cooling mechanisms (the MTA).

In parallel, based on an analysis of just the newly uncovered features, in the same 1960s, an alternative model (the MTD) was proposed that highlights the governing role of nucleation in the developed boiling heat transfer [2, 5–11]. Ultimately, the MTD led to adequate description of a vast array of experimental data on developed boiling heat transfer, outlined an effective way to enhance heat transfer successfully implemented since the 1980s in the form of high-performance boiling surfaces.

The principal difference between the MTD and MTA is particularly clearly manifested when comparing the characteristic lengths of the process. In the case of the MTA, this is the parameter of the macro-hydrodynamics of two-phase medium (e.g., the bubble detachment diameter or the internal diameter of the channel), the multiple change of which practically does not affect the superheat of the heating surface relative to the saturation temperature.

In the case of the MTD, this is the average effective radius (ER) of nucleation centers, which just reflects controlling role of nucleation. This linear scale not only contributes to universal description of heat transfer in developed boiling mode regardless of the geometry of the medium and the type of a boiling liquid but also serves as a tool for heat transfer enhancement. The ratio of two different scales (of the order of 10^5–10^6) would seem to emphasize the qualitative gap between the MTD and MTA.

2.1. Pumping effect of growing bubble

A special role in the prediction of the PEGB and development of the MTD was played by the discovery of local temperature pulsations of heating surface [12]. Establishment of coincidence of main cooling effect with onset of bubble growth (**Figure 1**, points **a** and **c**) has led to breakthrough in understanding of boiling phenomenon. The particular surprise was the disclosure of the secondary role of heat removal during the detachment of the bubble and its replacement by liquid mass that was considered as the main cooling effect according to the ideas existing at that time.

Simultaneously, a serious problem arose in terms of reconciling the identified pattern with a fairly firmly established fact of the predominant role of heat removal by the liquid phase, which led to the tendency of unjustified exaggeration of the role of the microlayer evaporation (MLE) [12]. On a qualitative level, a complete clarification of this problem was achieved by predicting the PEGB [5, 6].

The PEGB represents acceleration by vapor bubble of liquid jet at the initial stage of growth normal to the heating surface, accompanied by microcirculation in the boundary layer (**Figure 2**). According to the model [5–7], the PEGB is caused by the abrupt variability of the transverse momentum transfer by evaporation along the surface of the bubble, say, by the abrupt variability of the reactive force applied to the surface of the growing bubble. Let us look at the effect in combination with a typical cycle of local temperature pulsation.

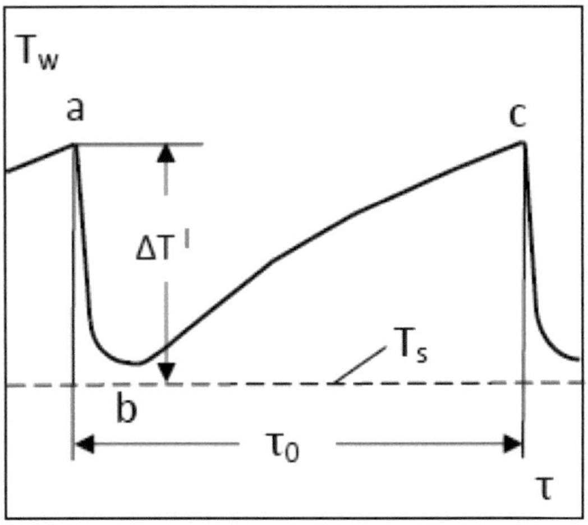

Figure 1. Typical cycle of local temperature pulsation of heating surface [12]: T_w—local superheat of heating surface; T_s—temperature of saturation; $\Delta T'$—maximum superheat; τ—time; τ_o—duration of the pulsation cycle.

After point b (**Figure 1**), the convective cooling effect is greatly reduced and the wall and adjacent liquid begin to warm up, mainly by transient conduction. At point c, wall overheating becomes sufficient to start the growth of the next bubble. The PEGB is again launched with another powerful short-term cooling effect, similar to sections a–b. In terms of collecting heat of the overheated liquid and transporting it first to the growing bubble and then to bulk liquid, the scheme seems almost ideal.

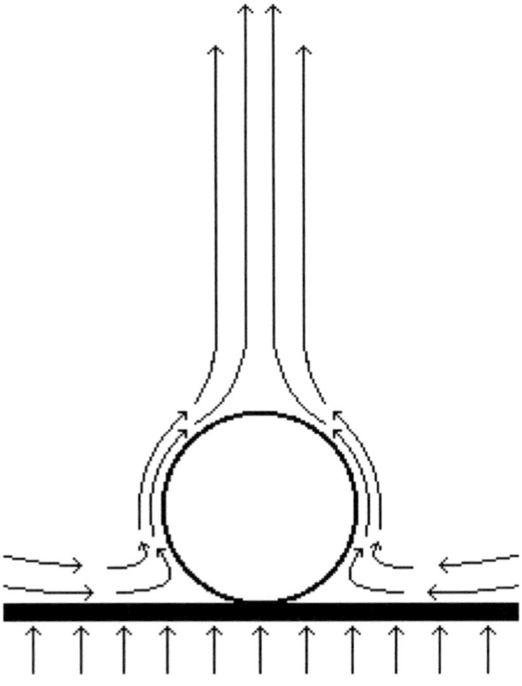

Figure 2. A model of pumping effect of growing bubble (PEGB) [5].

Intensity of the PEGB strongly depends on initial superheat of boiling surface. Therefore PEGB is much more intensive at relatively low pressures, small-sized nucleation sites and high surface tension (e.g. in liquid metals). "Switched on" simultaneously with the onset of bubble growth, the PEGB quickly reduces initial gradient of temperature due to that it arises and "cuts off" itself even if a bubble still remains on the wall. By the way, according to considerations [13], just this feature leads to quite impressive phenomenon of a bubble detachment against gravity force [14].

As a whole, the PEGB reconciles character of local temperature pulsation with prevailing role of liquid phase convection in the majority of boiling processes opening thereby a new line of attack on the boiling problem.

Over the past decades a number of experimental proofs were obtained directly confirmed existence and importance of the PEGB. Unfortunately, accidental rather strong manifestations of the effect with jet velocities in the range 1–5 m/s observed in some yearly experiments were left without proper interpretation [15, 16], for example, speeded-up liquid jet flow (5 m/s) penetrating through full-grown large preceding vapor bubble (**Figure 3**), the frame of which was published without any comment [15].

Real steps toward study of the PEGB during boiling on thin wires firstly were made in the works [17, 18]. Rather powerful manifestations of the PEGB were observed and recorded, including phenomenon of vapor bubble departure against gravity field. Diverse dynamical effects were studied including bubble-specific motion on micro-wires. Non-gravity nature of the observed phenomena was confirmed. Numerical model of bubble motion and jet flows through subcooled boiling on micro-wires was developed. At the same time, the interpretation of received data and evaluation of the role of the Marangoni effect in the PEGB became a subject of discussion [19, 20].

Very powerful manifestation of the PEGB is observed during laser drilling of nickel and copper [21]. The fixed velocities of the ejected liquid jets achieve 100–150 m/s. Though the authors link

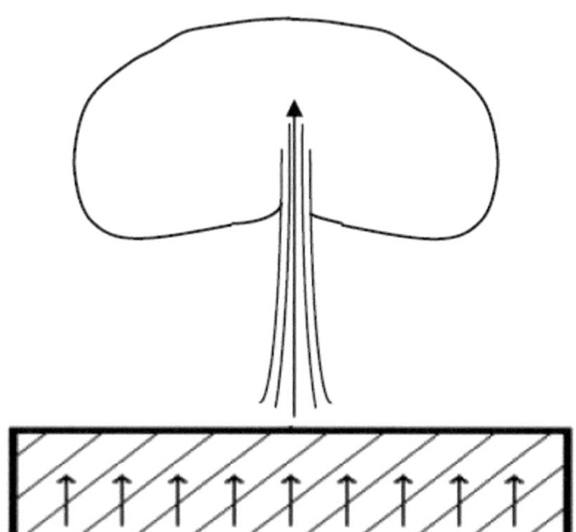

Figure 3. Liquid jet flow (5 m/s) penetrating through full-grown large preceding vapor bubble [15].

the effect to vapor bubble nucleation (to say, to the PEGB), this specific phenomenon requires further investigation. Tangible episodic manifestations of the PEGB were recorded also during boiling on down-facing heating surface [22].

The specific case of jet flow, containing the chain of micro-bubbles, was observed during experiments on subcooled boiling on micro-wires in micro-gravity [23]. Recorded velocities of micro-bubble jets (4–14 mm/s) are at 2–3 order lower than the jet velocities observed in aforementioned experiments (1–5 m/s and more). It may turn that the observed micro-bubble jet is the end result of numerous "micro-launches" of the PEGB, each of which detaches from the heating surface a single micro-bubble by the scheme [13].

In general, the studies are episodic. Especially when compared with the relatively minor MLE the study of which was the subject of numerous studies. It still is not examined the dependence of the effect on the properties of boiling medium and heating surface, heat flux, saturation pressure, and subcooling, the orientation of heating surface in the space. The most important in terms of understanding of the phenomenon relationships between the PEGB and local temperature pulsations are not investigated at all. In this connection, it still remains outside the field of view of researchers' specific type of thermal fatigue associated with cyclic thermal stresses generated by the PEGB [24].

The results of modern comparative studies of different boiling heat transfer mechanisms are also worthy of attention. In this context, it is of particular importance the actual confirmation of the status of the PEGB as the main cooling mechanism during boiling heat transfer by comprehensive review [25].

The review covers numerous experimental works, analytical and numerical studies. Having analyzed the experimental data obtained through various modern methods, such as micro-heater array, micro-heat flux sensors, and liquid crystal techniques, and comparing them with the results of numerical and analytical studies, the author makes an unambiguous conclusion about the impossibility to explain the observed pattern of heat inflow in a bubble during boiling by known to the author of the review heat transfer mechanisms, including MLE and contact line heat transfer.

The main outcome of the review is the fundamental conclusion about dominant role of heat transfer by liquid phase through "transient conduction and micro-convection." It also is concluded that "none of the proposed bubble heat transfer models described in the Introduction are consistent with the experimentally observed heat transfer signatures."

As follows from **Figure 2**, given in [25], characteristic of the basic heat transfer mechanism really is a brief description of the PEGB. The PEGB is an almost ideal mechanism for collecting heat accumulated by transient conduction in liquid boundary layer with its further transport through micro-convection to the almost whole surface of the bubble and then to bulk liquid.

Unfortunately, despite rather wide international publishing [2, 6, 7, 9–11], the PEGB turned to be unknown to the author of the review [25]. It also turned to be unknown an approximate analytical solution [2, 8, 13] just considering the combination of the transient conduction and microcirculation. These facts prevented the review to identify the real beneficiary of the study.

At the same time, the conclusion about the main role of "transient conduction and micro-convection" should not be taken as all-embracing. Studies of cooling mechanisms still do not adequately cover such processes as boiling at very high pressures or small gravitational accelerations in which the MLE can outperform the PEGB in importance.

In terms of refinement of the model [5–7], it deserves a serious attention the potential dynamic consequences of the rapid transition from one stable capillary state to another at the stage of nucleation (**Figure 4**), similar to the dynamic effect of "jumping droplets" [26].

In position 1, the nucleus meniscus still holds the wetting angle θ with the inner surface of the conical cavity, but the radius is almost equal to the radius of the mouth. During the nucleation (during overcoming the mouth), rapid transition of the same wetting angle θ to the base heating surface (position 2) occurs with associated dynamic effects, similar to the coalescence of the droplets. The transition is accompanied by change in the surface energy, corresponding thermal effect, and expansion work.

In fact, the emerging bubble is a capillary micro-heat engine that pushes the fluid from the heating surface, to say, promotes the PEGB. The relevant task is to determine the contribution of this micro-heat engine in the overall effect.

2.2. The MTD: heat transfer controlled by nucleation

The MTD is based on the fundamental fact of independence of heating surface superheat relative to the saturation temperature in developed boiling mode not only on the individual contributions of various cooling mechanisms but even on the number and composition of

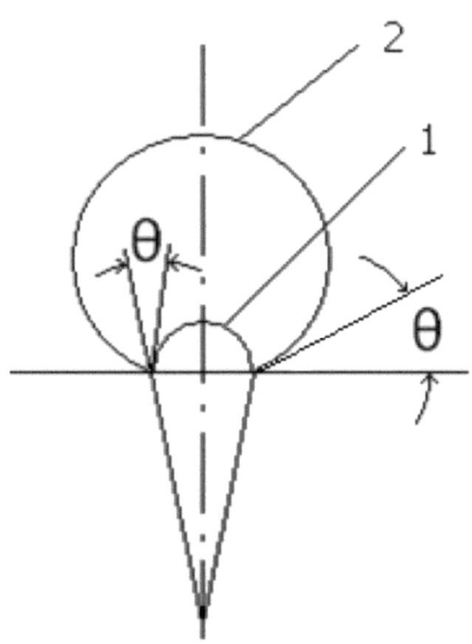

Figure 4. The scheme of the nucleation.

these mechanisms. Another important basic fact is the launch of the main cooling mechanisms by the onset of bubble growth and the short duration of their action.

In boiling of saturated liquid, one can distinguish four cooling mechanisms (**Figure 5**). Among them only the MLE [12] is linked to immediate evaporation on the boiling surface (**Figure 5a**). Other three mechanisms, bubbling [1] (**Figure 5b**), the PEGB or jet-like (**Figure 5c**) and the micro-membrane pumping (MMP) [27] (**Figure 5f**) are linked to liquid phase convection. With another approach, the number of cooling mechanisms could be greater. For instance, cooling mechanisms such as through pushing the liquid by growing bubble, through displacement of overheated liquid layer, or through drift liquid current subsequent to detached bubble sometimes are thought to be separate mechanisms. Here, these mechanisms are seen as the stages of the bubbling mechanism.

Less well-known MMP is a specific cooling mechanism due to permanent vibration of nuclei in potential centers synchronously with the local temperature pulsation of the heating surface. As the temperature increases, a nucleus surface ("micro-membrane") expands to critical profile, stops expanding when the nearest nuclei launches the PEGB with relevant cooling effect, and returns to the previous position with the temperature drop.

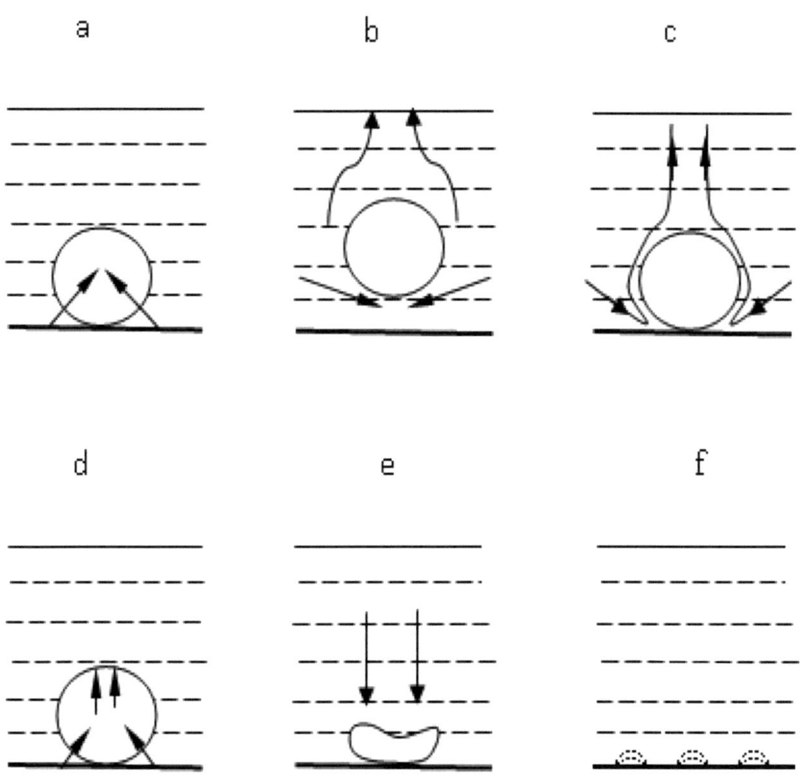

Figure 5. Schematics of cooling mechanisms: a—MLE; b—bubbling; c—PEGB; d—heat pipe-like; e—BCD; f—MMP.

Subcooling puts in operation two additional cooling mechanisms: heat pipe like (evaporation-condensation) [28] (**Figure 5d**), being an extra version of MLE, and bubble collapse-driven mechanism (BCD) (quasi-cavitational) [29] (**Figure 5e**) associated with the collapse of a bubble on the surface under the influence of influx of highly subcooled liquid.

As you can see, the main cooling mechanism (the PEGB) and its main assistants, the MLE, possibly the MMP and the BCD in the case of subcooled liquid, are all launched by the onset of bubble growth and have a short-term effect.

Data showing a virtually zero effect of a significant redistribution of the total heat flux between the various cooling mechanisms on the developed boiling heat transfer law are analyzed in the reviews [13, 30]. Here, we can confine ourselves to an impressive example of the BCD [29], which is absent altogether in the boiling of a saturated liquid and is quite intense in the case of surface boiling of a highly subcooled liquid. Despite this, these two processes reproduce the same developed boiling heat transfer curve.

Finally, based on these features of developed boiling, the MTD assumes control of the superheat by nucleation through multiple triggering short-run cooling actions of different cooling mechanisms. It also is assumed that onset of a bubble growth takes place at the instant the average temperature of the meniscus of critical size overcomes the temperature of thermodynamic equilibrium in the system nucleus-liquid-center.

It should also be clarified that the introduction of the MTD does not necessarily mean the inapplicability of the MTA in the analysis of developed boiling heat transfer, in general. We are talking only about extremely low efficiency of the latter in this particular case.

Let us take as an example an imaginary experiment with the process of developed boiling at a given heat flux. Changing within broad limits the acceleration of gravity and subcooling, we can set thousands of regimes of developed boiling with different compositions of cooling mechanisms and their various contributions to the overall heat transfer.

An accurate calculation of each such regime through the MTA is still an insoluble task. At the same time, when numerical models achieve such perfection, we will face a very peculiar situation: it turns out that thousands of complicated calculations result in the same overheating of the boiling surface relative to the saturation temperature. As for the MTD, it simply aims to identify the same overheating through the mechanism that sets it.

The MTD incorporates one-parameter model of boiling surface consisting of unlimited number of identical stable nucleation centers with the same ER characterized by unchangeable level of the superheat, triggering the growth of the first and following bubbles. The role of such a center may be played by conical recess with apex angle β satisfying the condition:

$$\frac{1}{2}\beta < \theta < 90^0 \tag{1}$$

The minimum curvature radius of the nucleus surface ρ_0 (the ER) in similar center is equal to the radius of the mouth [31].

The coverage area of the operating nucleation center is determined in a self-organized manner by the ability of the initially activated nucleation center to prevent by own cooling effect the activation of neighboring potential centers with the same ρ_0. If such zone of influence is reduced (e.g., with increasing heat flux), the former periphery overheats, and an additional center or centers with the same ER turn into operation there. Simultaneously, the new periphery is formed closer to the center. Of course, the process can also proceed in the reverse order. In such a framework, heating surface affects heat transfer by a single parameter, the ER.

Next, in the first place, within the framework of the MTD, an approximate analytical solution is made for the area adjacent to nucleation center [2, 8, 13]. The analysis approximates local temperature variation by the curve presupposing instantaneous drop in the wall temperature down to the saturation temperature at the onset of bubble growth (instantaneous start-up and shut-down of very intensive heat removal, e.g., by microcirculation (the PEGB) and immediate evaporation (the MLE)) and further warming-up of the wall through transient conduction up to the moment of onset of the next bubble growth (**Figure 6**).

The superheat ΔT_{eq} necessary for bubble growth onset, should be achieved at the meniscus of the nucleus in average. As critical nucleus is in the zone of temperature gradient, concomitant heating surface superheat $\Delta T'$ is much above ΔT_{eq}. Corresponding unsteady-state process is considered as warming-up of initially isothermal liquid semi-infinite space (with initial temperature equal to T_s) through transient heat conduction at suddenly posed boundary condition $q = Const$. The superheat ΔT_{eq} is determined by the relationship [18]:

$$\Delta T_{eq} = \frac{2\sigma T_s}{r \rho_0 \rho_g'} \tag{2}$$

where σ is the surface tension, r is the heat of evaporation, and ρ_g is the density of vapor.

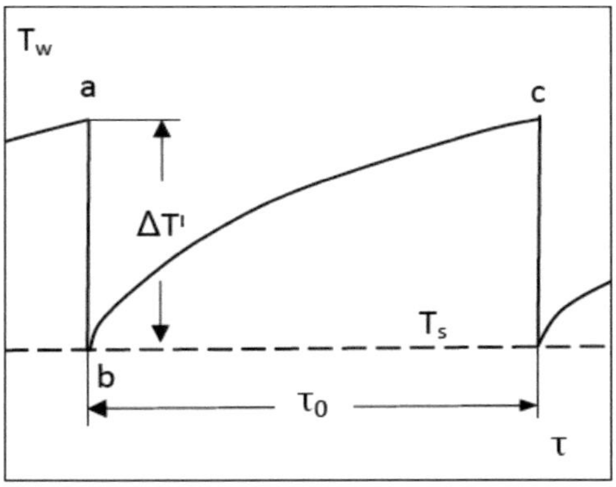

Figure 6. The first approximation.

On the basis of the general solution of the problem [32], the equations are obtained for the Nusselt number (Nu) and the rise time of the heating surface temperature (τ^*):

$$Nu = \frac{3\sqrt{\pi}}{2}\frac{\rho_0}{2\sqrt{\alpha\tau^*}}, \tag{3}$$

$$\frac{\rho_0}{2\sqrt{a\,\tau^*}}\frac{1}{ierfc\frac{\rho_0}{2\sqrt{a\,\tau^*}}} = \frac{K}{2}, \tag{4}$$

where h is the heat transfer coefficient (HTC), k is the thermal conductivity of the liquid phase, a is the thermal diffusivity of the liquid phase, and q is the heat flux.

$$Nu = \frac{h\rho_0}{k} \tag{5}$$

$$K = \frac{q\,\rho_0^2\,r\,\rho_g}{\sigma\,kT_s} \tag{6}$$

According to relevant comparison [2, 13], the analytical solution (3)–(4) predicts the order of HTC during boiling of nitrogen, water and sodium at atmospheric pressure. Taking in account the great difference between the liquids, such an outcome of approximate analysis still can be considered as a serious support of validity of the MTD. Important outcome of the solution is disclosure of the number K and characteristic length—the ER.

The theory is further refined through introducing some qualitative considerations of the periphery of the action zone of the center, given that it makes major contribution to the average superheat. The prolongation of liquid micro-convection by inertia after the termination of the action of PEGB also is taken into account. Based on some qualitative considerations, in addition to K, following modified Reynolds number is introduced:

$$\mathrm{Re}_{*,s} = \frac{C_p\,\sigma\rho\,T_s}{r^{3/2}\,\rho_g^2\,\nu} \tag{7}$$

where C_p is the heat capacity of the liquid, ρ is the density of the liquid, and ν is the kinematic viscosity of the liquid.

Finally the following correlation, Shekriladze and Ratiani, for developed boiling HTC is developed [2, 8–11, 13]:

$$Nu = 0.91 \cdot 10^{-2}\,K^{0.7}\,\mathrm{Re}_*^{\,0.25}, \tag{8}$$

An important outcome of the Eq. (8) is the disclosure of rather strong dependence of HTC on the characteristic length ($h \sim \rho_0^{0.4}$). Just this dependence marked the basic direction of boiling heat transfer enhancement by creating on the heating surface nucleation centers with large ER. With the exception of a misprint in article [8] (the coefficient of 0.88×10^{-2} instead of 1.22×10^{-2} in equation (10)), the presence in the part of publications of the other constant in the same Eq. (8) (1.22×10^{-2} instead of 0.91×10^{-2}) is due to the different records of Re^*: through specific work of expansion or through heat of evaporation (taking into account that $P(v_g\text{-}v) \approx 0.1r$ (P is the absolute pressure, v_g is the specific volume of the vapor, and v is the specific volume of the liquid)).

2.3. The ER, correlation of experimental data, and heat transfer enhancement

Disclosed by solutions (3)-(4), characteristic length especially clearly showed the basis of universality of the UC. Just the fact of the generation of control impulses by the nuclei of about 10 μm in size create the basis for the independence of wall superheat in the developed boiling mode from macro-hydrodynamics of two-phase medium, intensity of mass acceleration, the geometry and sizes of the heating surface, including microchannels. All of these parameters can affect the range of heat fluxes (beginning and end) of the developed boiling mode but not heat transfer law within the mode.

By the way, longstanding disregard of the MTD-UC was accompanied by a remarkable phenomenon: the concept and the term "characteristic length" left the scientific publications on boiling heat transfer for decades. The importance of knowledge or experimental determination of this parameter has ceased to be discussed at all.

As a result, an opportunity has been lost to stop the ordinary experimental practice to study boiling heat transfer without measuring the ER, a single parameter of the heating surface greatly affecting the HTC. If you try to invent an analogy to this situation, we could talk about the experiment on the hydrodynamics of the channel flow without measuring the cross-sectional dimensions of the channel.

The possibility of translating virtually all known correlations into the category of empirical relationships written in dimensionless form (due to the lack of the real characteristic length in them) was also missed.

In terms of the confirmation of the validity of the UC, it becomes particularly important a very few experimental studies including data on the ER and covering greatly differing liquids (sodium [33], water [34], refrigerants [35]) on the heating surfaces with highly different ER. Correlation of these data by the UC, borrowed from [2, 13, 36], is presented in **Figure 7**.

The correlation represents a fundamental confirmation of the validity of the MTD-UC and the role of the ER as the characteristic length. It is also obvious that the data presented in **Figure 7** cannot be described in a unified manner by the correlations that do not contain the ER (i.e., by all known correlations, other than UC). Incidentally, during boiling of sodium at $\rho_0 = 50\ \mu m$, HTC is 2.5 times higher than on commercial surface, other conditions are the same. The same enhancement factor for the refrigerants is 3.1.

Universal character of the Eq. (8), also can be demonstrated by the correlation borrowed from [22] (**Figure 8**) including, together with experimental data on pool boiling of Cesium and Hydrogen, the data on flow boiling of R11 and HCFC1 in a narrow passage and flow boiling of subcooled water in a microchannel.

To a certain extent, the problem of poor knowledge of boiling surfaces is also mitigated by the use in many experiments of commercial heating surfaces (mainly rolled pipes), which is characterized by the ER equal to 5 μm based on some indirect evidences [2, 8–11, 13, 34].

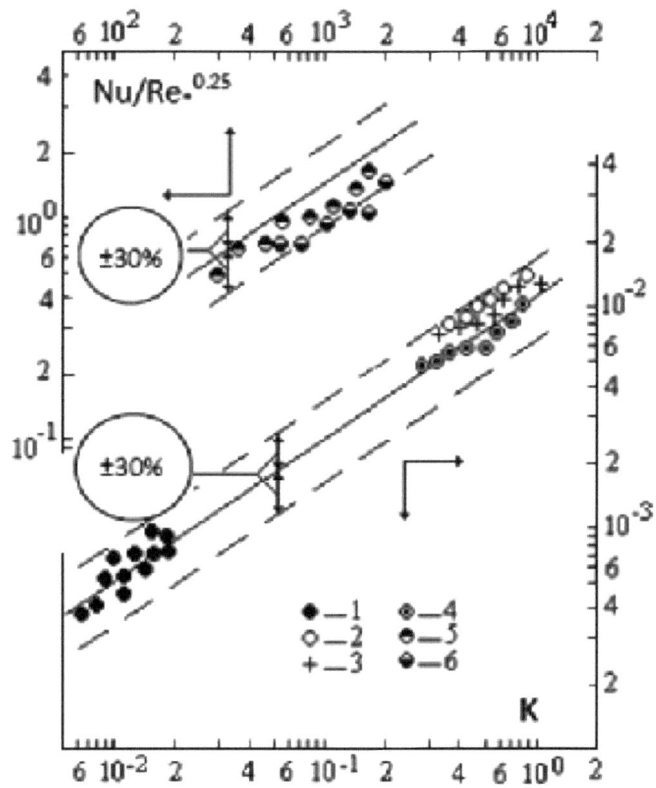

Figure 7. Comparison of Eq. (8) with experimental data on developed boiling on the surfaces with the known values of the ER: 1—sodium [33], ρ_0 = 50 *μm*; 2–4—water [34], ρ_0 = 5 *μm*; 5—R 12 [35], ρ_0 = 86 *μm*; 6—R 22 [35], ρ_0 = 86 *μm*; solid lines—Eq. (8).

Relying on a similar assessment of commercial surfaces, the validity of Eq. (8) is confirmed by a wide database on developed pool boiling of all groups of liquids including liquid metals and cryogens without matching different constants and powers to different surface-liquid combinations. The correlation covers the data on boiling of water, ammonia, ethyl alcohol, benzene, biphenyl, ethane, ethylene, R11, R12, R22, R113, R134a, R142, HCFC123, Na, K, Cs, Hg, CO, NO, BF_3, N, Ne, and H [2, 8–11, 13, 33].

Now let us look at the problem of boiling heat transfer enhancement. As mentioned, the MTD has determined the basic principle of boiling heat transfer enhancement: providing plenty of stable nucleation sites with large ER. At the same time, the UC has predicted the highest achievable enhancement factor ($h \sim \rho_0^{0.4}$). Over the past decades, enhanced boiling surfaces have been developed in direct using this basic principle. Outstanding achievements of this line of R&D have led to substantial progress in relevant technologies.

However, all the above circumstances have not prevented complete silencing of the MTD-UC-ER. The role of the boiling fundamentals turned out to be hidden by a rather simple way: the important issues to specify the scientific bases of development of the enhanced boiling surfaces and analyze the results in the same basic framework, they are generally avoided in the publications. A recent survey [40] may serve as an example of such an approach.

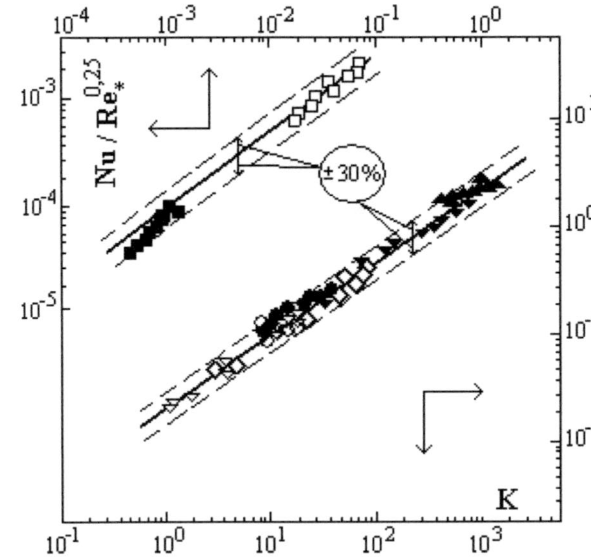

Figure 8. Correlation of experimental data on pool boiling HTC of cesium [36]: ■—1.82·10³ Pa; □—1.58·10³ Pa; pool boiling of hydrogen [37]: ∇—0.82·10⁵ Pa;▼—5.16·10⁵ Pa; ▲—8.50·10⁵ Pa; flow boiling in narrow passage [38]: ●—R11, 1.0·10⁵ Pa; o—HCFC123, 1.0·10⁵ Pa; flow boiling of subcooled water in a microchannel [39]: ◊—1.0·10⁵ Pa; solid lines—Eq. (8).

Finally, we should also address the issue of the limitations of the MTD. As it follows from the above correlations, developed boiling represents the most conservative basic regime of boiling heat transfer characterized by the dependence of HTC on restricted number of influencing factors. According to Eq. (8), together with the physical parameters of boiling area, developed boiling HTC depends only on two "external" factors—heat flux and the ER. As it follows from relevant analysis, such a conservatism of developed boiling heat transfer can be linked to the existence of a great (practically unlimited) number of stable nucleation sites with roughly uniform effective radii, short duration of each action of any cooling mechanism and prevailing contribution of heat removal by liquid phase convection.

According to the multi-factoring concept (MFC) [41], any failure to meet these conditions results in essential transformation of heat transfer regularities up to drastic increase of the number of influencing HTC factors. For instance, depending on concrete conditions, the circle of influencing HTC factors may be widened by the parameters of inter-phase hydrodynamics, intensity of body force, contact angle, subcooling, sizes, form, orientation, and thermal characteristics of the heating surface, micro-geometry, and distribution of nucleation sites, and prehistory of the process. Besides, multi-factoring may be accompanied by "passing on the baton" from the MTD to the MTA. As it follows from qualitative consideration, there can be distinguished two main types of multi-factoring:

- The first— connected with onset of dependence of effective radius (ER) on a degree of penetration of liquid into nucleation site (wetting-dependent multi-factoring (WDM)),
- The second—connected with transition to prolonged duration or uninterrupted regime of action of any intensive cooling mechanism (duration-dependent multi-factoring (DDM)).

The MFC opens up a promising path to the description of the observed in experiments diversity of boiling heat transfer curves, including boiling hysteresis (for more details, see [13, 22, 41]).

Finally, another very important aspect of the boiling heat transfer research should be noted. Due to the complete neglect of the above boiling fundamentals, a very important direction of boiling heat transfer research, numerical modeling, has lost consistent bases [42]. As a result, the direction still has not reached the level to be taken into account when analyzing the existing experimental data bases [43, 44].

3. Hard steps to recognition

The results of the first independent comparison of UC with experimental data were published only on the eve of the 50th anniversary of the publication of the correlation itself [43]. Despite the problems, with some applied methodological approaches and interpretations, the results of the comparison allow us to draw some important conclusions.

Using quite broad experimental database on heat transfer during pool boiling of 55 non-metallic liquids on copper heating surfaces, the authors identified the nine "most advanced" equations among tens and even hundreds of correlations published for more than half a century. Further, comparing these nine equations, they have identified three equations with the "low level" of the mean relative deviation (MRD): the equation of Gorenflo and Kenning updated in 2010 (the MRD 9.5%), the UC (10.8%), and the equation of Stephan and Preusser (12.1%) surpassing other equations in accuracy.

Unfortunately, in [43], important results of the comparison were not supported additionally by their comprehensive analysis. A thorough criticism of the corresponding part of the work was given in [45].

Here, we only note that the significance of the study would be greatly increased by clarifying the potentially appreciably higher accuracy of the UC.

The matter is that the UC participates in competition with other equations on unequal conditions. On the one hand, the UC is the only equation that includes the real characteristic length of the process (ρ_0). On the other hand, the standard boiling heat transfer experiment bypasses just the definition of this single parameter of heating surface influencing heat transfer.

In this connection, when processing the experimental data in the framework of the UC, it became necessary to characterize commercial heating surfaces on the basis of indirect estimates with a constant average value of ρ_0 [8]. For this reason, quite acceptable in itself, MRD of 10.8% can correspond to the UC only in the really unfeasible situation of the real constancy of ρ_0 of all experimental surfaces. Actually, of course, the numerous experimental surfaces deviated from the accepted value. As a result, in the still hypothetical situation, when the value of ρ_0 is known for each surface, the accuracy of the UC-based generalization can be noticeably higher. As shown in [46], in such a case, MRD can decrease almost two times.

More details about prolonged absurd situation with the characteristic length of boiling heat transfer, in general, are elucidated in [46]. At the same time, an important outcome of [46] is the first independent confirmation of the fundamental nature of the MTD-UC.

The conclusions drawn on the basis of the results of [43] are substantially strengthened by the results of the studies [44] devoted to the generalization of extensive experimental data on flow boiling. An experimental HTC database containing 2783 data points built from 26 open literatures for annular flow is covered. The database includes both macrochannel and mini-/microchannel data and covers wide range of working conditions. The annular flow database consists of seven working fluids, covering hydraulic diameters of 0.5–14.0 mm, mass velocities of 50–1290 kg/m^2 s, liquid-only Reynolds numbers of 240–55, 119, vapor qualities of 0.10–0.98, and reduced pressures from 0.01 to 0.77. In addition, 19 existing prediction methods for flow boiling are compared.

Really, the results of generalization of the experimental data showed fundamental characters of the MTD-UC and versatility of the UC, which outperformed in accuracy all the competitive equations. However, due to some sad missteps, very important work did not end with the adequate conclusions.

A critical mistake was the removal of the UC from the list of competing equations with subsequent exclusion from the published text of the UC-based generalization of the experimental data (necessarily preceding the obtaining of the UC-based so-called novel correlation). This step fundamentally contradicts the very logic of the MTD-UC, which reasonably claims to cover all classes of developed boiling heat transfer processes (see, for example, **Figure 8**).

As followed from the relevant analysis, the novel correlation [46] can hardly be regarded as a new result. It differs from the UC by two corrections of the opposite sign, each of which is noticeably smaller than the scatter of the experimental data (details of the examination are available in the "letter to the editor," submitted to Applied Thermal Engineering).

In any case, we are dealing with the fact that the correlation, ignored for 50 years, in 2017 wins in a wide competition for the best description of the channel flow boiling heat transfer.

Ultimately, independent confirmation of the successful competition of the UC with two different groups of equations in describing the two main classes of processes of boiling heat transfer certainly gives it a special status. In addition, if remembering the correlation by the UC and the experimental data on boiling heat transfer of liquid metals and highly different liquids (water, refrigerants, and sodium) on the surfaces with known different values of the ER, the UC will generally remain beyond any competition.

4. Concluding remarks

In the final part, there always is a desire, together with the conclusions, to talk about plans for the future. However, in this particular case, because of historical reasons, the accumulated problems are so vast that their representation would take a lot of space. Therefore, in this context, we recommend the reader to get acquainted with the publications of recent years [45, 46], where the tasks for the future are described in sufficient detail.

The real picture of boiling heat transfer research is rather gloomy. Half a century of total ignoring of the boiling fundamentals could not but lead to logical consequences. Tens and even hundreds of thousands of scientific publications, numerous editions, and conferences exist by themselves. In rare cases, when the accumulated knowledge should explain the reality in the form of experimental databases, the terms physical model, theory, numerical model, criteria, cooling mechanism, characteristic length completely disappear, and "calculation methods" come to the fore, the pure empiricism represented in dimensionless form.

Excessive tightening makes the recognition process increasingly painful. First, it turns out that the scientist, who established the key role of transient conduction and microcirculation, does not know that this is just the PEGB. Then others establish an excellent description of flow boiling heat transfer by the UC and by making minor corrections of the opposite sign transform the fundamental result of 1960s into the "novel correlation." The third and fourth forget to add to the examined databases the experimental data on heat transfer during boiling of liquid metals, generally leaving the UC out of competition. Developers of highly efficient boiling surfaces forget to indicate the basics of the success.

The author's systematic calls for a broad discussion on the problem remain unproductive for decades. Heat transfer research community still cannot gain will and carry out targeted actions to rid of the heavy burden of the past. The problem goes beyond the scope of the particular scientific discipline.

Author details

Irakli Shekriladze

Address all correspondence to: i.shekriladze@gtu.ge

Georgian Technical University, Tbilisi, Georgia

References

[1] Jacob M. Heat Transfer. V. New York: Wiley. 1949;**1**:505 p

[2] Shekriladze IG. Developed boiling heat transfer. International Journal of Heat and Mass Transfer. 1981;**24**:795-801

[3] Kruzhilin GN. Heat transfer from heating surface to boiling single-component liquid in conditions of natural convection. Izvestia AN SSSR. OTN. 1948;**7**:967-980

[4] Rohsenow WM. A method of correlating heat transfer data for surface boiling of liquids. Transactions of the American Society of Mechanical Engineers. 1952;**74**:969-976

[5] Shekriladze IG. On the mechanism of nucleate boiling. Bulletin of the Academy of Sciences of Georgian Soviet Socialistic Republic. 1966;**41**:392-396

[6] Shekriladze IG. Mechanism of steam bubble formation. NASA TM X-59398; 1967. 14 p

[7] Shekriladze IG, Mestvirishvili SA, Rusishvili JG, Zhoszholiani GI, Ratiani VG. Studies in the mechanism of boiling and enhancement of evaporative cooling coefficients. Heat Transfer-Soviet Research. 1980;**12**:91-95

[8] Shekriladze IG, Ratiani GV. On the basic regularities of developed nucleate boiling heat transfer. Bulletin of the Academy of Sciences of Georgian Soviet Socialistic Republic. 1966;**42**:145-150

[9] Shekriladze IG. Developed boiling heat transfer of cryogenic liquids and refrigerants. In: Afgan N, editor. Heat and Mass Transfer in Refrigeration and Cryogenics. Washington: Hemisphere; 1987. pp. 580-591

[10] Shekriladze IG. Mechanisms of heat removal in the process of developed boiling. Heat Transfer-Soviet Research. 1990;**22**:445-463

[11] Shekriladze IG. Frozen pathways to breakthrough in boiling heat transfer theory. Proceedings of 11th International Heat Transfer Conference. 23-28 August 1998; Kyongju, Korea; 2: 473-478

[12] Moore FD, Mesler RB. The measurement of rapid surface temperature fluctuations during nucleate boiling of water. American Institute of Chemical Engineers Journal. 1961;**7**(5):620-624

[13] Shekriladze IG. Boiling heat transfer: Mechanisms, models, correlations and the lines of further research. The Open Mechanical Engineering Journal. 2008;**2**(1):104-127

[14] Subbotin VI, Kaznovski SP, Korotaev SK. Investigation of the dynamics of vapor bubbles in boiling of water on thin wires under natural convection. Atomnaia Energia. 1970; **28**:9-13

[15] Van Stralen SJ, Zijl W, De Vries DA. The behavior of vapor bubbles during growth at subatmospheric pressures. Chemical Engineering Science. 1977;**32**:1189-1196

[16] Afgan ON. Boiling Liquids Superheat. Oxford: Pergamon Press; 1976. 186 p

[17] Wang H, Peng XF, Wang BX, Lee DJ. Jet flow phenomena during nucleate boiling. International Journal of Heat and Mass Transfer. 2002;**45**:1359-1363

[18] Wang H, Peng XF, Christopher DM, Garimella SV. Jet flows around microbubbles in subcooled boiling. Journal of Heat Transfer. 2005;**127**:802

[19] Shekriladze I. Comments on the paper jet flow phenomena during nucleate boiling. International Journal of Heat and Mass Transfer. 2003;**46**:2711-2712

[20] Shekriladze I. Discussion: "Dynamics of bubble motion and bubble top jet flows from moving vapour bubbles on microwires". Journal of Heat Transfer. 2006;**127**:1260-1261

[21] Yilbas BS, Sami M. Liquid ejection and possible nucleate boiling mechanisms in relation to laser drilling process. Journal of Physics D: Applied Physics. 2005;**30**:1996-2005

[22] Shekriladze I, Machavariani E, Gigineishvili G, Rusishvili J, Shekriladze D. Investigation of duration- dependent multifactoring during boiling on down-facing heating surface. Proceedings 14th International Heat Transfer Conference. 8-13 August 2010; Washington DC, USA; 1:735-744

[23] Munro T. Heater geometry and heat Flux effects on subcooled thin wire nucleate pool boiling in microgravity. Digital Commons & USU, Utah State University. 2012:1-89

[24] Shekriladze IG. Nucleate boiling heat transfer: Temperature pulsations or local thermal shocks. Materials Performance and Characterization. 2014;**3**(4):69-85. DOI: 10.1520/MPC20130114

[25] Kim J. Review of nucleate pool boiling bubble heat transfer mechanisms. International Journal of Multiphase Flow. 2009;**35**:1067-1076

[26] Steyer A, Guenoun P, Beysens D. Spontaneous jumps of a droplet. Physical Review Letters. 1992;**68**:64-66

[27] Shekriladze IG. Heat transfer in two-phase areas with intensive evaporation and condensation. Doctor of Science in Technology Thesis. The Bauman Moscow Highest Technical School. 1982:282 p

[28] Robin TT, Snyder NW. Bubble dynamics in subcooled nucleate boiling based on the mass transfer mechanism. International Journal of Heat and Mass Transfer. 1970;**13**:305-318

[29] Nesis YI, Sologub IS. Temperature fluctuations in subcooled boiling at a single site. In: Kipenie i kondensatsia. Riga: RPI Press; 1984. pp. 5-13

[30] Shekriladze IG. Boiling heat transfer: An overview of longstanding and new challenges. Journal of ASTM International. 2012;**9**(1):1-41

[31] Griffith P, Wallis SD. The role of surface conditions in nucleate boiling. Chemical Engineering Progress (Symposium Series). 1960;**56**:49-60

[32] Luikov AV. Heat Conduction Theory. Moscow: Visshaia Shkola Press; 1967. 599 p

[33] Marto PL, Rohsenow WM. Effects of surface conditions on nucleate pool boiling of sodium. Journal of Heat Transfer. 1966;**88**:149-157

[34] Shoukri M, Judd RL. Nucleation site activation in saturated boiling. Journal of Heat Transfer. 1975;**97**:96-102

[35] Chumak LV, Malaia LV, Vinichenko IV. Enhancement of heat transfer of cryogens on the pipe surface. Kholodilnaia tekhnika. 1979;**2**:31-34

[36] Subbotin VI, Ovechkin DM, Sorokin DN. Heat transfer during pool boiling of cesium. Teploenergetika. 1968;**6**:63-66

[37] Grigoriev VA, Pavlov YuM, Ametistov EV. Boiling of Cryogenic Liquids. Moscow: Energia Press; 1973. 243 p

[38] Bao ZY, Fletcher DF, Haynes BS. Flow boiling heat transfer of freon R11 and HCFC123 in narrow passages. International Journal of Heat and Mass Transfer. 2000;**43**:3347-3358

[39] Liu D, Garimella SV. Flow boiling heat transfer in microchannels. Journal of Heat Transfer. 2007;**129**:1321-1332

[40] Attinger D, Frankiewicz C, Betz AR, Ganguly R, Das A, Kim CJ, et al. Surface engineering for phase change heat transfer: A review. MRS Energy & Sustainability: A Review Journal. 2014:1-40

[41] Shekriladze IG. Developed boiling heat transfer – Forty years of the model of "the theatre of director". Proceedings 13th International Heat Transfer Conference. Sydney, Australia; 13-18 August 2006. 13 p. DOI: 10.1615/IHTC13.p28.370

[42] Dhir VK, Warrier GR, Aktinol E. Numerical simulation of pool boiling: A review. Journal Heat Transfer. 2013:**135**(6-17):061502. DOI: 10.1115/1.4023576

[43] Gorenflo D, Baumhögger E, Herres G, et al. Prediction methods for pool boiling heat transfer: A state-of-the-art review. International Journal of Refrigeration. 2014;**43**:203-226. DOI: 10.1016/j.ijrefrig.2013.12.012

[44] Yuan S, Cheng WL, Nian YL, Yuan S, Zhong Q, Fan YF, Hel J. Evaluation of prediction methods for heat transfer coefficient of annular flow and a novel correlation. Applied Thermal Engineering. 2017;**114**:10-23. DOI: 10.1016/j.applthermaleng.2016.11.170

[45] Shekriladze IG. Boiling heat transfer theory: To overcome historical deadlock. Proceedings 15th International Heat Transfer Conference. Kyoto, Japan; 10-15 August 2014; 6067-6981. DOI: 10.1615/IHTC15.pbl.008817

[46] Shekriladze IG. Tabooed universal characteristic length and misled boiling heat transfer research. Journal of Physical Science and Application. 2015;**5**(5):334-344

Chapter 6

Calculations of Heat Transfer in the Furnaces of Steam Boilers According to the Laws of Radiation of Gas Volumes

Anatoly N. Makarov

Additional information is available at the end of the chapter

http://dx.doi.org/10.5772/intechopen.75529

Abstract

The laws of heat radiation from black body and the laws of Stefan-Boltzmann (Jožef-Ludwig) , Max Planck, and Wilhelm Wien are fundamental laws of physics. All in all, a little more than 30 fundamental laws of physics, studied by pupils and students worldwide, were disclosed. Scientific disclosure of fundamental laws influences mainly power technology, fuel, and energy resource saving. In the late nineteenth century, the laws of heat radiation from gas volumes and the laws of Makarov were disclosed. Since the radiation laws from blackbody are fundamental laws of physics, then the laws of heat radiation from gas volumes are fundamental laws of physics. The effect of using laws of heat radiation from gas volumes on fuel saving and reduction of development pressure on the environment in many countries of the world is shown.

Keywords: physics, scientific discovery, laws, Nobel prize, heat radiation, gas volumes, combustion chamber

1. Introduction

Radiant heat transfer is the main kind of heat transfer in furnaces and combustion chambers and accounts for 90–98% of the total heat transfer in steam boiler fireboxes [1–3].

Since the late nineteenth century and throughout the twentieth century, heat transfer in torch furnaces, fireboxes, and combustion chambers was calculated based on the law that was experimentally established by Stefan in 1879 in studying radiation from solid bodies, which was then theoretically substantiated by Boltzmann in 1884. In the late nineteenth to the early twentieth century, solid lumped fuel (coal, peat, and wood) was fired in furnaces on fire grates, and the first descriptions of heat transfer processes were essentially descriptions of problems and calculation of radiant heat transfer between two arbitrarily located surfaces (a fuel bed and a heating surface) on the basis of Stefan-Boltzmann's law. In 1924, Kirpichev gave an analysis of methods for solving this problem that had been developed by different researchers [4], and Stefan-Boltzmann's law is presently formulated as follows:

$$q = c_s \varepsilon_{red}\left[\left(\frac{T_1}{100}\right)^4 - \left(\frac{T_2}{100}\right)^4\right]\varphi_{12}, \quad (1)$$

where q is the density of heat flux radiated from the fuel to the heating surface (W/m^2), cs is the black body emissivity factor (W/(m^2 K$_4$)), ε_{red} is the reduced emissivity, T_1 and T_2 are the temperatures of fuel bed and heating surface (K), and φ_{12} is the view factor for radiation from the fuel bed on the heating surface.

In the twentieth to the twenty-first centuries, flaring of gas, liquid, pulverized fuel in furnaces, fire boxes, and combustion chambers was widespread. Fuel flaring is characterized by volume emission, a three-dimensional radiation model [1–6]. In torch, gas volume emits 10^{15}–10^{30} particles of atoms. Radiation of each particle and atom on the calculated area should be considered. The calculation of heat radiation on the calculated area of all the atoms in the gas volume and the torch requires the solution of triple integral equations [7–10]. The solution of triple integral equations to determine the average path length of beams from the emitting particles, atoms, and angular radiation coefficients of the gas volume on the calculated area in the twentieth to the twenty-first centuries has not been found [7, 8]. The laws of radiation from gas volumes were not disclosed.

It is considered that the problem of calculating heat transfer in torch furnaces, fireboxes, and combustion chambers was solved with the appearance of computers and the use of numerical simulations of integral equations of heat transfer [11, 12]. However, long-term analytical and experimental studies of heat transfer have shown that the results of the numerical solution of integral equations of heat transfer on computers are not valid [9]. The method uses the laws of heat radiation of a blackbody, solid bodies, and Stefan-Boltzmann law (1); however, gas volume radiation is not subject to the laws of Stefan-Boltzmann [9, 10]. This method uses the Stefan-Boltzmann law and a large mass of approximate values of the temperatures and optic coefficients of surface and volume zones, and the accuracy of calculations is 40–80% [9–13].

Paradoxical cases are observed using the existing calculation methods. The torch power can be increased by additionally heating the air supplied to the burner. For example, with air heated from 20 to 600°C, the torch power increased by 17%, and its temperature rose from 1300 to 2000°C, i.e., by a factor of 1.5 [14]. According to expression (1), the density of heat flux radiated from the torch to the calculated zone should increase by a factor of 5, and the heating rate of articles being processed should also increase by a factor of 5, which is in contradiction with the energy conservation law. Under the real conditions of heating furnace operation, with air subjected to preheating and with the torch power increased by 17%, the heat flux density and the heating rate increase by 12–15%, i.e., in direct proportion to the growth of torch power and not to the fourth power of temperature [14].

In the twentieth century, the torches and emitting gas volumes remained a "black box" despite the applied enormous intellectual resources to solve the problem. Formulas for determining the main parameters of heat radiation from gas volumes, torches, formulas for determination average beam path length from quadrillions of radiating atoms, and the local angular coefficients of radiation from radiation flux densities on the calculated area were not available. The solution to the problem has stalled.

2. Laws of radiation from spherical and cylinder gas volumes

At the end of the twentieth century, in 1996–2001 the laws of heat radiation from gas volumes [13, 15, 16] and the laws of heat radiation from gas isothermal isochoric concentric spherical (**Figure 1**) and coaxial cylinder gas volumes (**Figure 2**) were dislosed, the volumes, that the torches, gas volumes of the furnaces, fireboxes, and combustion chambers are currently modeled by [17–19].

The laws are called Makarov's laws with the goal of adherence to the age-old scientific traditions and copyright [13]. Based on the scientific discovery, geometric, physical, and mathematical models of gas volume and torch as a source of heat radiation have been developed. In the gas volumes formed during flare combustion of the fuel, spherical or cylindrical gas volumes are inscribed. Radiating gas atoms are simulated by emitting quadrillions of spheres, uniformly filling the spherical and cylindrical gas volumes.

The statement of the scientific disclosure is as follows. "The average path length of beams from quadrillions of radiating particles of each isochoric isothermal concentric spherical or coaxial cylindrical gas volumes to the calculated area is equal to the arithmetic mean distance from the symmetry axis of volumes to the calculated area and the angular coefficients, flux densities of radiation from gas volumes on the calculated area are equal. The flux density of radiation from the central spherical or central cylindrical gas volume of a small diameter on the calculated area is equal to the sum of the fluxes of the radiation fluxes from all the concentric spherical or coaxial cylindrical volumes on the calculated area at the radiation power released in the

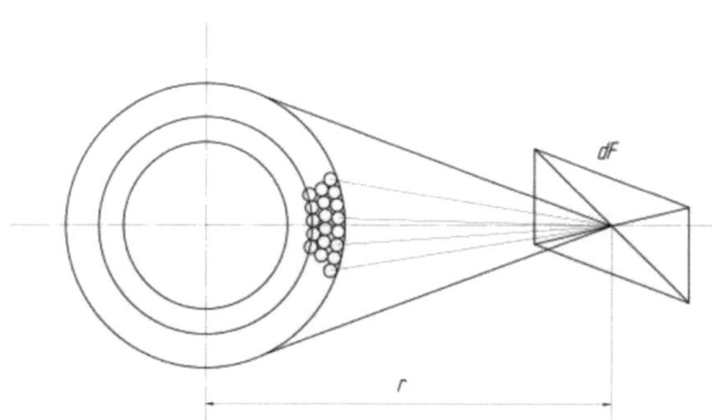

Figure 1. The radiation from isothermal isochoric concentric spherical gas volumes on the calculated area dF.

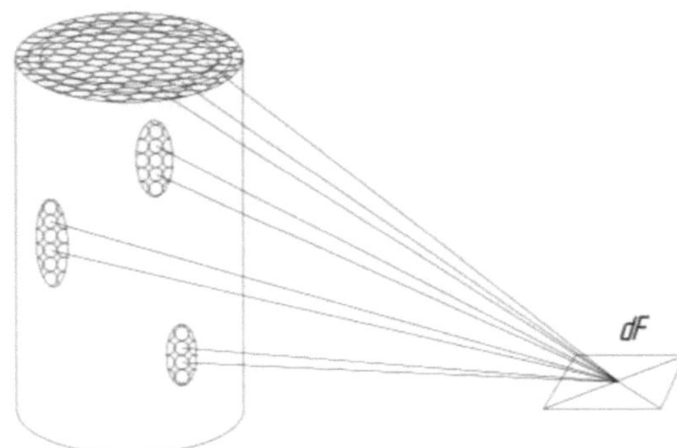

Figure 2. The radiation from isothermal isochoric coaxial cylinder gas volumes on the calculated area dF.

volume of a small diameter, equal to the sum of the radiated powers released in all spherical or coaxial cylindrical gas volumes radiating on the calculated area.

The density of the heat flux incident from the cylindrical or spherical gas volumes to the calculation area is directly proportional to the power, the local angular emission factor of the gas volume to the calculation area and inversely proportional to the absorption coefficient of the gas medium, the average path length of the rays from the emitting particles of the gas volume to the calculation area and the area of calculation area."

Mathematical notation of the laws is as follows:

$$l_1 = l_2 = l_3 = \ldots = l_i = \left(\sum_{i=1}^{n} \frac{l_i}{n}\right) = l \quad (2)$$

where l_1, l_2, l_3, and l_i are the average beam path lengths from the first to the ith cylindrical or spherical gas volumes to the calculated area dF and l is the arithmetic mean distance from the axis of symmetry of the cylindrical volumes or the center of symmetry of the spherical volumes to the calculated area dF:

$$\varphi_{F_1dF} = \varphi_{F_2dF} = \varphi_{F_3dF} = \ldots = \varphi_{F_idF} \quad (3)$$

where $\varphi_{F_1dF}, \varphi_{F_2dF}, \varphi_{F_3dF}, \varphi_{F_idF}$ is the local angular coefficient of radiation from a surface of the first, the second, the third, and the ith coaxial cylindrical or concentric spherical gas volumes on the calculated area dF, respectively:

$$q_{F_1dF} = q_{F_2dF} = q_{F_3dF} = \ldots = q_{F_idF} \quad (4)$$

where $q_{F_1dF}, q_{F_2dF}, q_{F_3dF}, q_{F_idF}$ is the density of the radiation fluxes incident from the first to the ith coaxial cylindrical or concentric spherical gas volumes on the platform dF:

$$q_{F_1dF} = \sum_{i=1}^{n} q_{F_idF} \quad (5)$$

where $q_{F_i dF}$ is the density of radiation flux incident from the central cylindrical or spherical gas volumes of a small diameter on the calculated area dF:

$$q_{FdF} = \frac{\varphi_{F_i dF} \cdot P_F \cdot e^{-kl}}{F_0} \tag{6}$$

where $\varphi_{F_i dF}$ is a local angular coefficient of radiation from the central cylindrical or spherical gas volumes of a small diameter on the calculated area dF, P_F is the radiation power of the central cylindrical or spherical gas volumes, and F_0 is the area of the calculated platform dF.

Mathematical notation of the laws of heat radiation from gas volumes and the laws of Makarov is obvious and grounded in a similar manner to the statement and mathematical notation of Newton's third law of motion in texts on physics for students of secondary schools and technical universities:

"The force with which two bodies act upon each other are equal in magnitude and opposite in direction":

$$F_1 = -F_2, \tag{7}$$

where F_1 is the force with which body 1 acts on body 2 and F_2 is the force with which body 2 acts on body 1.

Laws of heat radiation from gas volumes possess the compactness and the accuracy of the description of physical phenomena in a similar manner to the fundamental laws of physics. For example, a fundamental law of physics, Ohm's law, describes the relationship between the current I flowing in the conductor and the voltage U applied to the conductor and the conductor resistance R:

$$I = \frac{U}{R} \tag{8}$$

Similarly, the law of heat radiation from gas volumes characterizes the dependence of flux density of heat radiation q of gas volume from the angular coefficient of the radiation φ, radiated power P, and the average beam path length rays l of gas volume. For the calculation of parameters of heat radiation from gas volumes (6) φ, P, and l analytical expressions, formulas were derived [16–19].

A unique, natural harmony of heat radiation from quadrillions of particles of spherical and cylindrical gas volumes is disclosed, namely, that the average beam path length from these particles is equal to the arithmetic mean distance from the symmetry axis of volumes to the calculated area.

Complex, a triple integration of no solution within the gas volume to determine the average beam path length is reasonably replaced by computing actions of elementary mathematics and analytic geometry; this produces the same result, which would have been gotten in triple integration.

The uniqueness of the scientific discovery is that the flux densities of radiation and angular radiation coefficients of spherical, coaxial, and cylindrical gas volumes to the calculated area are equal and it is sufficient to hold a single integration of trigonometric functions within the height of the cylindrical gas volume of a small diameter, located on the axis of symmetry to define them [15–22].

Heat radiation from cylindrical gas volumes of diameter 2, 5, and 10 m and more in calculations can be equivalently modeled by heat radiation from cylindrical gas volumes of an infinitely small diameter and the axis of their symmetry. Scientific discovery of heat radiation from gas volumes provides researchers and designers with great opportunities for improvement of electric arc and torch furnaces, fireboxes, and combustion chambers.

With the discovery and development of the laws of geometrical, physical, and mathematical models of torch, the radiating gas volumes and torches as sources of heat radiation become an investigated physical phenomenon, not a "black box." The formulas for calculating the density of the radiation flux from the gas volume, the torch on the calculation area (6), for determining the local angular coefficients of gas volumes on the calculation area [15–22], for determining the mean path length of quadrillion rays (2) from gas volumes on the calculation area were obtained. Basing on the scientific discoveries of the laws of heat radiation from gas volume, the theory of thermal radiation from the gas volume and the new concept of calculating heat transfer in torch furnaces, fire chambers, and combustion chambers were developed [19]. The theory of thermal radiation of the gas volume includes the output 14 of the formulas for calculating the coefficients and fluxes of the radiation of the flame on the heating surface in Vivarelli, mutually perpendicular coils and arbitrarily located planes.

In accordance with the new concept and the theory, cylinder gas volumes, from which the calculation of radiation fluxes on the calculated areas and heating surface is performed, are inscribed in torches.

Radiation fluxes from torch, heated surfaces, and combustion products are determined for each calculated area taking into account multiple reflections and torch for each calculated area platform determined by taking into account multiple reflections and absorptions. The calculations of heat transfer in steam boiler boxes [9, 10, 18–20], torch heating furnaces [7, 8, 15–17], and combustion chambers of gas turbine installations [19] are made with the use of the new concept.

The calculations allow to determine rational energy modes of electric arc and torch furnaces, fireboxes, and combustion chambers in which fuel consumption reduces and operational life increases. In 15 years since the first publication of the author of scientific discovery in printing, the theory of thermal radiation of the gas volume and the new concept of calculating heat transfer in torch furnaces, fire chambers, and combustion chambers have been tested by time; the results of calculations are confirmed by the results of experimental studies on existing kilns, furnaces, and combustion chambers; and the accuracy of calculations does not exceed 10%. Since the radiation laws of a blackbody and the laws of Stefan-Boltzmann and Planck, these wines belong to the fundamental laws of physics, and the laws of radiation by gas volumes are both fundamental laws of physics.

http://dx.doi.org/10.5772/intechopen.75529

The laws of heat radiation, the theory of heat radiation from gas volumes, and the new concept of calculation in electric arc and torch furnaces, fire boxes, and combustion chambers were published in the form of text [19], which is used for teaching university students. The method for calculation that had existed until the scientific discovery had not allowed to calculate and to manage rational heat transfer in torch furnaces, since the error of calculations was 20–50%, so the efficiency of fuel energy in torch furnaces is 25–45% at the present time. The use of scientific discovery and its base-developed theory allows to determine the rational parameters of the torch (capacity, length, expansion angle) and its spatial position to the heating surface (vertical, horizontal, inclined at a certain angle).

Rational position of products and torches and burners will increase consumption efficiency of fuel energy half-twofold from 25–45 to 65–75% and decrease fuel consumption twofold over the coming years all over the world.

3. Calculation of heat transfer in steam boiler furnaces

3.1. Calculation of heat transfer in the firebox of a TGMP-204 steam boiler

A unified procedure for calculating heat transfer in electric arc and torch furnaces, fireboxes, and combustion chambers has been developed proceeding from the discovered regularities pertinent to heat transfer in torch gas layers [19]. The resulting integral heat fluxes consisting of radiant fluxes falling on the heating surfaces from the torch, wall and arch lining, combustion products, and convective fluxes are all calculated according to this procedure. Innovative designs of torch furnaces and fireboxes have been developed proceeding from the discovered regularities, and the use of which makes it possible to obtain a higher output from fireboxes, more uniform steam generation in tubes, more uniform heating of articles, and smaller consumption of fuel.

The distribution of integral radiant fluxes over the boiler firebox walls and bottom surfaces was calculated taking the TGMP_204 boiler as an example used as part of a 800 MW power unit, and the firebox of which has the shape of a rectangular parallelepiped of height $H_f = 46$ m, width $a = 20.66$ m, and depth $b = 10.40$ m. The firebox rear wall has an aerodynamic nose in its upper part. The boiler operates on fuel oil and is equipped with 36 double-flow vortex burners with a throughput capacity of 5.2 t/h each, which are installed in opposite directions in three tiers on the front and rear walls of the firebox. The burners installed on each wall are placed in a common duct through which air and recirculating gases are supplied. The air excess factor in the furnace $\alpha = 1.03$, and the gas recirculation ratio $r = 0.14$. With these values of air excess factor and gas recirculation ratio, the average values of particle diameter dp, density ρ, concentration μ, and the medium attenuation coefficient k are determined from the formulas given in [1–3]. The calculated values of these parameters were found to be $d_p = 0.278$ μm, $\rho = 2 \times 103$ kg/m^3, $\mu = 0.06$ g/m^3, and $k = 0.162$.

The distribution of temperature along the height of steam boiler fireboxes was investigated [23–25]; the results of temperature measurements are reported in many publications, e.g., in

[12]. The torch fills the entire firebox chamber and has the shape of a straight elliptical cylinder; the isotherms shown in **Figure 4** divide it along the height into six volume bodies.

Five volume bodies with ellipses in their bases and vertices and with a parabolic generatrix (the torch vertical parts) have the shape of elliptical paraboloids resting on the sixth volume body having the shape of a truncated ellipsoid of revolution (the horizontal part of torch 7).

The own radiation from gas volumes and the radiation power decrease along the firebox height in accordance with temperature variation along the flame height. Below, we denote the power releasing in the flame horizontal part by *Ph* and the powers releasing in five vertical volume zones (from the bottom to top) by *P*1–*P*5. The power releasing in the torch is determined from the expression as follows:

$$P_{tr} = Q_i^r B_r, \tag{9}$$

where Q^r_i is the fuel heating value equal to 41 MJ/kg and *Bf* is the fuel flow rate (kg/h). During the operation of 36 burners with a throughput of 5.2 t/h, the fuel flow rate will be $B_f = 187.2 \times 103$ kg/h, and the torch power will be $P_{tr} = 2155$ MW.

Introducing the assumption that the radiating volumes are isothermal within the confines of their volume zones and taking into account that the volume bodies have identical bases and different heights, we can write the following proportion for determining the power releasing in each of the six volume zones:

$$P_r{:}P_1{:}P_2{:}P_3{:}P_4{:}P_5 = T_r^3 h_r{:}T_1^3 h_1{:}T_2^3 h_2{:}T_3^3 h_3{:}T_4^3 h_4{:}T_5^3 h_5, \tag{10}$$

where T_h and T_1–T_5 and h_h and h_1–h_5 are the temperatures and heights of the corresponding zones.

The following values of powers releasing in the torch volume zones were obtained from expression (10): MW: $P_h = 645$, $P_1 = 905$, $P_2 = 216$, $P_3 = 172$, $P_4 = 129$, and $P_5 = 87$.

For calculating the integral radiation fluxes falling from the torch on the waterwall surfaces and for reducing the calculation error, the torch should be decomposed as follows: three straight circular cylinders are inscribed into the straight elliptical cylinder representing the torch vertical part (see **Figure 3**) and these circular cylinders will model the radiation from the flame vertical part both over its height and volume, i.e., over the firebox width and depth. The way in which the radiation fluxes from the cylinder are distributed over the heating surface does not depend of the cylinder diameter; therefore, the torch vertical part should be subdivided into 15 small-diameter cylinders representing linear radiation sources by 3 in each vertical volume zone.

After that, we determine the density of integral radiation flux from each *j*th cylindrical source in the horizontal and vertical volume zones falling on the *i*th elementary area on the wall surface:

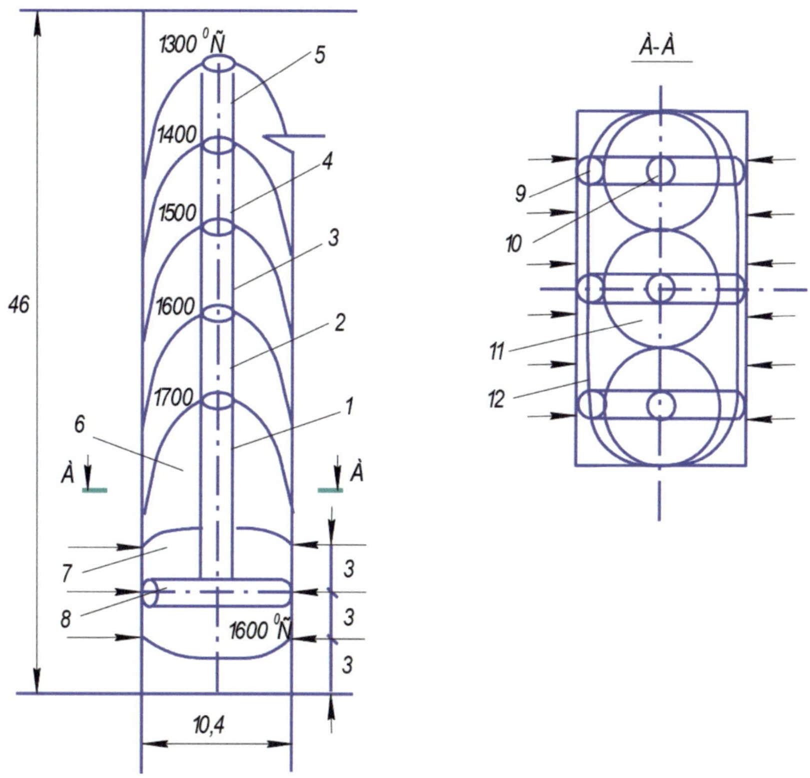

Figure 3. Schematic design of the regenerative soaking pit with a two-tier unit of regenerators (a) and the distribution of integral heat fluxes falling on the lateral surfaces facing the soaking pit longitudinal symmetry axis over the ingot height (b). (*1*) chamber; (*2*) cover; (*3–5*) rear, lateral, and front walls, respectively; (*6*) regenerator units; (*7* and *8*) air and gas regenerators; (*9*) mixing chamber; (*10*) technological holes; (*11*) ingots; and (*12*) torch.

$$q_{inj} = \frac{\varphi_{ji} P_j e^{-kl}}{F_i}, \tag{11}$$

where *j*th is the cylindrical radiation source on the *i*th area, which is determined from the analytical expressions given in [19]; *Pj* is the power of the *j*th cylindrical source (MW); *Fi* is the surface area of the *i*th elementary area (m^2); and *l* is the average beam length (m).

The density of incident integral radiation flux from the torch on the *i*th elementary area is determined as the sum of integral heat flux densities from all cylindrical sources, i.e.,

$$q_{iin.tr} = \sum_{j=1}^{18} q_{itrj}. \tag{12}$$

The distribution of integral radiation fluxes falling on the front and rear walls of a TGMP-204 steam boiler firebox is almost the same and is characterized by curves *1* and *5* (**Figure 4a**),

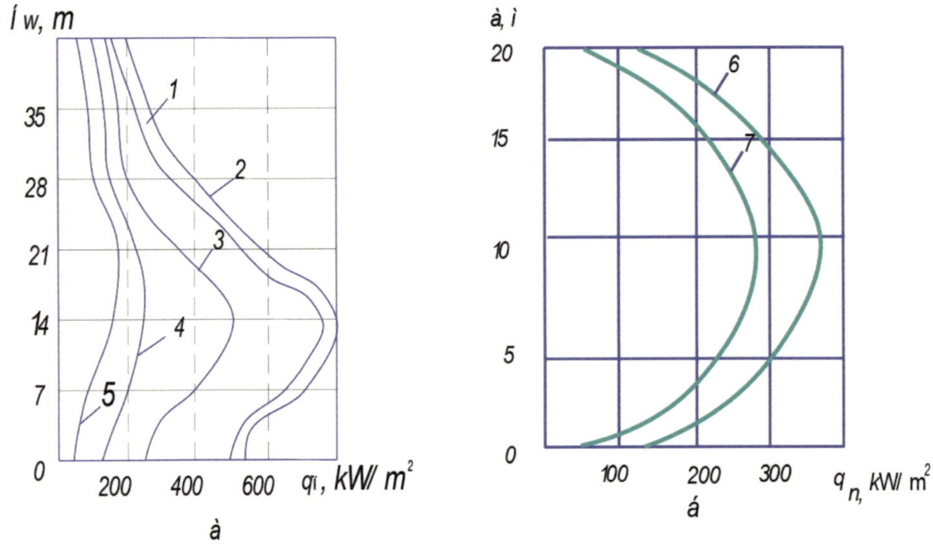

Figure 4. Distribution of the integral radiation flux density over the walls (a) and bottom (b) of the TGMP-204 boiler firebox. (*1*) Results of calculation over the front wall symmetry axis; (2) measurement results for the same; (*3–5*) results of calculation over the lateral wall vertical symmetry axis, over the lateral wall height at a distance of 4 m from the lateral wall vertical symmetry axis and over the front wall height at a distance of 8 m from the front wall vertical symmetry axis; and (*6* and *7*) results of calculation over the bottom's major symmetry axis and over the horizontal line parallel to the bottom major symmetry axis at a distance of 4 m from it.

and so is the distribution of integral radiation fluxes falling on the firebox right- and left-hand walls, which is characterized by curves *3* and *4*. The calculation results coincide with the results from measurements of integral radiation flux densities along the frontal wall vertical symmetry axis [1], testifying that the developed torch mathematical model adequately reflects the real conditions.

The maximal integral radiation fluxes are observed on the front and rear walls at a height of 12–16 m (4–5 m above the top tier of burners). The waterwall surfaces at a height of 12–16 m experience the maximal radiation from the horizontal volume zone and from the torch's first vertical volume zone. The densities of integral radiation fluxes reach here 780 kW/m^2 on the front and rear walls and 520 kW/m^2 on the lateral walls. The integral radiation fluxes observed on the walls at the firebox bottom level are a factor of 1.7–1.9 smaller and equal to 460 and 270 kW/m^2, respectively. This is because the wall lower belt is located at a considerable distance away from the first vertical volume zone and the more so from the second to the fifth volume zones. The density of integral radiation fluxes falling from the torch on areas lying at a height of above 20 m decreases along the wall height in a similar manner.

The densities of integral radiation fluxes on the walls under the ceiling are equal to 150 and 110 kW/m^2, respectively, on the front and lateral walls along the vertical symmetry axis. Such decrease of integral radiation fluxes on the wall surfaces under the ceiling is due to the fact that the wall upper belt is situated at a considerable distance away from the torch horizontal and first vertical volume zones, in which 72% of the torch power is released, whereas only around 10% of the torch power is released in the nearby fourth and fifth vertical volume zones.

The variation of integral radiation flux densities over the wall perimeter is also essentially nonuniform in nature. The densities of integral fluxes radiated from the torch in the front and rear wall "hot belt" situated at a height of 12–16 m vary from 780 kW/m^2 at the wall vertical symmetry axis to 180 kW/m^2 at their periphery; i.e., they drop by a factor of 4.3. This is because the vertical symmetry axes of the front and rear walls are situated at the shortest distances from the linear sources and from the central and peripheral cylinders by which the torch is modeled.

The densities of integral fluxes radiated from the torch at the vertical symmetry axis and at the periphery of lateral walls differ from each other to a significantly lesser extent. The densities of integral fluxes radiated from the torch at a height of 12–16 m are equal to 520 and 290 kW/m^2, respectively, at the vertical symmetry axis and at the periphery of lateral walls. The flux densities in the underceilng zone of lateral walls differ from each other to a still lesser extent and are equal to 120 and 95 kW/m^3, respectively, at the vertical symmetry axis and at the periphery. This is because the distance from the cylindrical sources of radiation to the vertical symmetry axis of lateral walls differs insignificantly from the similar distance to the periphery of lateral walls [19].

Figure 4b shows the calculated distribution of integral radiation flux densities over the bottom surface. The 250 kW/m^2 isorad forms an elliptic hot spot in the bottom surface center with the sizes along the major and minor axes equal to 12 and 8 m. The integral radiation fluxes at the bottom periphery do not exceed 125 kW/m^2.

As is well known [23], the surface density of deposits inside the tubes increases with the density of integral radiation flux falling on the waterwall surfaces. With a heat flux equal to 200 kW/m^2, the surface density of deposits inside the tubes is 0.1 kg/m^2, whereas at 500 kW/m^2, it is equal to 0.3 kg/m^2; that is, the density of deposits inside the tubes grows in proportion to the heat flux. Hence, reducing the densities of integral heat fluxes and making them more uniform over the perimeter and height of steam boiler firebox walls are presently a topical problem.

An experimental confirmation of the results obtained from the performed calculations aimed at determining the distribution of integral radiation flux densities over the surfaces of the TGMP-204 steam boiler firebox can be found in [1], as well as in [24, 25].

In [19], the experimentally determined distribution of integral radiation fluxes over the height of the left-hand lateral waterwall of a TGMP-204 KhL steam boiler firebox is presented for the case of boiler operation in the mode of power unit maximal load equal to 800–820 MW1. The difference between the maximal local incident radiation fluxes in the firebox of a TGMP-204 steam boiler obtained by calculation (780 kW/m^2) and by measurements (870 kW/m^2) [25] does not exceed 12%, which confirms that the developed torch mathematical model adequately reflects the real conditions.

Nonuniform distribution of the densities of integral fluxes radiated from the torch over the wall perimeters and heights gives rise to similar nonuniformity of steam generation and deposits in the tubes. Engineers who develop and design steam boilers, as well as researchers in this field, should apply additional efforts on improving the designs of fireboxes aimed at decreasing the heat fluxes over the wall perimeters and heights and at making them more uniform, which will lead to similar reduction and more uniform distribution of deposits in the tubes. The shortcoming of the considered firebox of a TGMP-204 boiler is that the waterwall heating surfaces in the furnace lower part experience high heat loads, which result in a high growth rate of deposits inside the tubes. High heat loads entail a growth of temperature of waterwall tube metal surfaces and facilitate the occurrence and development of high-temperature corrosion in these surfaces. All the abovementioned factors have a negative effect on the service life of waterwall heating surfaces and, hence, result in less reliable operation of the entire boiler.

The use of the proposed calculation procedure allows one to get a more comprehensive idea about heat transfer and to design new improved fireboxes for steam boilers. This improvement is achieved through changing the shape of firebox and waterwall surfaces and the inclination angle of oppositely arranged burners, which leads to rational distribution of heat loads [26, 27].

In [27], a firebox for firing gas and oil fuel was proposed. This firebox consists of a bottom, arch, walls, waterwalls repeating the firebox inner surface, and oppositely placed burners built in the walls. The firebox is made in the form of two truncated cones with their larger bases facing each other. The firebox bottom serves as the base of the lower cone; the lateral walls of the upper cone are matched with vertical walls forming the shape of a straight cylinder. The burners are inclined to a horizontal plane at an angle of 5–10°.

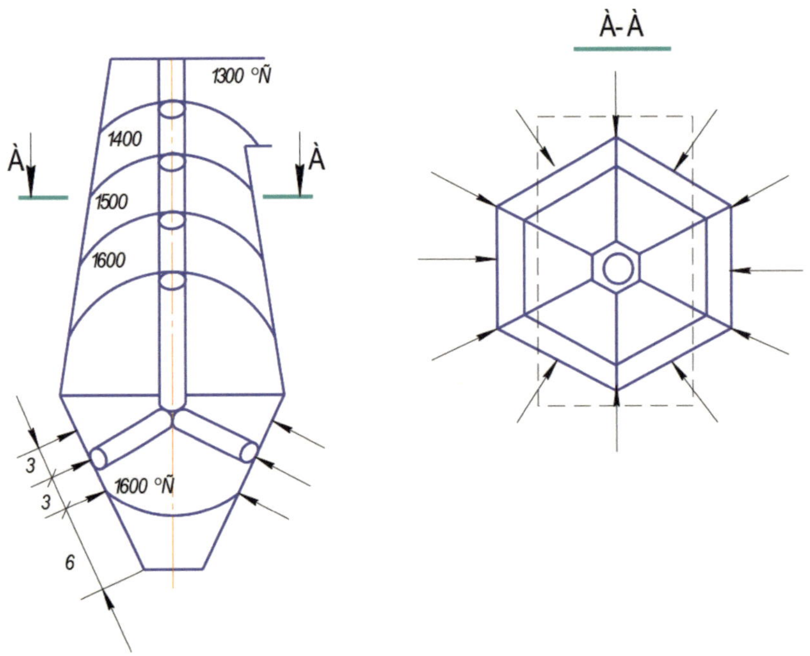

Figure 5. Firebox of the TGMP-204-I steam boiler (the dashed line in view *A–A* depicts the section of the TGMP-204 boiler furnace).

Figure 5 shows the innovative firebox of a TGMP-204-I steam boiler with rational distribution of heat loads intended for use as part of an 800 MW power unit. This furnace has the form of two truncated hexagonal pyramids with a common base having the shape of a regular hexagon with a radius of 9.6 m. The firebox is fitted with 36 vortex double-flow burners with a throughput of 5.2 t/h each. The burners are installed in three tiers uniformly over the perimeter of walls at the corners and in the middle of each face of the lower inverted pyramid. The firebox has a height of 46 m, and its volume is approximately equal to that of the TGMP-204 steam boiler. The air excess factor in the TGMP-204-I boiler firebox $\alpha = 1.03$, the gas recirculation ratio $r = 0.14$, the concentration of particles $\mu = 0.06$ g/m^3, the particle diameter $dp = 0.278$ μm, the particle density $\rho_p = 2 \times 10^3$ kg/m^3, and the attenuation coefficient k in the medium = 0.162.

Figure 5 shows the expected distribution of temperature over the height of the TGMP-204-I boiler firebox. Owing to the sloped walls, the firebox has a larger volume in the zone of burners; the maximal expected temperature in the active fuel combustion zone is 1600°C, and the temperature of combustion products leaving the firebox chamber is 1300°C. The torch is modeled by a vertical cylinder with five volume zones and by 12 inclined cylinders (**Figure 5**). The torch power P_{tr} = 2155 MW. The procedure described above was used to calculate the densities of heat fluxes falling from the torch on the waterwall surfaces (**Figure 6**).

Owing to the fact that all waterwall surfaces and all faces of the truncated pyramid are situated at the same distance from the torch and from the cylindrical sources of radiation by which the torch is modeled, the integral radiation flux density along the wall symmetry axis and at the wall periphery is characterized by one curve shown in **Figure 6**. The integral radiation flux density has the same distribution over all waterwall surfaces and over all faces of the upper and lower truncated pyramids. The heat flux density in the zone of maximal heat loads at a height of 12–16 m in the TGMP-204-I boiler firebox is by 200 kW/m^2 lower than it is in the

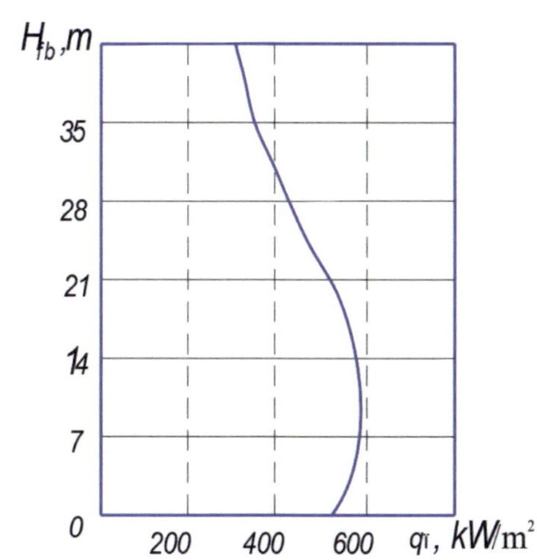

Figure 6. Distribution of integral radiation flux density over the walls of the TGMP-204-I boiler furnace.

TGMP-204 boiler firebox. As a result, more uniform steam generation is obtained in all tubes of waterwall surfaces over the firebox perimeter.

Owing to lower heat loads, the waterwalls operate at lower temperature, which result in a lower rate of their corrosion. This, in turn, results in a longer service life of waterwall heating surfaces and in a longer interval between their outages for cleaning the boiler with acid.

The proposed design of the steam boiler firebox has also other advantages over the existing ones. The new firebox has a larger volume in its part opposite to the burners, and the torch has a lower temperature in this volume, which results in less intense generation of nitrogen oxides. In addition, the waterwall surfaces in the firebox upper part are situated closer to the torch axis, due to which the working fluid absorbs heat more intensely and the heat fluxes are distributed along the firebox height more uniformly (see **Figure 6**). It is expected that the formation of deposits inside the tubes will be less intense in the proposed steam boiler firebox, that its operational costs will be reduced, that the firebox will have higher efficiency, and that the waterwall surfaces and working fluid will absorb heat more intensely [26].

Comparison between the distribution pattern of integral radiation fluxes over the TGMP-204 boiler firebox walls and that of the TGMP-204-I boiler firebox with the similar capacity testifies that the latter has certain advantages over the former: the heat loads of the waterwall surfaces are more uniformly distributed over the firebox height, the maximal heat loads in the burner zone are decreased by 25%, the torch has a lower temperature, and a smaller amount of nitrogen oxides is generated.

3.2. Calculation of heat transfer in the firebox of a TGMP-314 steam boiler

Here, we examine calculations of heat transfer in the furnace of a type TGMP-314 steam boiler with modeling of the flare by radiative zones and large cylindrical gas volumes.

The TGMP-314 steam boiler furnace for a 300 MW power-generating unit is in the shape of a rectangular parallelepiped (**Figure 7a**, **b**) with height Hw = 35 m, width a = 14 m, and depth b = 7 m. The boiler runs on fuel oil and has 16 burners with a combined capacity Bb = 67 torch-mounted counter to one another in two rows of eight burners each, at heights of 3 and 6 m from the bottom surface. The minimum heat of combustion of the fuel oil is Q^r_1 = 41 MJ/kg. The attenuation coefficient for radiation in the medium in the furnace is k = 0.162 [19]. The distribution of isotherms along the height of the furnace with oppositely mounted burners is shown in **Figure 7a**. In terms of its height and perimeter, the flare fills the entire furnace and is in the shape of a right elliptical cylinder. The isotherms divide it height-wise into seven volumes. Six of these volumes, with ellipses at their base and top, and a parabola as a lateral generator, are elliptical purebloods which lie on the seventh volume, which is a truncated ellipsoid of rotation. An ellipsoid of revolution formed by the 1900°C isotherm is contained within the truncated ellipsoid.

Figure 7c shows the calculated radiative fluxes incident on the shielding surfaces of the furnace walls of a TGMP-314 steam boiler furnace. The distribution of the integrated radiative flux along the vertical axis of symmetry of the front wall is characterized by graph 1, which

http://dx.doi.org/10.5772/intechopen.75529

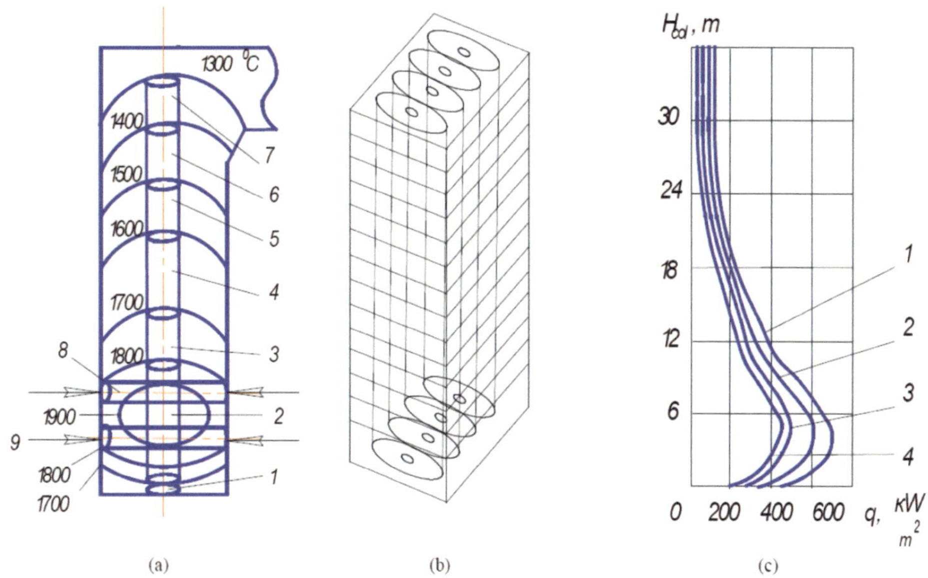

Figure 7. A type TGMP-314 steam boiler furnace with modeling of the flare by radiating cylindrical volumes: (a) distribution of isotherms and cylindrical radiation sources, (b) dividing the furnace into large cylindrical radiating volumes, and (c) the distribution of the integrated radiative flux along the height of the front and side walls: (*1–7*) arrays of vertical cylindrical sources; (*8* and *9*) arrays of horizontal cylindrical sources; (*10*) measurement and computational results along the vertical axis of symmetry of the front wall; (*11*) calculated results along the vertical axis of symmetry of the side wall; and (*12* and *13*) calculated results along the heights, respectively, of the side wall (at a distance of 2 m from its vertical axis of symmetry) and the front wall (at a distance of 4.7 m from its vertical axis of symmetry).

also illustrates the distribution of the integrated radiative flux along the vertical axis of symmetry of the rear wall.

The measured and calculated integrated radiative fluxes along the vertical axis of symmetry of the front wall differ by less than 10% [28, 29]. This confirms the adequacy of the model developed here for the flare in type TGMP-314 steam boiler furnaces. The maximum heat release zone, located at a height of 2–5 m from the bottom of the furnace, is characterized by maximal integrated radiative fluxes of the flare onto the front and rear walls at a level of 680 kW/m^2. At heights of 2–5 m on the side walls, the radiative fluxes along the axis of symmetry are 590 kW/m^2; at a distance of 2 m from the axis of symmetry, they fall to 440 kW/m^2.

On the periphery of the front wall at a height of 2–5 m, the radiative fluxes from the flare fall to 250 kW/m^2. The heat release in the flare decreases along the height of the furnace; the integrated radiative fluxes of the flare on the shielding surfaces also decrease.

The nonuniform distribution of the radiation fluxes from the flare along the perimeter and height of the screen surfaces of the walls gives rise to similar nonuniformity of vaporization and deposits in the tubes [2, 3]. The high heat loads in the bottom part of the front and back walls increase the temperature of the screen tubes and promote high-temperature corrosion in them. As a result of this, the problem of lowering the maximum temperature of the flare and equalizing the radiation flux densities on the screen surfaces along the height and perimeter of the firebox walls of steam boilers is of high priority. The importance of finding a solution to this problem also increases because the temperature in the fuel combustion zone affects thermal nitrogen oxide NOx emissions the most [30].

An innovative modification of the shape of the firebox of a steam boiler was proposed in order to obtain a rational distribution of the heat loads along the screen surfaces and to reduce nitrogen oxide emissions. The bottom part remains a rectangular parallelepiped; it is expedient to make the top part (**Figure 8a**) a truncated four-sided pyramid [26].

We shall examine heat transfer in such a firebox for the example of the previously studied boiler of the TGMP-314 power plant [19]. To suppress nitrogen oxide and organize two-step combustion of fuel, the nozzles are installed to feed air in amounts of 16 units (8 units per tier at heights 5.5 and 11 m). Oil is fed into a burner with air deficiency á = 0.8–0.85; the remaining air required for complete combustion of the fuel equals Äá = 0.2–0.25 and is fed through a nozzle higher up on the flare. The implementation of the two-step combustion of the fuel makes it possible to lower the temperature of the flare in the active burn zone and the nitrogen oxide emissions by 25–60%.

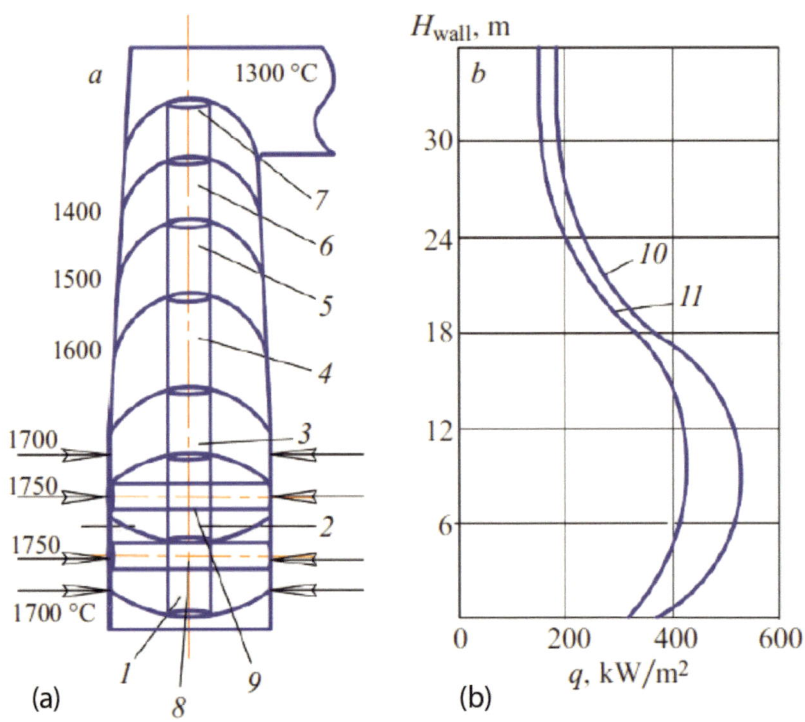

Figure 8. Firebox of TGMP-314 steam boiler with innovative modification: (a) distribution of the isotherms and cylindrical sources of radiation which are used to model the flare (same as in **Figure 7**) and (b) radiation flux density distribution along the height of the front wall 1 and side wall 2.

In order that the implementation of two-step combustion of fuel not to increase the temperature of the combustion products at the outlet of the firebox, the radiative heat transfer was organized from the flare to the screen surfaces in the middle and top parts of the firebox. To this end, the front, back, and side walls are built to slope at 4–6° relative to the vertical axis starting from the height 15 m, forming a truncated pyramid of height hp. = 20 m. The walls converge onto the axis of the flare, the radiative heat transfer from the flare to the screen surfaces increases, the heat fluxes along the height of the screen surfaces equalize, and the temperature of the flare decreases. The expected distribution of the temperatures in the firebox of a TGMP-314 steam boiler modernized in this manner is shown in **Figure 8**. As a result of the two-step combustion of fuel, the temperature in the zone of active combustion decreases to 1750°C. As a result of the improvement of the radiative heat transfer, the temperature of the combustion products leaving the firebox does not exceed 1300°C with the walls converging on the flare. The heat transfer in the modernized firebox of the TGMP-314 steam boiler was calculated.

The computational results obtained from expression (12) are displayed in **Figure 8b**. It is evident that the highest values of the maximum radiation flux densities—546 kW/m^2—lie on the front walls at 10 m on the vertical axis. The radiation flux densities on the vertical axis of the front and side walls equal 352 and 306 kW/m^2, respectively, at a height of 18 m, 246 and 205 kW/m^2 at 24 m, and 175 and 148 kW/m^2 at 30 m. Comparing the radiation flux densities of the flare along the screen surfaces of the front side in the fireboxes of conventional and modernized TGMP-314 boilers (**Figure 8b**) shows that in the firebox of the modernized boiler the highest radiation flux densities on the screen surface decreased by 24% from 680 to 546 kW/m^2 in the bottom part, increased by 40% from 252 to 352 kW/m^2 at height 18 m, and increased by 47% from 167 to 246 kW/m^2. In the top part at height 30 m, the heat fluxes from the flare in the firebox of the modernized boiler increased by 41% compared with the conventional boiler: from 126 to 178 kW/m^2. Similar variation of the radiation flux density distributions also occurs along the height of the screen surfaces of the side walls.

Altering the configuration of the firebox of a steam boiler from a rectangular parallelepiped in the bottom part to a truncated rectangular four-sided pyramid whose walls incline at angle 5–6° to the vertical plane inside the firebox in the central and top parts made it possible to increase the heat fluxes on their screen surfaces and reduce the no uniformity of the heat flux distribution along the height and perimeter of the firebox.

These changes ease the operating conditions of the tubes in the bottom part, lower the maximum gas temperatures in the interior volume of the firebox and the gas temperature at the exit from the firebox, reduce the production of nitrogen oxides in the firebox, and increase the service life of the screens.

4. Conclusions

The scientific discovery of the laws and the development of the theory of heat radiation from gas volumes is a contribution to the foundation of modern physics, as it allows calculating and managing the transfer of heat around the world in tens of thousands of electric arcs and torch furnaces, steam boiler boxes, and combustion chambers of gas turbine units, reducing energy consumption and saving millions of tons of fuel, reducing emissions of pollutants and anthropogenic load on the environment, and improving the quality of life in many countries. The laws and the theory of heat radiation of the ionized and non-ionized gas volumes and the laws of Makarov were included in the text [19], in the amount of fundamental knowledge on the quantum nature of radiation, and are in line with the laws of heat radiation from absolutely black body and with even more than 30 fundamental laws of physics.

Author details

Anatoly N. Makarov

Address all correspondence to: tgtu_kafedra_ese@mail.ru

Tver State Technical University, Tver, Russia

References

[1] Blokh AG. Heat Radiation in Boiler Units. Leningrad: Energiya; 1967. 326 p

[2] Blokh AG, Zhuravlev YA, Ryzhkov LN. Radiant Heat Transfer: A Handbook. Moscow: Energoatomizdat; 1991. 432 p

[3] Nevskii AS. Radiant Heat Transfer in Furnaces. Moscow: Metallurgiya; 1971. 440 p

[4] Kirpichev MV. About Heat Transfer in Boilers. Moscow: Energiya; 1924. 128 p

[5] Siegel R, Howell Y. Thermal Radiation Heat Transfer. Moscow: Mir; 1975. 934 p

[6] Ozirik MN. Radiative Transfer and Interaction with Conduction and Convection. Moscow: Mir; 1976. 616 p

[7] Makarov AN. Modeling of a torch and calculation of heat transfer in furnaces, fire boxes, combustion chambers. Part I. Calculation of radiation from solids and gas volumes by the laws of radiation from solid bodies. International Journal of Advanced Engineering Research and Science. 2016;**3**(12):44-48. DOI: 10.22161/ijaers/3.12.9

[8] Makarov AN. Modeling of a torch and calculation of heat transfer in furnaces, fire boxes, combustion chambers. Part II. Calculation of radiation from gas volumes by the Laws of radiation from cylinder gas volumes. International Journal of Advanced Engineering Research and Science. 2016;**3**(12):49-54. DOI: 10.22161/ijaers/3.12.10

[9] Makarov AN. Laws of heat radiation from surfaces and gas volumes. Word Journal of Engineering and Technology. 2015;**3**:260-270. DOI: 10.4236/wjet.2015.34027

[10] Makarov AN. Laws of radiation from large gas volumes and calculation of heat transfer in steam boiler boxes. Mechanical Engineering Research. 2015;**5**(1):42-55. DOI: 10.5539/mer.v5n1p42

[11] Alekseenko SV, Burdukov AP, Dekterev AA, Markovich DM, Shtork SI. Physical and mathematical modeling of aerodynamics and combustion in the combustion chamber of the power plant. Teploenergetika. 2011;**9**:67-72

[12] Maydannyk MN, Verbovetsky EH, Dekterev AA, Chernetsky MY. Mathematical modeling of fire box and rotary flue of furnace P-50 R in joint combustion of solid and gaseous fuels. Teploenergetika. 2011;**6**:37-42

[13] Makarov AN. Theory of radiative heat exchange in furnaces, fire boxes, combustion chambers is replenished by four new laws. Science Discovery. 2014;**2**:34-42. DOI: 10.11648/j.sd.20140202.12

[14] Mastryukov BS. Thermal Engineering Calculation of Industrial Furnaces. Moskow: Metallurgia; 1972. 287 p

[15] Makarov AN. Laws of heat radiation from spherical gas volumes. Part I. Laws formulation. International Journal of Advanced Engineering Research and Science. 2017;**4**(3):74-79. DOI: 10.22161/ijaers.4.3.11

[16] Makarov AN. Laws of heat radiation from spherical gas volumes. Part II. Modeling of heat radiation from volume bodies by radiation from spherical and cylindrical gas volumes. International Jornal of Advanced Engineering Research and Science. 2017;**4**(3):80-87. DOI: 10.22161/ijaers.4.3.12

[17] Makarov AN. Calculation of heat transfer in torch furnaces by gas volume radiation laws. World Journal of Engineering and Technology. 2016;**4**:488-503. DOI: 10.4236/wjet.2016.43049

[18] Makarov AN. Flare temperature and nitrogen oxide emission reduction in the TGMP-314I steam boiler firebox. Power Technology and Engineering. 2016;**50**(2):200-203. DOI: 10.1007/s10749-016-0683-x

[19] Makarov AN. Heat Exchange in Arc and Torch Metallurgical Furnaces and Power Stations: Studies. A Grant for Students of Higher Education Institutions. St-Petersburg: Lan'; 2014. 384 p

[20] Makarov AN. Regularities of heat exchange in gas layers of a torch and screens of fire chambers of boilers. Part I. Geometrical and physical модкль torch as source of the thermal radiation. Teploenergetika. 2014;**9**:26-32

[21] Makarov AN. Regularities of heat exchange in gas layers of a torch and screens of fire chambers of boilers. Part II. Laws of radiation of gas layers and the method of calculation of heat exchange developed on their basis in furnaces, fire chambers, combustion chambers. Teploenergetika. 2014;**10**:24-31

[22] Makarov AN. Regularities of heat exchange in gas layers of a torch and screens of fire chambers of boilers. Part III. Examples of calculation of heat exchange in torch furnaces and fire chambers of boilers. Teploenergetika. 2014;**11**:46-54

[23] Davidzon MI. About influence of heat flux density on formation of intrapipe deposits. Teploenergetika. 2001;**1**:72-73

[24] Kulikov AA, Kurbeer BA, Akintsev IY. Optimization of the mode of a fire chamber of the boiler TGMP-204 HL of the unit of 800 MW of the Surgut GRES2. Teploenergetika. 1988;**3**:25-28

[25] Abryutin AA, Chuprov VV, Antonov AY. A comparative research of summary heat exchange in fire chambers of boilers of the power unit of 800 MW with hearth and wall configurations of torches in case of combustion of fuel oil. Teploenergetika. 1988;**3**:31-36

[26] Makarov AN, Voropayev VV, Krivnev EI. Patent 2285200 The fire chamber for combustion of oil-gas fuel. Inventions. 2006;**28**:36-37

[27] Makarov AN, Shevchenko MN. Patent 2400668 A fire chamber for combustion of oil-gas fuel. Inventions. 2010;**27**:39-40

[28] Gerasimov VV, Pieces IA, Duda IM, Nezhentsov YN, Yablokov IV, Sivtsov AI, Terentyev AD. Test results and trial maintenance of power units with load of 330 MW. Express information of the "Maintenance and Repair of Power Station" series. Teploenergetika. 1988;**1**:1-12

[29] Abryutin AA, Antonov AY, Ousman YM, Shtilman of this year, Levin MM. Features of heat exchange in a fire chamber of a powerful black oil boiler unit with hearth configurations torches. Teploenergetika. 1981;**9**:17-26

[30] Kozlov YV, Zroychikova TV, Belov VA. Methods of reliability augmentation of furnace screens of boilers. Teploenergetika. 2003;**5**:25-29

Chapter 7

Conjugate Heat Transfer of an Internally Air-Cooled Nozzle Guide Vane and Shrouds

Lei-Yong Jiang, Yinghua Han and Prakash Patnaik

Additional information is available at the end of the chapter

http://dx.doi.org/10.5772/intechopen.74104

Abstract

The accurate evaluation of the aero-thermodynamic working environments of gas turbine critical components is essential in the development of advanced gas turbine engines, such as cooling flow arrangement and service life assessment. In this chapter, as a technology demonstration, the conjugate heat transfers of the internally air-cooled nozzle guide vane (NGV) and shrouds of a gas turbine engine at flight conditions are numerically examined. The simulations are performed with a high-fidelity CFD model and adequately defined boundary conditions. The effect of the non-dimensional distance from wall surfaces, y+, on the wall temperature is studied. The characteristics of the complex three-dimensional flow and temperature fields are revealed, and the heat fluxes between the hot gas, NGV body, and cooling airflow at selected cross sections are presented and discussed. It is clear that the traditional one-dimensional semiempirical approach is no longer suitable. Generally, the Mach number is higher, and the temperature is lower on the NGV suction side than on the pressure side. It is found that two high-temperature zones occur on the NGV pressure side and the temperature over the middle section is relatively low. These findings are related to where the cooling holes and outlets are located and consistent with the field observation of NGV damages.

Keywords: conjugate heat transfer, gas turbine, nozzle guide vane, internal air cooling

1. Introduction

Conjugate heat transfer (CHT) means the combined heat transfer processes between and in solid and fluid, where conduction usually dominates in solid and convection generally governs in fluid. The CHT analysis is based on solid physics and governed by a set of mathematical differential equations with well-defined boundaries.

The CHT analysis has now become an essential tool in the development and maintenance of thermal systems from aerospace, telecommunication, medicine, and atmosphere/ocean interaction to food processing. It can provide realistic 3D heat transfer analyses for practical complex devices, and this is why it has been extensively used in a wide range of applications to replace the traditional 1D semiempirical approach in CHT analyses.

In this chapter, as a technology demonstration, the conjugate heat transfer of the internally air-cooled nozzle guide vane (NGV) and shrouds of a gas turbine engine at flight conditions is numerically analyzed. Due to the continuous demand for high efficiency [1], the NGV is always exposed to high-pressure and high-temperature working environments and the field variables surrounding and inside the NGV change drastically. A higher temperature reduces the fatigue strength of the material, while a larger temperature gradient increases the thermal strain excursions and causes higher stresses. Both of them can reduce the NGV fatigue life. Therefore, for reliable life analysis of engine critical components, accurately assessing their aerothermodynamic working environments is essential.

Due to the high-pressure and high-temperature conditions, experimental measurements are particularly difficult inside the gas turbine engine. Computational fluid dynamics (CFD) analyses are normally used and continuously validated for calculating the aero-thermal loads for gas turbine engines. Mazur and coworkers [2] performed the conjugate heat transfer analysis between the NGV/shrouds and flow for a power generation gas turbine combustor. They found that based on the calculated thermal loads, stress, and creep strain analyses, it was feasible to assess the nozzle remaining lifetime. However, the shroud thermal boundary conditions were not given in their paper. Charbonnier and coauthors [3] validated the heat transfer predictions with the measured data for a film-cooled linear cascade. They found that the CFD predictions were in good agreement with the measurements of the vane surface Mach number, film cooling row discharge coefficients, wall heat flux, and adiabatic film cooling effectiveness. A conjugate heat transfer solver was developed by Heidmann and coworkers [4], where no grid was required inside the solid regions. They used this code to study the heat transfer of the middle span section of a film-cooled turbine vane with a few blade materials. The results indicated that the heat transfer in a film-cooled vane was complex and the simple 1D conduction analysis from the vane surface to the plenum surface was no longer valid.

In spite of being a common practice in the industry, the detailed CFD analysis of internally air-cooled nozzle guide vanes and shrouds with high-fidelity geometry and well-defined boundary conditions is rare. There may be three reasons. Firstly, the realistic geometry of real-world turbine NGV/shrouds is complex, particularly for the internal cooling flow channels. Secondly, fine mesh near the walls is required in order to accurately calculate heat transfer [5], and thus the mesh size can be considerably large even for a single NGV and shrouds. Thirdly, it can be difficult to find reliable boundary conditions, where both the engine combustor flow field and secondary air-cooling flow have to be considered. The significant effect of secondary air-cooling flow on the end walls of an air-film-cooled NGV was found by Harsqama and coworkers [6].

As a technology demonstration, a numerical investigation has been performed to study the conjugate heat transfer of the internally air-cooled nozzle guide vane of a practical gas

http://dx.doi.org/10.5772/intechopen.74104

turbine engine at flight conditions. The CFD model with high-fidelity geometry, complex flow features, temperature contours around/inside the NGV, and heat fluxes at typical NGV cross sections are presented and described. Finally, a number of highlights are given.

2. High-fidelity CFD model of air-cooled NGV

2.1. Computational domain and mesh of internally air-cooled NGV and shrouds

A conventional can-annular gas turbine combustor is chosen for this study. An annular chamber accommodates 6 combustion cans, and 60 NGVs are located at the exit of the combustor. The high-fidelity geometry of the NGV and shrouds is retained in the present work. For example, the curved three-dimensional internal/external surfaces are maintained.

Figure 1 illustrates the computational domain and mesh of an internally air-cooled NGV with internal and external shrouds. It is a 6° sector in the engine combustor coordinates, and the periodic boundaries of the geometry are rotated twice to avoid cutting through the NGV body. The flow domain starts upstream of the flat plates of the internal and external shrouds and is extended downstream of the shrouds by half of the NGV chord length. Both flow and solid meshes are shown in **Figure 1(a)**, and in **Figure 1(b)**, only the solid mesh is illustrated. The mesh at one middle cross section is given in **Figure 1(c)**, where the thick black lines represent the NGV body cross section which separates the flow and solid region meshes. In order to properly solve the conjugate heat transfer between the fluid and solid regions, fine grids were created inside the NGV body and flow regions around the NGV. The NGV CAD geometry file was generated based on laser-scanned cloud of points, which resulted in difficulties in mesh generation. A number of meshes were generated and attempt runs were carried out to improve numerical simulation quality. Eventually, a mesh size of 6.3 million cells was used for all simulations. Fine grids were also created in the vicinity of the NGV internal and external walls, fins, holes, and cooling air slot, as shown in **Figure 1(b)**. In the flow regions away from the NGV walls, coarse nodes were generated. At the flight conditions, the nondimensional wall boundary parameter, y+ was in a range of ~0–25 on the cooling slot, fins, and NGV internal walls. It varied from 5 to 35 at the NGV external walls and from 5 to 130 at the shroud walls.

As illustrated in **Figure 1(a)**, and **1(b)**, the hot gas mixture, the main flow, from the engine combustor runs over the NGV body. The cooling air coming from the combustor annular chamber enters the NGV thin-wall cooling slot through the cooling air inlet above the flat plate of the internal shroud, travels through six holes facing the NGV leading edge and one hole looking downstream into the cooling chamber inside the NGV body, then flows out of the NGV body at the four rectangular air exits, merges together with the main flow, and finally exits the flow domain. There are twelve cooling fins on either side of the cooling chamber to increase the heat transfer between the cooling air and NGV internal surfaces. The cooling air in the NGV body absorbs heat from the NGV internal surfaces and decreases its external surface temperature through impingement and convection.

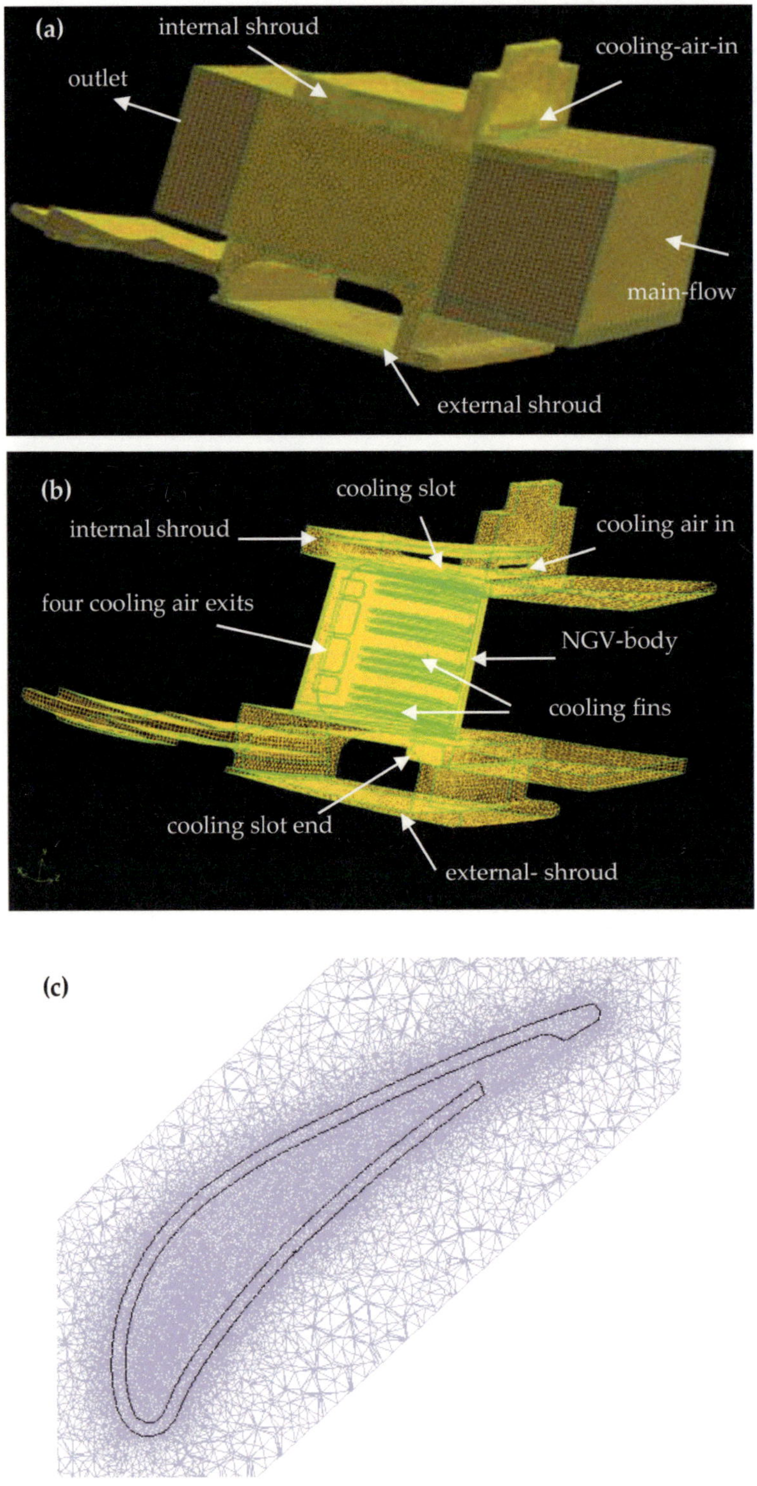

Figure 1. Computational domain and mesh: (a) flow and solid mesh, (b) solid mesh, and (c) mesh at one middle cross section.

Case	Size	y+	Δt	Δt/t
Mesh-1	21,200	~200	14.9	1.3%
Mesh-2	282,000	~60	8.0	0.7%
Mesh-3	180,000	~10	9.7	0.8%
Mesh-4	265,200	~0.5	~0	~0

Table 1. Variation of hot wall temperature with y + value.

As indicated in **Figure 1**, the high-fidelity geometry of the NGV and shrouds is maintained in the present work. For example, the curved 3D internal/external surfaces between the NGV body and shrouds are retained.

2.2. The effect of the nondimensional distance from the wall (y+) on the wall temperature

The main objective of this work is to adequately compute conjugate heat transfer among the external combustion mixture, NGV body, and internal cooling airflow. Therefore, the mesh independence effect in the numerical solutions is examined by simplified 2D simulations. The computational domain of these 2D simulations contains a flat metal plate, and its thickness is equal to the mean value of the NGV body. A hot gas flow with the velocity and temperature corresponding to the combustion mixture flows over one side of the plate, and a cooling flow with the velocity and temperature equal to those in the NGV simulation, runs along the other side of the plate.

The mesh size and result of four test cases are summarized in **Table 1**, where rectangular cells are generated for all meshes. In Mesh-1 and Mesh-2, grids are distributed uniformly in both coordinates, while for Mesh-3 and Mesh-4, in order to maintain reasonable mesh size, fine grids are created near the walls and gradually expand away from the walls.

As indicated in **Table 1**, the hot wall temperature for y + ≈ 0.5 (Mesh-4) is taken as the best result, Δt represents the deviation of wall temperature from the Mesh-4 case, and Δt/t denotes the difference in wall temperature between Mesh 4 and other cases normalized by that of the Mesh-4 case. The comparisons suggest that in the range of y + values (0.5–200), the maximum deviation is limited to about 1.3%. The temperature values of Δt and Δt/t at the cooling side wall are almost identical as those in **Table 1**, and thus they are not presented here. As stated before, y + varies from ~0 to 35 at the NGV external walls and in a range of 5–130 at the shroud walls in the present work, and thus the mesh used in the simulations is fine enough for wall temperature calculation.

3. Numerical simulations

In the present study, steady compressible turbulent thermal flow fields are resolved. A commercial software code, ANSYS Fluent, is applied to all the numerical simulations. In the following subsections, the numerical approach, boundary conditions, and solution methods are discussed briefly.

3.1. Numerical approach

The Favre-averaged conservation differential equations for mass, momentum, and energy can be readily found in the literature, such as [7, 8], and thus they are not presented here. The viscous heating and radiation heat transfer items in the energy equation are not considered in the present study since their effect on the flow field around the NGV is negligibly small.

The SST k-ω model (the shear-stress transport model) is selected to model turbulent transfer items in the flow. Major improvements are observed in the prediction of separation and adverse pressure gradient flows in comparison with the standard k-ω and k-ε models [9]. This model has been successfully used in the heat transfer simulation for the film cooling of a linear cascade, and the results are in good agreement with the experimental measurements [3].

In engineering applications, it is difficult to keep y + values to a prescribed value for all wall-adjacent cells throughout the domain for complex devices. To keep numerical solutions insensitive to wall mesh, an automatic wall treatment is formulated for the SST model in ANSYS Fluent [8]. According to local y + values, it shifts automatically from the viscous sub-layer treatment to the fully turbulent region formulation, while in the buffer layer between the viscous sub-layer and fully turbulent region, a blending function is used.

In the present work, the thermal properties of air are utilized for the combustion gas mixture because the overall fuel-air ratio in gas turbine combustors is low [1]. The polynomials as a function of temperature are applied to compute the specific heat of air. The molecular viscosity and thermal conductivity of air at the average temperature over the whole domain are used in simulations. The polynomials are also used to compute the specific heat and conductivity for Hastelloy X alloy.

3.2. Boundary conditions

The boundary conditions at the inlet are taken from a previous benchmark simulation of the gas turbine combustor at the flight conditions. The simulation is similar to those carried out in [10], where the two-phase, compressible, reacting flow of a 60° sector of the can-annular combustor with ten NGVs (no internal cooling flow passages) attached is resolved. The average total temperature and pressure obtained from this simulation at the matching location are defined as the inlet boundary conditions for the main flow (see **Figure 1**). As stated earlier, the NGV cooling air comes from the annular chamber of the combustor. The average temperature and pressure values at the ten NGV cooling air slots from the simulation are specified as the cooling air inlet conditions for the cooling flow (see **Figure 1**). The cooling airflow rate is 2.3% of the total inlet airflow. To consider the effect of the flow tangential velocity on the pressure distribution, a radial equilibrium pressure distribution is applied to the domain exit plane. The Reynolds number calculated with the chord length and inlet flow conditions is 1.0×10^5.

The thermal boundary conditions on the shroud outer surfaces are estimated from the secondary airflow information (flow temperature and Mach number) of [11]. A 2D simulation is carried out, and in this simulation, the stream-wise cross sections of the internal/external shrouds remain the same, the NGV body is removed from the computational domain, and two additional flow zones beside the external and internal shrouds are included in the domain. The wall temperatures obtained from the simulation are utilized at the outer surfaces of the NGV/shroud assembly. The turbulence kinetic energy and specific dissipation rate at the inlets of the NGV/shroud assembly are calculated with an estimated turbulence intensity of 10% and inlet hydraulic diameters. At the side boundaries of the 6° sector, periodic boundaries are specified.

3.3. Solution methods

A pressure-based coupled solver with a second-order accuracy scheme is used to solve the flow field. At convergence, the normalized residuals are less than 4×10^{-5} for all velocity components and energy and around 6×10^{-4} for turbulent parameters. The mass flow rate monitored at the four rectangular outlets of the internal cooling airflow remains unchanged at least for the first three digits. All these ensure that the simulation reaches steady conditions. A 4-node LINUX cluster, with 64-bit, 2.6 GHz, 8-core, and 64 GB RAM per node, is used to perform all the simulations.

4. Results and discussion

In the following subsections, some of the simulation results around and inside the NGV/shrouds at the flight conditions are given and discussed.

4.1. Flow field of the NGV/shrouds

The Mach number contours at the section passing through an upstream hole and the downstream hole of the cooling airflow passage are shown in **Figure 2**. **Figure 2(a)** represents the whole cross section, and **Figure 2(b)** is a plot zoomed near the NGV. The temperature contours at the same cross section are illustrated in **Figure 3**, and the local temperature contours around the NGV are enlarged in **Figure 3(b)**. In this figure and other figures in the chapter, the temperatures are normalized by the maximum temperature in the computational domain. Lines in the plots denote the NGV and internal cooling slot metal walls.

Figures 2 and **3** clearly display the compressible flow features. In the inlet section, the flow velocity and temperature are more or less constant. The flow acts noticeably differently on the pressure and suction sides as it approaches the NGV. A stagnation point is formed at the NGV leading edge of the pressure side, as shown in **Figure 2**. After the leading edge, the flow gradually accelerates, and the Mach number or velocity remains relatively low till reaching the trailing edge. More complex interactions among the main flow, cooling flow, and NGV geometry happen near the rectangular opening. The flow rapidly speeds up from the end edge of the rectangular opening, and behind the trailing edge, a small recirculation or stagnation zone is observed. The static temperature at the pressure side is relatively high except for the small downstream region, as shown in **Figure 3**, and this is because the flow velocity is relatively low except for the small region from the end of the rectangular opening to the trailing edge.

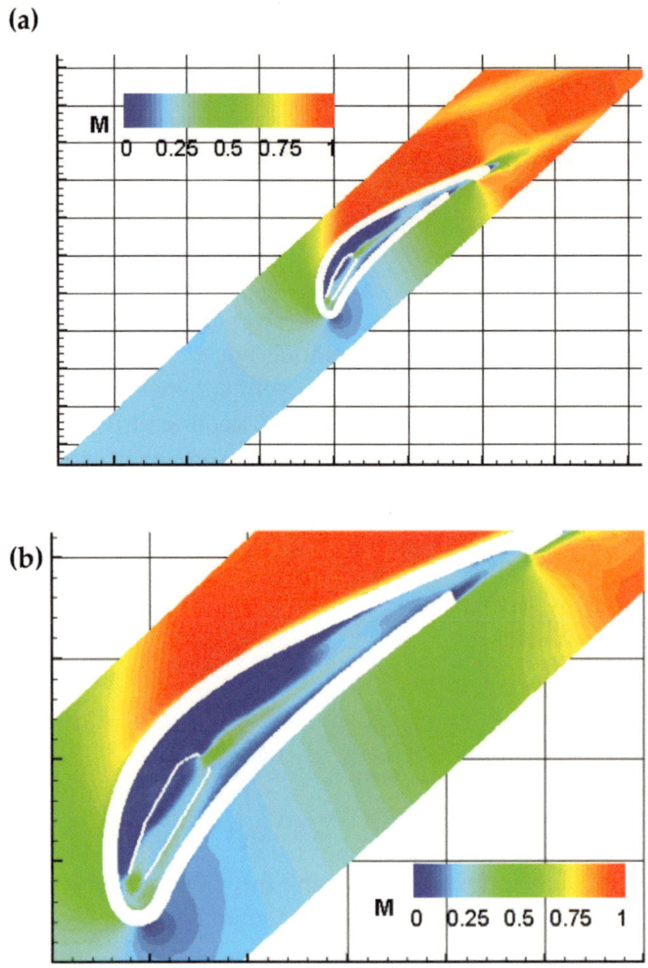

Figure 2. Mach number contours at the section across an upstream hole and the downstream hole of the cooling slot. (a) the whole section and (b) the zoomed local region.

In contrast, the flow velocity accelerates quickly along the suction surface after the leading edge, stays high in most of the downstream region, and rapidly decelerates at the trailing edge. The maximum Mach number is slightly higher than 1.0 in the flow, and two local supersonic spots occur in **Figure 2**. Because of the rapid flow acceleration along the suction side, the temperature drops quickly from the leading edge and remains relatively the same level at most downstream regions. In short, as illustrated in **Figure 2**, the pressure side wall is subject to higher temperature than the suction side wall, which agrees the observation in [2].

At this section in the cooling chamber, the flow velocity is fairly high around the leading edge, middle, and downstream portion. However, the cooling air velocity is low in a large region around the suction side. It is anticipated that the cooling chamber and fin geometries are the reasons for these observations.

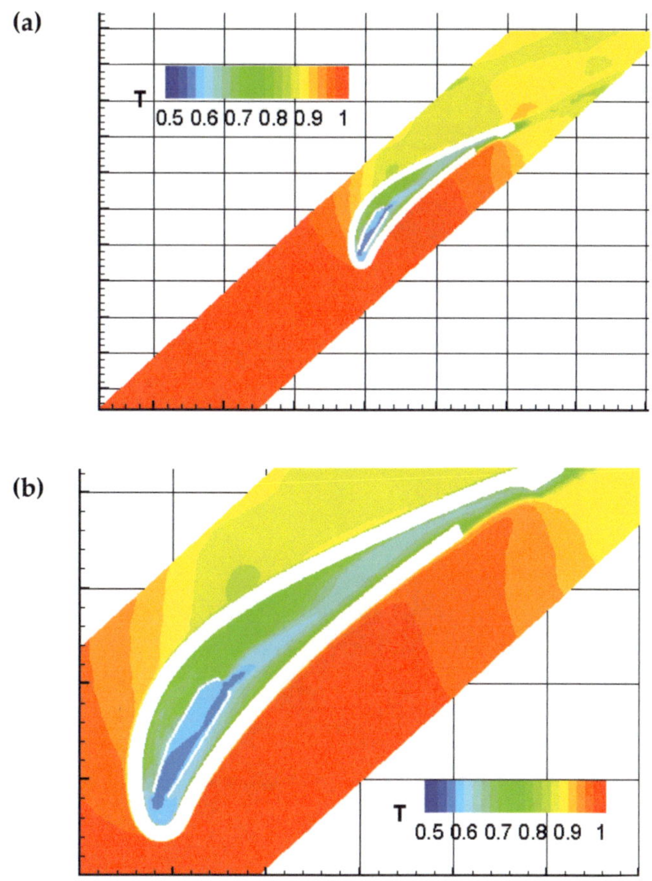

Figure 3. Temperature contours at the section across an upstream hole and the downstream hole of the cooling slot. (a) the whole section and (b) the zoomed local region.

The velocity vector plot is presented in **Figure 4** at the same cross section as **Figures 2–3**. **Figure 4(a)** shows a whole picture of velocity vectors around the NGV section wall, **Figure 4(b)** displays vector details around the upstream cooling hole of the cooling slot, and **Figure 4(c)** is a plot zoomed around the downstream hole. The vector length indicates its magnitude, and the color stands for its Mach number. For the present design, both impingement and convection cooling are utilized to cool the NGV walls. As shown in **Figure 4(a)** and **(b)**, the cooling air directly impacts on the inner surface of the NGV near the leading edge. In all three plots of **Figure 4** the convection cooling is clearly observed, i.e., the cooling air runs along the NGV external/internal surfaces.

Table 2 gives the mass flow rate through each cooling hole as a percentage of the total cooling air. The upstream holes in **Table 2** are listed as 1 to 6 beginning from the internal shroud side. At the four upstream holes (3–6) and the downstream hole, the mass flow rates are at the same level, and at upstream holes 1–2, the mass flow rates are close to each other. The maximum difference in the cooling flow rate is 3.2% between holes 2 and 6.

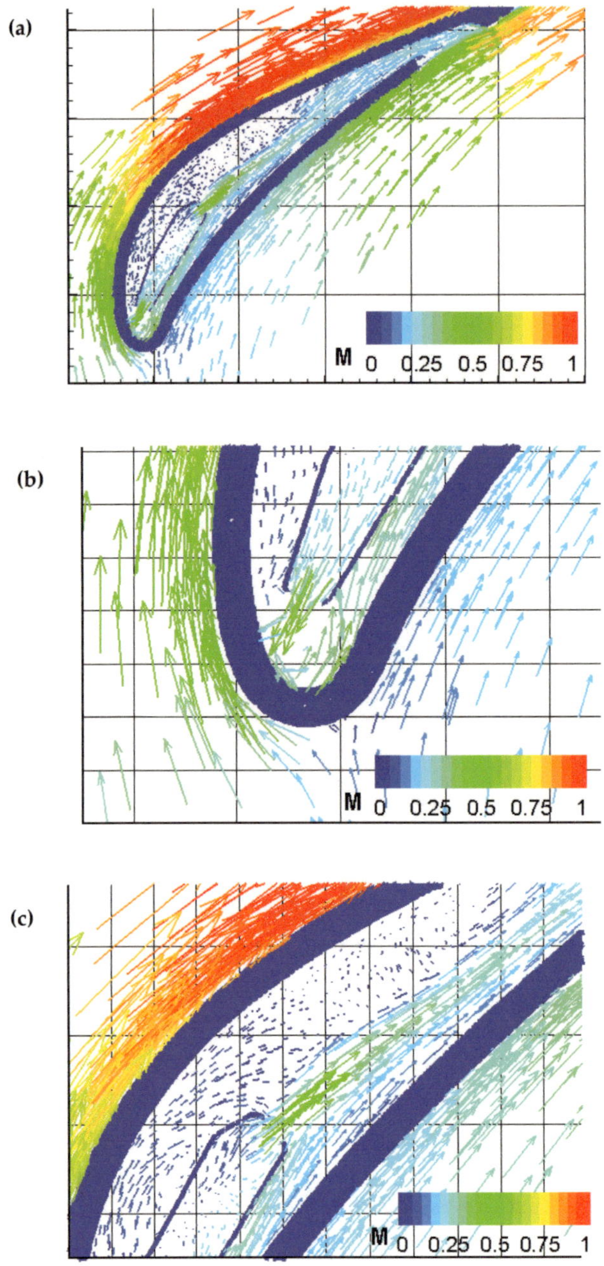

Figure 4. Velocity vectors colored by Mach number levels: (a) the section across an upstream hole and the downstream hole, (b) the section across an upstream hole, and (c) the section across the downstream hole.

4.2. Temperature distributions of NGV/shroud walls

Figure 5 shows the normalized temperature contours of the NGV and shroud walls. A legend scale of T = 0.5–0.875, instead of 0.5–1.0, is used in **Figures 5–8** for clarity. The high

	No.	Mass flow rate
Downstream hole	1	14.1%
Upstream holes	1	12.9%
	2	12.4%
	3	15.1%
	4	14.7%
	5	15.1%
	6	15.6%

Table 2. Mass flow rates through cooling holes.

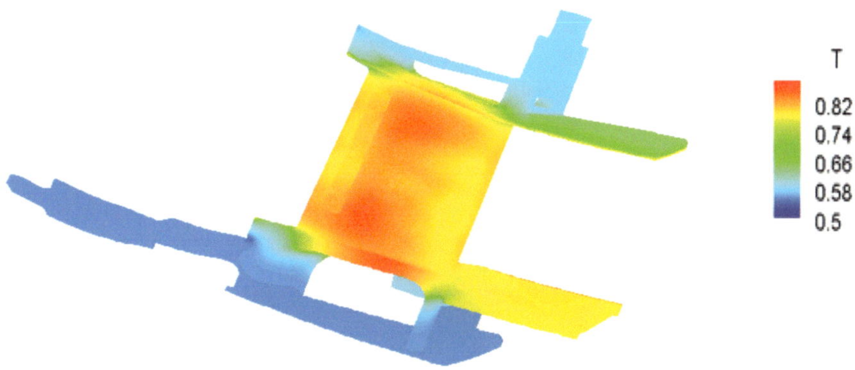

Figure 5. Temperature contours of the NGV and shrouds.

temperature is found in two local areas, one upper and one lower, on the NGV pressure side near the rectangular cooling air outlets as well as the upper middle region of the external shroud, with a maximum value of 0.87. At the shroud outer surfaces, the wall temperature is comparatively low.

The temperature contours at the NGV external and internal surfaces are presented in **Figures 6–7**, respectively, and the temperature distributions over the fin surfaces are shown in **Figure 8**. **Figures 6–8** illustrate the temperature distributions at both pressure and suction sides. At these surfaces, the temperature variation is significant. As mentioned earlier, the higher temperature occurs on the external walls of the pressure side than that on the suction side, and two local high-temperature areas around the rectangular outlets are observed on the pressure side. These findings are consistent with the temperature distributions over the NGV internal and fin surfaces These results clearly imply that the semi-empirical 1-D heat transfer analysis is not suitable for the practical complex 3-D heat transfer phenomena.

In **Figure 9**, the metal temperature contours at three sections with two, (a) and (b), cutting through the hot areas of the NGV pressure side, and one, (c), through the middle portion, with a temperature scale of 0.7–0.86. The metal temperature variations at these sections are obvious and in agreement with the findings in **Figures 5–8**. The maximum variation of metal

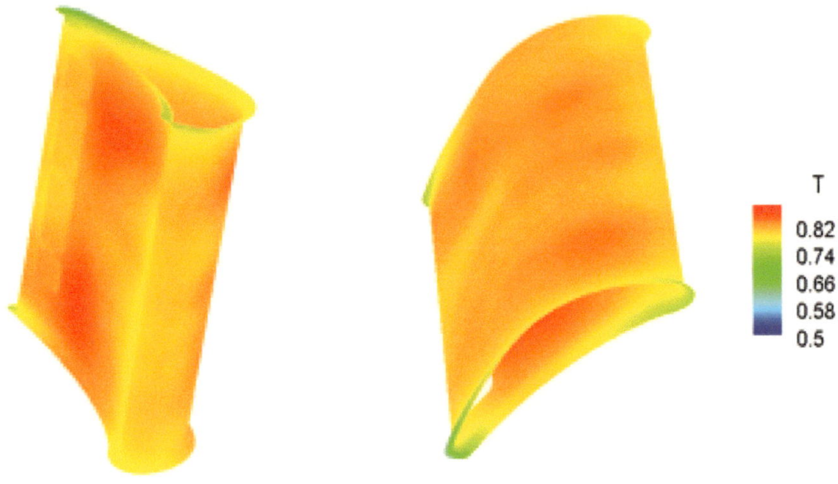

Figure 6. Temperature contours at the external surfaces of the pressure and suction sides.

Figure 7. Temperature contours at the internal surfaces of the pressure and suction sides.

wall temperature in **Figure 9** is 0.12. The Biot number at the NGV internal/external surfaces is only 0.006 in the present heat transfer case, much less than 1. This means that the resistance of the heat conduction in the NGV solid regions is much less than that of the heat convection through its surfaces. As a result, the temperature distribution in the NGV body is mainly determined by the complex heat convection among the NGV body, external main flow, and internal cooling flow.

The temperature distributions at the two sections across the centers of the hot spots on the pressure side observed in **Figures 5–9** is shown in **Figure 10**, and the outlines of the air-cooling slot, seven holes, and NGV main body are included in the figure. The upper section passes

http://dx.doi.org/10.5772/intechopen.74104

Figure 8. Temperature contours at the cooling fin surfaces.

through the middle between the two rectangular outlets of the cooling flow and does not cut through any cooling hole. The lower section just passes through the edges of a rectangular outlet and a cooling hole. It is anticipated that the cooling airflow is restricted to some extent at local areas of these two sections. This may be the reason why the temperature around these local areas is higher than the other areas. The temperature distributions at the middle section across an upstream hole and the downstream hole of the cooling slot are shown in **Figure 11**. This geometrical configuration may explain why the temperature at the middle of the NGV and fins is relatively lower than in the neighboring areas.

As stated before, the heat transfer is complex inside and outside the NGV/shrouds, and it is no longer governed by the semiempirical 1D convection-conduction-convection analysis, i.e., from the main hot flow to the NGV metal body and then to the cooling airflow in the cooling chamber. For instance, if only velocity or Mach number is considered (**Figure 2**), it is anticipated that the heat flux at the pressure side is lower than that at the suction side. However, according to the temperature contours in **Figure 3**, the heat flux at the pressure side tends to be higher than the suction side.

Figure 12 provides the heat flux through the NGV internal and external surfaces at the three sections, as shown in **Figures 9–11**. In this figure, the heat flux values are normalized by the mean maximum heat fluxes across the NGV internal and external surfaces, where the metal cross sections are shown in dark blue thick lines. The data points in red stand for the heat flux across the NGV external surface, while the green dots represent the heat flux across the internal surface. The positive sign of heat flux means the flux flows toward the cooling chamber, and for the negative sign, it is in the opposite direction. The characteristics of the heat flux patterns at the three sections are similar. Owing to the 3D nature of the NGV/shroud heat transfer, the heat fluxes across the NGV internal and external surfaces are not equal at all. Around the NGV leading edge, the heat flux across the internal surface is larger than that for

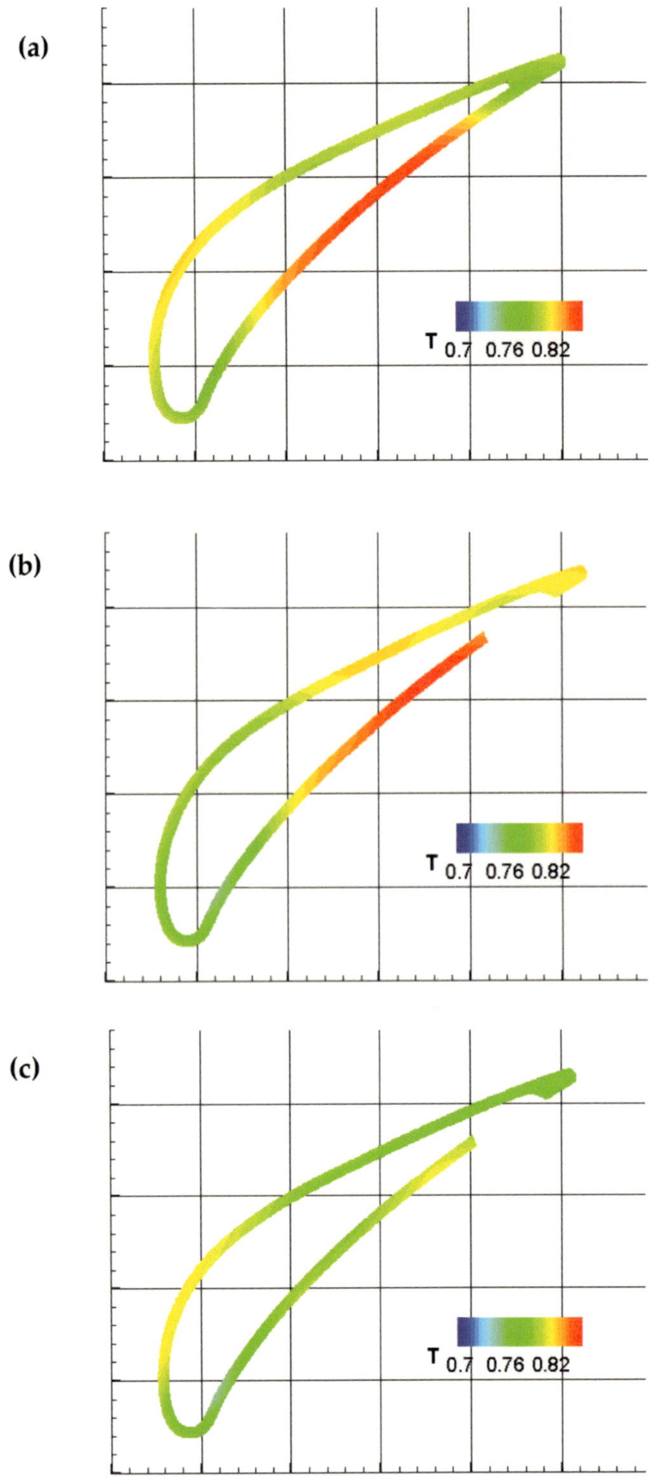

Figure 9. Metal temperature contours: (a) and (b) across hot spots and (c) across the middle of the NGV.

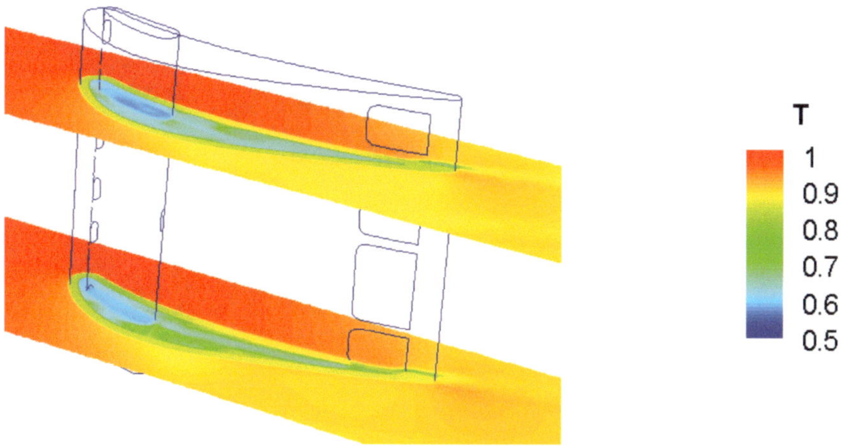

Figure 10. Temperature contours at the two sections across the high-temperature spots on the NGV external surfaces.

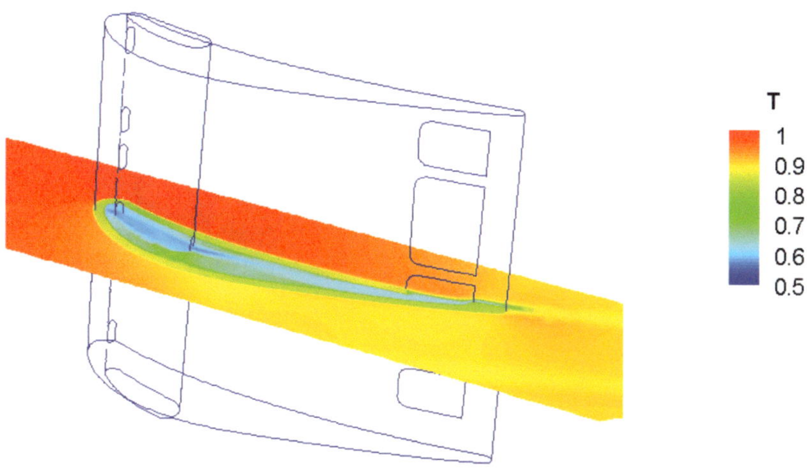

Figure 11. Temperature contours on the section across the downstream hole and an upstream hole of the cooling slot.

the external surface, in particular at the lower and middle cross sections (**Figure 12(b)** and **(c)**). Generally, the flux is high near the NGV leading edge and steadily decreases along the trailing edge. Local variations and bumps of the heat flux profiles are observed, which are determined by the local flow conditions and geometry.

Around the trailing edge, there is a substantial negative heat flux at the external surface. Note that the area with negative flux is a portion of or close to the cooling air exiting passage (**Figure 1**) and the cooling air runs over the tiny trailing edge region. Displayed in **Figure 13** are the flow/metal temperature distributions around the NGV trailing edge at these three sections: (a) for the upper section, (b) for the lower section, and (c) for the middle section. It is obvious that the flow temperature around the trailing edge region is lower than the neighboring metal temperature at all three cross sections. Therefore, the heat flux alters its direction, i.e., the negative heat flux is found at these small regions.

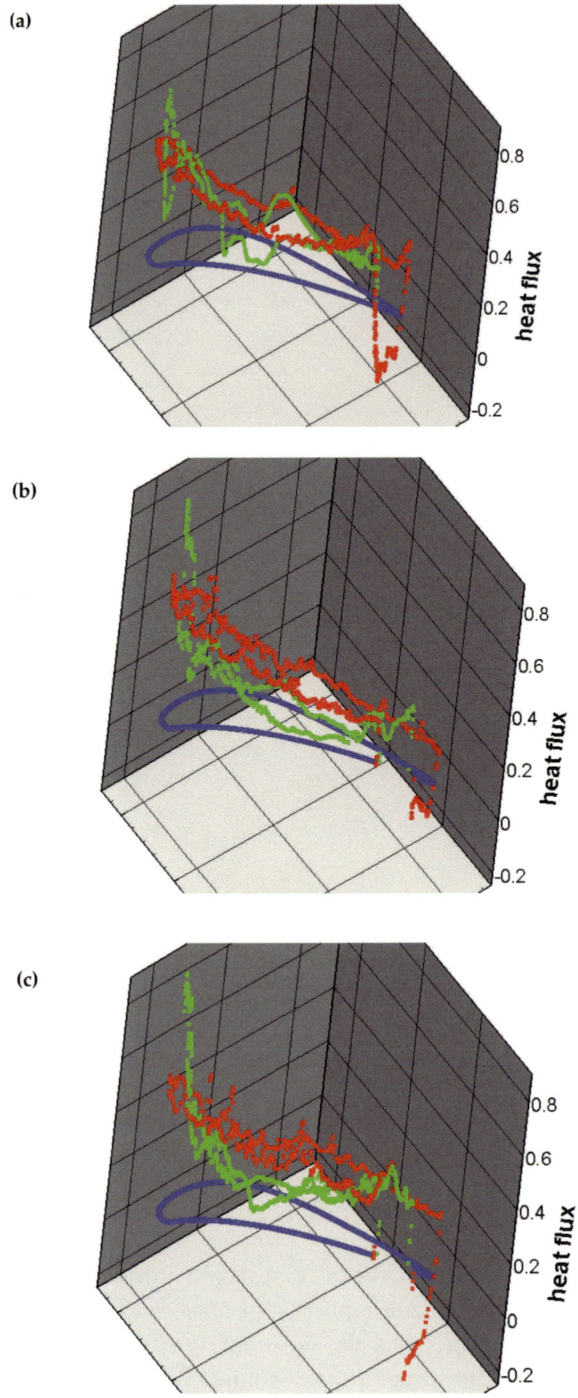

Figure 12. Normalized heat flux (red for external surface and green for internal surface): (a) at the upper section, (b) at the lower section, and (c) at the middle section.

http://dx.doi.org/10.5772/intechopen.74104

(a)

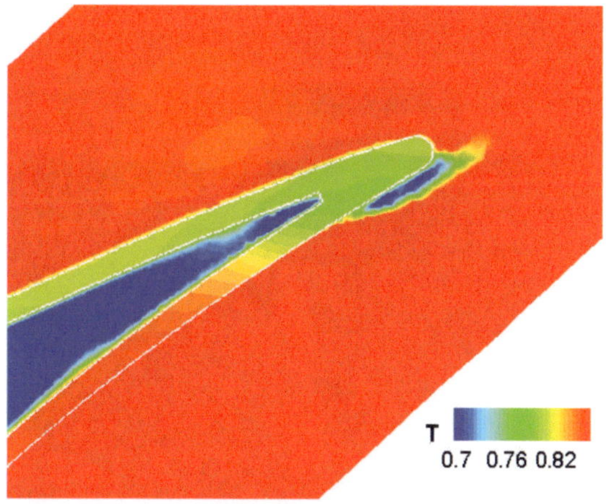

(b)

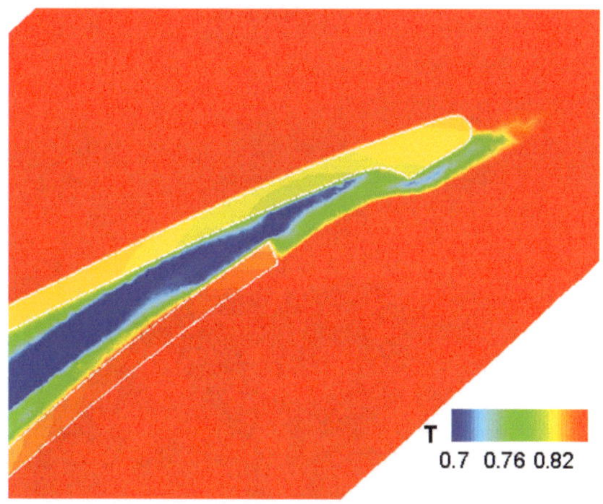

c)

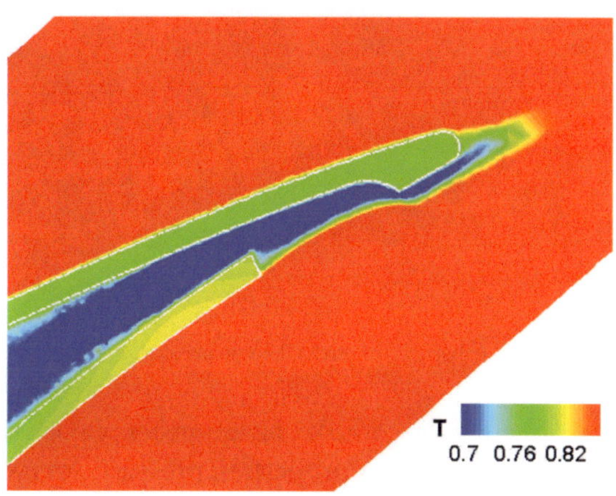

Figure 13. Temperature contours near the NGV trailing edge: (a) at the upper section, (b) at the lower section, and (c) at the middle section.

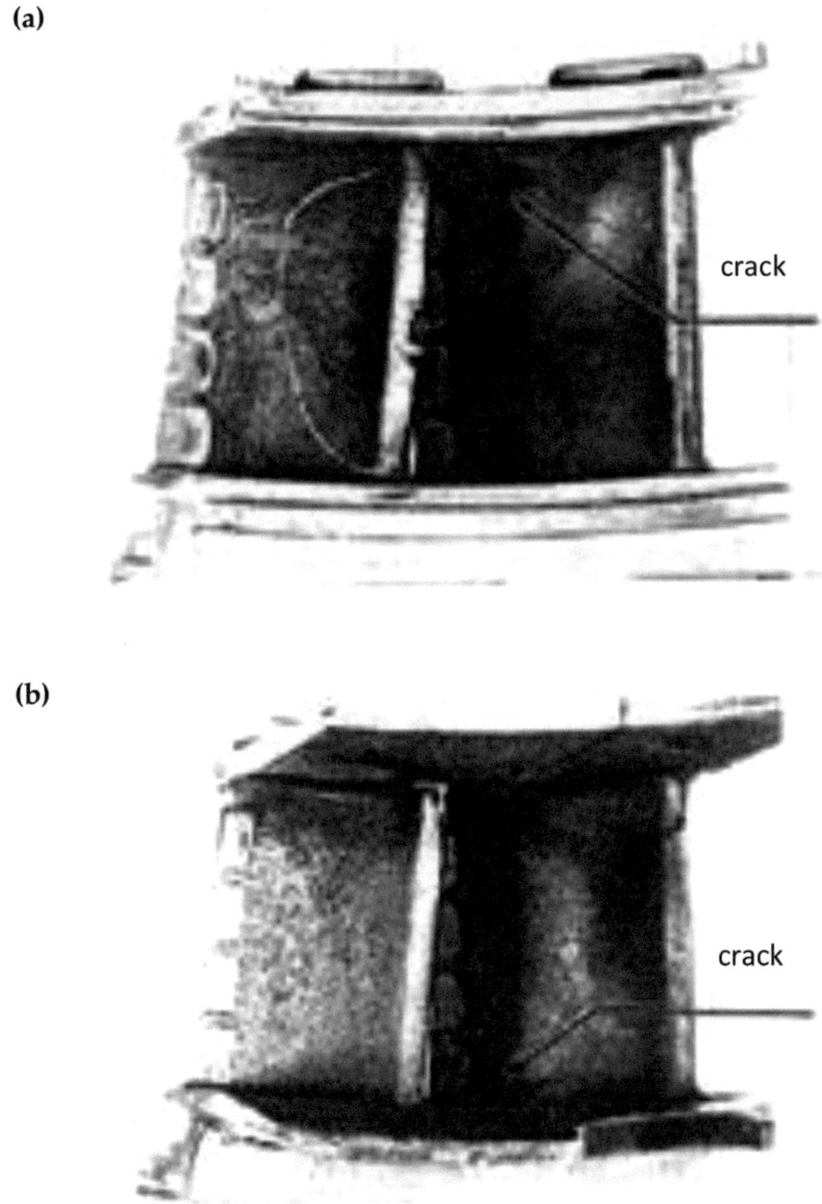

Figure 14. Cracks at the upper and lower portions of the NGV pressure side, starting from the rectangular cooling air exits.

There are many factors affecting the complex heat transfer phenomena observed in **Figure 13**: the external flow parameters (local gas properties, pressure, temperature, and velocity), the cooling flow variables (local air properties, pressure, temperature, and velocity), and the local geometries of the NGV and cooling flow passage (shape and location of the cooling holes, outlets, fins, shrouds, etc.). Moreover, as shown in **Figures 1**, **2–4**, **7–8**, and **10–11**, the cooling air enters the cooling chamber through the internal shroud, crosses the seven cooling holes, flows over the 24 cooling fins, and eventually exits the chamber through four rectangular outlets. It is expected that complex secondary flows can be generated because of the jet flows issuing from the cooling holes and the interactions among the cooling flow and cooling fins [12]. The detailed analysis of the secondary flow effect is outside of the scope of the present study, and it can play an important role in the NGV cooling [6].

The photos of damaged NGVs after a certain service time are illustrated in **Figure 14**, and there are cracks observed at the pressure side. A long crack from the rectangular outlet edge to the NGV leading edge is located at the upper portion of the NGV body as shown in **Figure 14(a)**, and a short crack also beginning from the rectangular outlet edge occurs at the NGV lower surface shown in **Figure 14(b)**. These field observations are in good agreement with the above-predicted results. As shown in **Figure 9**, the cross sections (a) and (b) are located in the upper and lower regions of the NGV body, close to or agreeing with the crack locations. The high metal temperature area is found from the middle of the span to the cooling rectangular outlet on the NGV pressure side. The high temperature decreases the metal strength, and the rectangular geometry and temperature variation increase local metal material stress. Consequently, these rectangular edges in the upper and lower regions at the pressure side are more vulnerable and prone to cracking.

As a technical demonstration case, the obtained results provide essential information for the design or life analysis of the NGV/shroud assembly. Moreover, the results recommend that the cooling flow can be optimized by rearranging the holes, outlets of the cooling flow passage, and/or altering the numbers and sizes of cooling holes. It is clear that the detailed heat transfer analysis of the NGV/shroud assembly helps to achieve this goal.

5. Conclusions

In this chapter, the flow field and conjugate heat transfer of an internally air-cooled nozzle guide vane and shroud assembly of a gas turbine engine at flight conditions are numerically studied. The high-fidelity CFD model is created, and the simulations are performed with properly defined boundary conditions. The characteristics of the complex flow and temperature fields are discussed. The Mach number is higher, and the temperature is lower on the NGV suction side than those on the pressure side. The results reveal that the NGV surface temperature varies considerably and the thermal stress could be excessively high. Two high-temperature spots on the NGV pressure side are observed, which can significantly reduce the NGV service life. These findings are closely correlated with the locations of the cooling holes and outlets of the cooling flow passage, and consistent with the field observations of damaged NGVs. As a technology demonstration case, the findings provide essential data for the design or service life assessment of the NGV/shrouds, for example, improving the cooling effectiveness.

Acknowledgements

The authors are grateful to the Air Defence System Program, National Research Council of Canada for supporting this research project. The authors would also like to express their thanks to Dr. Francois Fortin for the valuable comments and suggestions during the preparation of this manuscript.

Author details

Lei-Yong Jiang*, Yinghua Han and Prakash Patnaik

*Address all correspondence to: lei-yong.jiang@nrc-cnrc.gc.ca

Aerospace, National Research Council of Canada, Ottawa, Ontario, Canada

References

[1] Lefebvre AH, Ballal DR. Gas Turbine Combustion. New York: Taylor and Francis Group; 2010

[2] Mazur Z, Hernandez-Rossette A, Garcia-Illescas R, Luna-Ramirez A. Analysis of conjugate heat transfer of a gas turbine first stage nozzle. Applied Thermal Engineering. 2006;**26**:1796-1806

[3] Charbonnier D, Ott P, Jonsson M, Kobke T, Cottier F. Comparison of Numerical Investigations with Measured Heat Transfer Performance of a Film Cooled Turbine Vane. ASME IGTI Paper, GT2008-50623.2008

[4] Heidmann JD, Kassab AJ, Divo EA, Franklin Rodriguez F, Steinthorsson E. Conjugate Heat Transfer Effects on a Realistic Film-Cooled Turbine Vane. ASME IGTI Paper, GT2003-38553. 2003

[5] Jiang LY, Manipurath S, Bourque G, Houde M. Flow-field of a triple-walled gas-sampling probe with sub-cooled boiling effect. Flow Measurement and Instrumentation. 2005;**18**:156-116

[6] Harsqama SP, Burton CD, Chana KS. Measurements and computations of external heat transfer and film cooing in turbines. Proceedings of the 10th International Symposium on Air Breathe Engines (ISABE). 1991. pp. 1276-1284

[7] Ferziger JH, Peric M. Computational Methods for Fluid Dynamics. New York: Springer-Verlag; 2000

[8] ANSYS Fluent, Fluent 15 Documentation. Lebanon, NH, USA. 2015

[9] Menter FR. Two-equation Eddy-viscosity turbulence models for engineering applications. AIAA Journal. 1994;**32**(8):1598-1605

[10] Jiang LY, Corber A. Assessment of combustor working environments. International Journal of Aerospace Engineering. 2012;**2012**:6. Article ID: 217463. DOI: 10.1155/2012/217463

[11] Snedden G, Roos T, Naidoo K. Detailed disc assembly temperature prediction: Comparison between CFD and simplified engineering methods. ISABE Paper 2005-1130. 2005

[12] Samimy M, Breuer KS, Leal LG, Steen PH. A Gallery of Fluid Motion. UK: Cambridge University Press; 2003

Chapter 8

Photonic Metamaterials: Controlling Nanoscale Radiative Thermal Transport

Alok Ghanekar, Yanpei Tian and Yi Zheng

Additional information is available at the end of the chapter

http://dx.doi.org/10.5772/intechopen.72805

Abstract

We discuss concepts of radiative thermal diodes demonstrating dynamic control and modulation of radiative heat transfer. These concepts are analogous to electronic diodes and display high degree of asymmetry in radiative heat transfer. Change in optical properties of vanadium dioxide (VO_2) upon phase transition are exploited to influence thermal radiation. The first concept is based on a simple multi-layer structure containing a layer of VO_2 to attain dynamic optical response in the far-field regime. The active terminal of the diode changes from highly reflecting to highly absorbing upon phase transition of VO_2. In the second concept, a near-field thermal diode is considered that utilizes period gratings of VO_2. Radiative heat transfer across the near-field gap is modulated by altering tunneling of surface waves when phase change in VO_2 occurs. For minimal temperature difference of 20 K, rectification ratios have been reported and they are maximum in existing literature for comparable operating temperatures and configurations.

Keywords: metamaterials, near-field, far-field, thermal diode, radiative thermal transport

1. Introduction

Thermal diode [1], thermal transistors [2], thermal memory element [3] and similar thermal analogues of electronic devices have been topic of theoretical as well as experimental works. While earlier research has been on conduction (phonon) based devices [4–8], more recent studies have been focusing on radiation (photon) based thermal rectifiers [9–12]. Thermal rectification has numerous applications in thermal management, thermal logic gates [13–15] and information processing [16].

Analogous to electrical diode, thermal diode is a rectification device wherein magnitude of heat flux strongly depends on the sign of applied temperature bias. To quantify rectification, one can employ the widely used definition of rectification ratio, i.e., $R = (Q_f - Q_r)/Q_r$ where

Q_f and Q_r refer to forward and reverse heat flux, respectively [17]. Alternatively, rectification coefficient can be defined as $\eta = (Q_f - Q_r)/\max(Q_r, Q_f)$. There are numerous studies pertaining to near-field and far-field thermal radiation based rectification devices that exploit temperature dependent properties of a phase change materials such as vanadium dioxide (VO_2) and $La_{0.7}Ca_{0.15}Sr_{0.15}MnO_3$ (LCSMO) [11, 18, 19]. A number of studies deal with far-field thermal radiation [20, 21] while several others focus on modulation of radiative heat transfer in the near-field regime [18, 19, 22–26]. Ben-Abdallah and Biehs introduced a VO_2 based simple far-field radiative thermal diode, while Prod'homme et al. [27] proposed a far-field thermal transistor that uses a VO_2 base between a blackbody collector and a blackbody emitter. Zhu et al. [28] showed that temperature dependent optical properties of SiC can be used to attain negative differential conductance. Van Zwol et al. [22] proposed that one can take advantage of the phase transition from crystalline to amorphous state in AIST (an alloy of Ag, In, Sb, and Te) driven by a current pulse to obtain a large contrast in heat flux. In far-field limit, rectification is due to the change in emissive properties of a phase change material. In near-field limit, the difference in the coupling strength of polaritons or tunneling of surface waves between structures leads to thermal rectification. In general, it is observed that a higher rectification can be achieved in the near-field regime than in the far-field. However, it is challenging to develop such devices operating on the principle of near-field radiative transfer.

Spectral control has been studied to affect radiative heat transfer in both the far-field as well as near-field. Customization of absorption/emission spectra is often achieved by the use of multilayer thin film structures [29], nanoparticles [30, 31], dielectric mixtures [32, 33], photonic crystals [34, 35], 1-D/2-D gratings [36] and metamaterials [37, 38]. Absorbers that utilize Fabry-Perot cavities [39, 40], Salisbury screens [41] and Jaumann absorbers [42] and ultra-thin lossy thin films bounded by transparent substrate and superstate [43–45] have been investigated for decades. Quite notably, Nefzaoui et al. [46] proposed using multilayer structures consisting of thin films (e.g., Si, HDSi and gold) to obtain thermal rectification. Kats el al. [47] have theoretically and experimentally demonstrated that a thin-film of VO_2 on sapphire shows strong modulation of absorbance upon phase transition, particularly, at wavelength of 11.6 μm. Taylor et al. [48] recently proposed an emitter consisting a dielectric spacer between VO_2 film and a reflecting substrate to achieve dynamic radiative cooling upon phase transition of VO_2. Fabry-Perot resonance was achieved at 10 μm wavelength. We will see that, by tuning the resonance at right wavelength, maximum rectification can be achieved in the proposed design.

VO_2 has often been used in thermal rectification devices, because its phase-change from an insulator to a metal can be switched reversibly within a short time (~100 fs) [49]. The common devices use either a bulk VO_2 solid or its thin-film form. Here, we will see two cases of radiative thermal diodes with record rectification ratios achieved when compared to similar operating conditions and configurations.

2. Far-field thermal diode

A typical far-field thermal diode has two planar components separated by a distance much larger than thermal wavelength. The active component is made of a phase-change solid, whereas the passive component stays inert. **Figure 1** illustrates the vertical structure of the proposed thermal diode. The active component contains a tri-layer structure consisting of VO_2, potassium bromide (KBr) and gold thin films on a substrate. Thicknesses of VO_2 and KBr layers can be tuned to maximize rectification. The thickness of gold layer is fixed at 1 μm to block radiation from the substrate. For a given temperature bias, maximum (far-field) radiative heat transfer would be possible when both sides are blackbodies, while minimum heat transfer would take place when at least one side is a highly reflective mirror. Ideally, the active component should exhibit a transition from blackbody to reflective surface upon the reversal of a temperature bias which induces the phase change. This is exactly what the design attempts to achieve. Therefore, the passive component is chosen to be a blackbody. Any material other than a blackbody would not yield the maximum rectification. Structures 1 and 2 are at temperature $T_1 = T_C + \Delta T$ and $T_2 = T_C - \Delta T$, respectively. The mean temperature is chosen to be the phase transition temperature of VO_2 ($T = 341$ K). When $T_1 > T_2$ (referred to as forward bias), VO_2 layer is in its metallic phase; and when $T_1 < T_2$ (reverse bias), VO_2 layer becomes insulating with its optical axis aligned along the vertical direction, i.e., z-axis.

Phase transition of VO_2 is not abrupt [49, 50] and a complete insulator-metal transition does not occur until 350 K [26]. Rectification ratio depends on temperature bias as the temperature

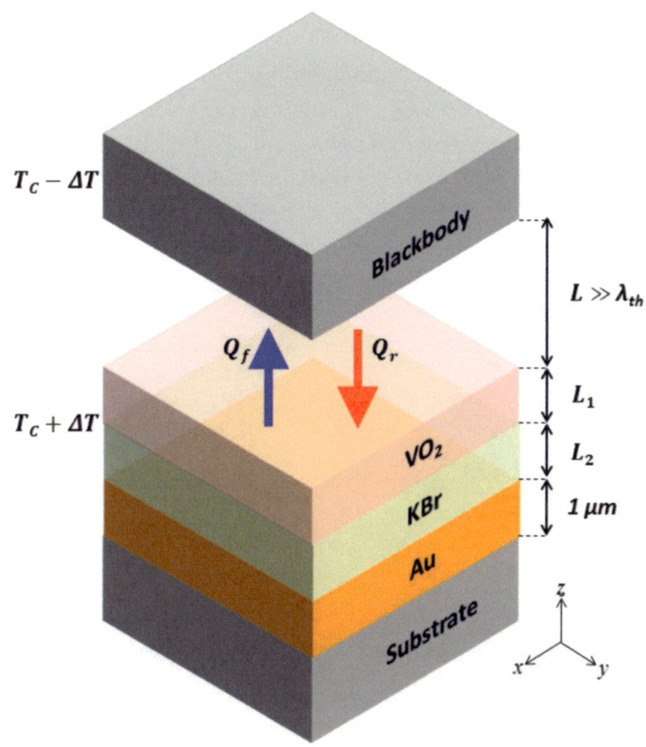

Figure 1. Schematic of a far-field thermal diode with a high rectification ratio. The active component has a tri-layer structure consisting of VO_2, KBr and gold thin films on a substrate with thicknesses L_1, L_2 and 1 μm, respectively. The passive component is a blackbody. $T_c = 341$ K is the phase transition temperature of VO_2.

dependence of radiative heat transfer is essentially nonlinear. Here, rectification values are calculated at a minimal temperature bias of 20 K i.e., $\Delta T = \pm 10$ K. Although transition of VO_2 exhibits a thermal hysteresis of about 8 K as presented in Refs. [49, 51], the phase transition is reversible. As we are concerned with heat flux values at 10 K above and below the critical temperature of VO_2, hysteresis behavior is ignored for the sake of simplicity. A multilayer structure can be designed to attain high absorbance or reflectance based on its dimensions and material properties. Multilayers with constituent thicknesses much smaller than the incident wavelength of light have been studied before [52]. Here, we see that in a VO_2 based multilayer structure, the dramatic change in the optical property of VO_2 upon phase-change facilitates an extensive variation in the surface reflectivity.

Concept shown in **Figure 1** has variable dimensions of VO_2 (L_1) and KBr (L_2) layer. These dimensions can be optimized by running Genetic Algorithm to maximize rectification ratio. When lower and upper bounds on L_1 and L_2 are kept at 25 nm and 2 μm, respectively, optimal dimensions can be found to be $L_1 = 25$ and $L_2 = 880$ nm, both are practical values. Further discussion will be focused on the design with these dimensions. **Figure 2** shows spectral heat flux ($dq/d\lambda$) of the proposed thermal diode in forward and reverse direction with temperature bias 20 K ($\Delta T = 10$ K). Forward heat flux is significantly higher than reverse flux as is clear from **Figure 2**. A comparison is shown for heat flux across blackbodies at temperatures 331 and 351 K, respectively. Inset in **Figure 2** displays angle-averaged emissivity of the active component in both scenarios. When VO_2 is metallic, the structure on the active component has high emissivity near the thermal wavelength ($\lambda_{th} = 1.27\hbar c/k_B T = 8.5$ μm for 341 K). As a significant portion of blackbody radiation falls within this range, this gives rise to a high heat flux in forward bias. However, when VO_2 is insulating, the structure has very low emissivity in the broad spectrum. The tri-layer structure behaves like a highly reflecting mirror resulting in very low heat flux. Consequently, high contrast in heat flow is achieved leading to a high rectification ratio of 11.3 ($\eta = 0.918$). In order to highlight the diode-like characteristics, heat flux across the device has been plotted against temperature difference in **Figure 3**. For comparison, simple case of bulk VO_2 is also shown, it has a rectification coefficient of $\eta = 0.49$.

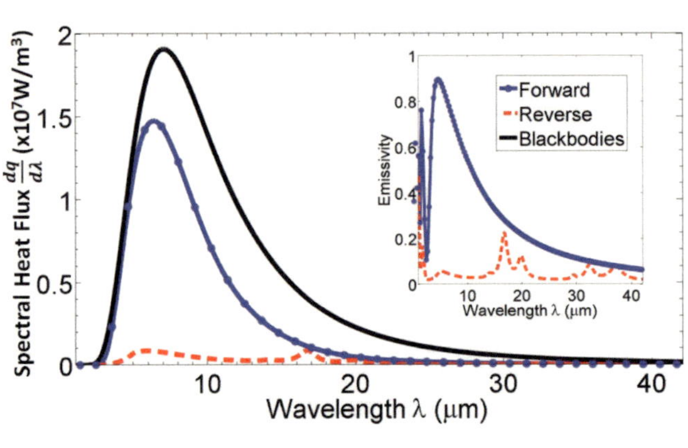

Figure 2. Spectral heat flux across the optimized thermal diode in forward and reverse bias scenarios. Spectral heat flux between blackbodies at temperatures 331 and 351 K is shown for reference. Inset shows hemispherical emissivity of the active component of the diode for the forward and reverse bias.

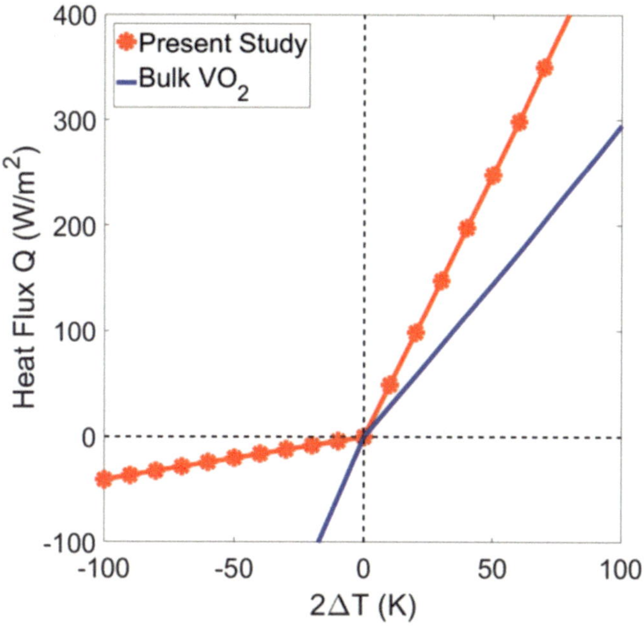

Figure 3. Heat flux plotted against temperature difference for thermal diode with bulk VO_2 and present structure.

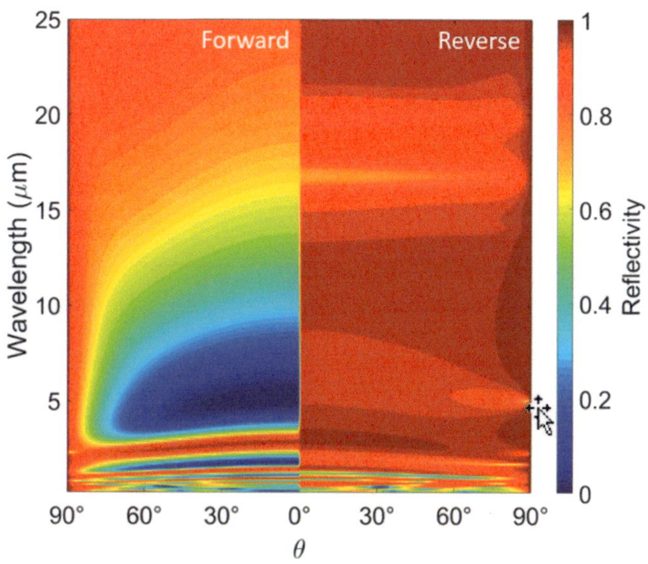

Figure 4. Angle dependent reflectivity of the active component of thermal diode plotted against wavelength and angle of incidence under forward and reverse bias.

Angle dependent spectral reflectivity of the active component of the thermal diode is plotted in **Figure 4** for the forward and reverse bias cases. When VO_2 is metallic, the tri-layer structure acts like a wide-angle antireflection coating for wavelengths between 4 and 10 μm. The dark spot in **Figure 4** corresponds to Fabry-Perot type of resonance that occurs around $\lambda = 4n_{KBr}(\lambda)L_2 = 5.3$ μm [47]. High absorption/emission in this wavelength region favors radiative heat transfer as thermal wavelength falls within this range. In reverse bias, the structure is highly reflective in a broad range of wavelengths giving rise to a very low absorption. Note that for thermal wavelength of 8.5 μm, Fabry-Perot resonance occurs (for metallic VO_2) when thickness of KBr layer is $L_2 = \lambda_{th}/4n_{KBr}(\lambda_{th}) = 1.4$ μm. This configuration, however, would not necessarily achieve maximum rectification as the structure may not be purely reflecting when VO_2 is its insulating phase.

Contrasting reflective properties of the structure are due to constructive and destructive interferences of electromagnetic waves generated by partial reflections at interfaces. As an electromagnetic wave travels through the media, it is partially reflected at each interface leading to multiple reflections from each layer. This causes interference of electromagnetic waves due to each partial reflection. Effective reflection coefficient of the structure is the phasor sum of these reflection coefficients due to (an infinite number of) individual reflections. When VO_2 is metallic, phasor sum of partial reflections results in destructive interference in the wavelength range of 4 μm to 10 μm. As a result, the structure is highly absorptive in the range. When VO_2 is insulating, individual reflections add up to a large value making the structure highly reflective for a broad range of the spectrum.

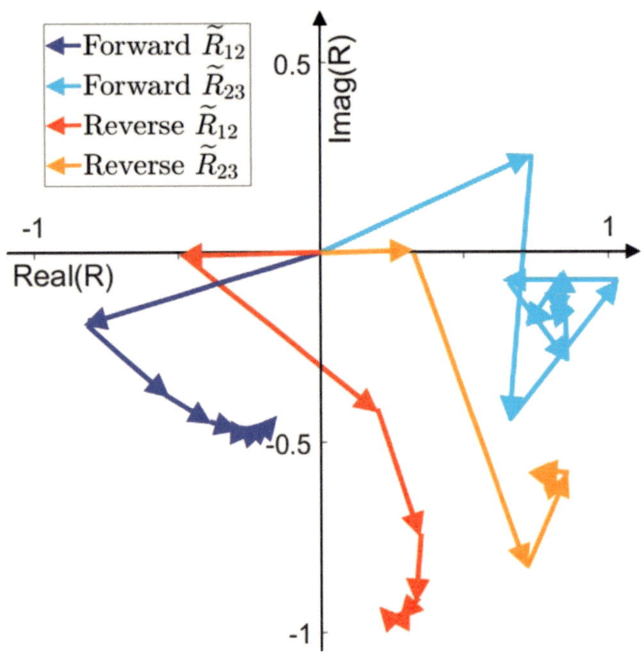

Figure 5. Effective reflection coefficient at air-VO_2 interface $(\tilde{R}_{1,2})$ and VO_2-KBr interface $(\tilde{R}_{2,3})$ as phasor sum of reflection coefficients due to each reflection for TE polarized incident plane wave of wavelength $\lambda_{th} = 8.5$ μm and angle of incidence 10°.

http://dx.doi.org/10.5772/intechopen.72805

Figure 5 shows phasor diagram of partial reflections at air-VO_2 interface and VO_2-KBr interface for TE polarized incident ray of wavelength $\lambda_{th} = 8.5$ μm and angle of incidence 10°. $\tilde{R}_{1,2}$ is the effective reflection coefficient at air-VO_2 interface and $\tilde{R}_{2,3}$ is the effective reflection coefficient at VO_2-KBr interface due to multiple reflections within KBr layer. They can be expressed as geometric series whose terms are relative amplitudes of partial waves due to first, second and third reflection and so on. For both metallic as well as insulating VO_2, the magnitude of $\tilde{R}_{2,3}$, $|\tilde{R}_{2,3}|$, is large. However, when VO_2 is in metallic phase, each partial reflection results in a phase-shift such that partial waves add up destructively leading to a small value of $|\tilde{R}_{1,2}|$ and low reflectivity, especially in the wavelength range centered around thermal wavelength. On the other hand, in reverse bias (insulating VO_2) phasors add constructively, giving rise to highly reflective surface properties for a broad range of wavelengths. A similar phenomenon can be observed for TM polarization as well. As KBr is transparent and has a negligible extinction coefficient for most of infrared region, much of the absorption takes place within the VO_2 layer. Transparent layer of KBr mainly influences the reflective properties by altering the phase of the light propagating through the media. Potentially, any other material transparent to infrared light such as magnesium fluoride or intrinsic silicon can be used in this concept. However, optimal dimensions of such a device might be different.

3. Near-field thermal diode

In a near-field radiative thermal diode the two terminals are separated by a distance less than thermal wavelength [18]. The active side has a phase change material and its counterpart has fixed material properties. **Figure 6** introduces two concepts of thermal diode that consist of two structures at a distance of $L = 100$ nm. In both concepts, structure 1 (active side) contains top layer of phase change material VO_2 at temperature $T_1 = 341\ K + \Delta T$. On the passive side, structure 2 has its temperature $T_2 = 341\ K - \Delta T$. Mean temperature is chosen to be the phase transition temperature of VO_2 at 341 K. When $T_1 > T_2$ (forward bias), VO_2 layer is in metallic phase; when $T_1 < T_2$ (reverse bias), VO_2 layer is in insulator phase with its optical axis aligned along the distance between them. Proposed configurations will be discussed in more details along with results.

In order to explain the results of the calculations for the proposed designs shown in **Figure 6**, rectification ratio against gap for four different configurations has been plotted in **Figure 7** for the temperature difference of 20 K ($\Delta T = \pm 10$ K). Before discussing the main design (**Figure 6(a)**) we consider the simplistic case which consists of bulk VO_2 on active side and bulk BN on passive side (case I). As insulator VO_2 and metallic VO_2 have different optical properties, one expects to observe some degree of asymmetry in heat flow. Such a configuration exhibits weak rectification which gradually increases at smaller gaps when surface waves become dominant. Next we consider using thin films. In case II, the active side has 1 μm layer of VO_2 over 1 μm thick gold and the passive side has 1 μm layer of BN on the top of 1 μm gold. 1 μm thick layer of gold is sufficient to block radiation from the substrate and to support the top layer as free standing thin layer is not practical [53]. This design shows an increased near-field rectification,

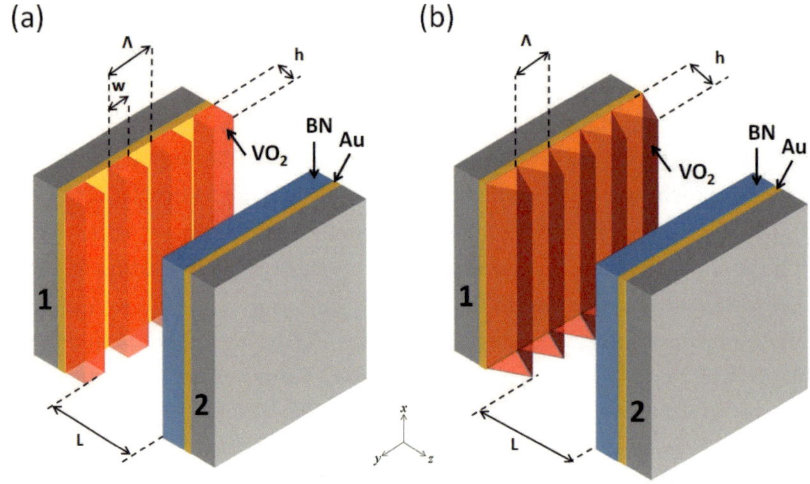

Figure 6. Schematics of near-field thermal diodes. (a) Active side has top layer of 1-D rectangular grating made of VO_2 of height h, width w, period Λ and filling ratio ϕ on a gold layer deposited on a substrate. (b) Rectangular grating is replaced by a triangular one of height h and period Λ. Passive counterpart of both designs consists of a BN layer on gold on the top of a substrate.

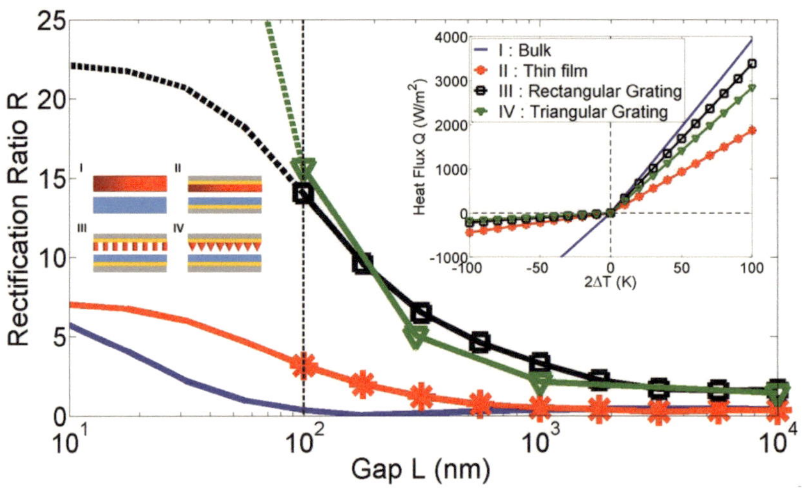

Figure 7. Gap-dependent rectification ratio for different thermal diode configurations. Passive structures are (I): Bulk BN, and (II), (III), (IV): 1 μm layer of BN over 1 μm layer of gold on a substrate. Active structures are (I): Bulk VO_2, (II): 1 μm layer of VO_2 on the top of 1 μm layer of gold on a substrate, (III): Rectangular 1-D grating of VO_2 with thickness $h = 0.5\mu m$, period $\Lambda = 50$ nm and filling ratio $\phi = 0.3$, (IV): Triangular grating of VO_2 with height $h = 0.5\mu m$ and period $\Lambda = 50$ nm. Inset figure shows heat flux as a function of temperature difference at 100 nm separation to highlight diode-like characteristics of different configurations.

far-field rectification being about the same. Rectification is stronger in near-field because of the asymmetry in heat flux due to different levels of tunneling of surface waves across the two interfaces. We now consider case III where thin film of VO_2 is replaced by a 1-D grating structure of height $h = 0.5$ μm, period $\Lambda = 50$ nm and filling ratio $\phi = 0.3$. The passive side remains the same as in case II. This configuration shows a significant enhancement in rectification and rectification value reaches around 14 at the gap of 100 nm. **Figure 7** also shows dependence of rectification ratio on gap for an alternative design (case IV) depicted in **Figure 6(b)**. This design has a 1-D triangular structure of height $h = 0.5$ μm and period $\Lambda = 50$ nm on active side. While this structure shows a similar trend as case III, a sharp increase in rectification ratio can be seen at smaller gaps. Rectification ratio of 16 is reached at 100 nm gap. Although results for the distances smaller than 100 nm may not be accurate completely as gap becomes comparable to the grating period (displayed using dashed lines in **Figure 7**). The trend is noteworthy for the triangular structure (case IV) as rectification ratio keeps increasing to higher values for shorter distances. Numerical methods such as Wiener chaos expansion method [54] can be employed for calculation of near-field heat flux for distances shorter than the grating period. **Figure 7** inset displays heat flux versus temperature difference for the four cases studied. Difference between slopes for forward and reverse bias is obvious. Diode-like characteristics are apparent especially for designs based on rectangular and triangular gratings.

Different materials such as SiO_2, SiC, BN, gold and polystyrene and structures (thin film or bulk) can be used on passive side and strongly influence rectification. When 1 μm layer of BN is used over reflecting gold surface on passive side, it results in maximum rectification for case II. The choice of passive structure used here may not remain optimal when active structure is modified. In order to emphasize the effect of 1-D gratings on active side, we have used the same structure on passive side for cases II, III and IV.

To illustrate why grating structure enhances the thermal rectification, we plot energy transmission coefficient $\xi(\omega, k_\rho)$ across the interfaces of thermal diode for case II (**Figure 8(a)** and **(b)**) and Case III (**Figure 8(c)** and **(d)**) at a gap of 100 nm. Here $k_\rho c/\omega$ is normalized parallel wavevector (for details on $\xi(\omega, k_\rho)$. For both cases, transmission coefficient is close to unity for forward bias well beyond light line ($k_\rho c/\omega = 1$). Transmission is high for two prominent frequencies that are close to characteristic wavelengths of BN (7.6 and 9.8 μm), as shown by dashed lines. Since metallic VO_2 does not support surface phonon polariton in the infrared region [26], near-field radiative transfer is mainly attributed to the symmetric and antisymmetric surface phonons supported by the BN layer ($k_\rho c/\omega \gg 1$ in **Figure 8**). In addition, there exists a secondary contribution due to Fabry-Perot modes and frustrated modes ($k_\rho c/\omega \approx 1$) in the near-field regime. High energy transmission in forward bias is due to tunneling of surface waves across interfaces. In reverse bias, although both BN and insulator VO_2 support surface phonon modes, they do not overlap (see **Figure 9**) and near-field radiative transfer is dominated by non-resonant surface waves. In addition, surface phonons of insulator VO_2 occur in frequency range where BN has low extinction coefficient ($\kappa \approx 0$), and vice versa. As a result, tunneling between BN and insulator VO_2 is much weaker than that of BN and metallic VO_2, that leads to thermal rectification. Observe that when thin film of VO_2 is replaced by a 1-D rectangular grating, transmission coefficient for reverse bias is reduced as seen in **Figure 8(b)** and **(d)**. Reduction in transmission coefficient comes from the presence of grating which suppresses the tunneling of surface waves supported by insulator VO_2 and BN. Tunneling between BN and metallic VO_2 however is relatively unchanged (**Figure 8(a)** and **(c)**). Consequently, a higher rectification is achieved. Resulting difference in spectral heat fluxes can also be observed.

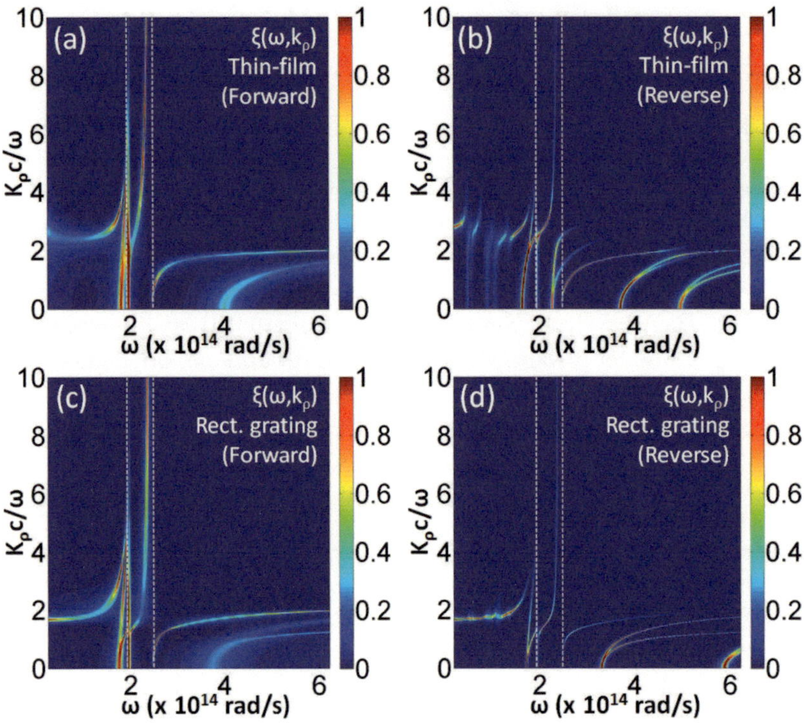

Figure 8. Coefficient of energy transmission $\xi(\omega, k_\rho)$ across the two interfaces of thermal diode plotted against angular frequency ω and normalized parallel wavevector $k_\rho c/\omega$ for (a) case II: Forward bias, (b) case II: Reverse bias, (c) case III: Forward bias, and (d) case III: Reverse bias.

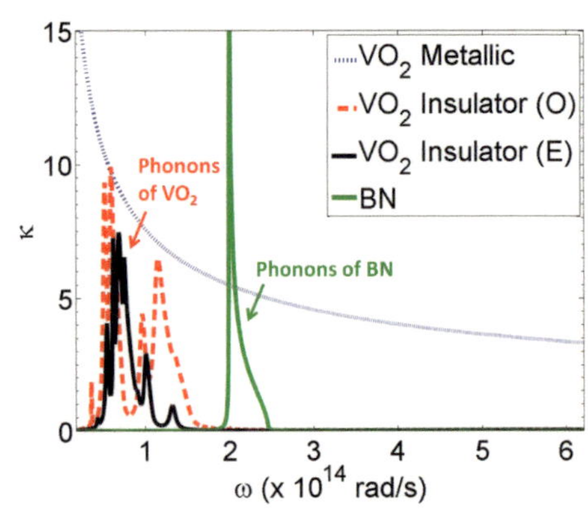

Figure 9. Extinction coefficient of metallic VO_2, insulator VO_2 (ordinary and extraordinary modes) and BN.

Spectral heat flux $dq/d\lambda$ in forward and reverse bias for the cases II and III at gap of 100 nm is plotted in **Figure 10**. For the structure with thin film of VO_2, majority of spectrum (solid lines) shows higher heat flux in forward direction than the reverse, which explains the rectification ratio being more than 1. The two prominent peaks that can be seen in the

http://dx.doi.org/10.5772/intechopen.72805

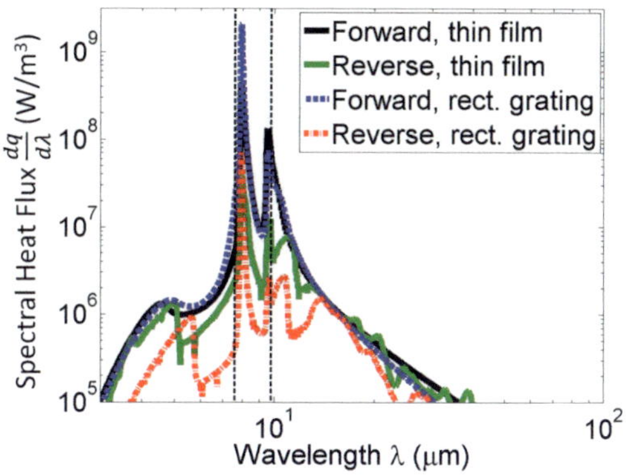

Figure 10. Comparison of spectral heat fluxes in forward and reverse bias for case II with a VO_2 thin film and case III with a rectangular grating structure. (Horizontal axis starts at 3 μm.)

spectrum are due to the frustrated modes of BN that happen around the characteristic wavelengths of BN at 7.6 and 9.8 μm. It is apparent that tunneling of surface waves across interfaces of insulator VO_2 and BN is weaker as the peaks have a reduced intensity. When the 1 μm thin film of VO_2 is replaced by the grating structure (dashed lines), forward heat flux shows little change. However, reverse heat flux is considerably reduced. It gives rise to an increased rectification.

One can analyze possible parameters that can influence the rectification for the design using 1-D rectangular grating (case III) for a gap of 100 nm in **Figure 11**. Filling ratio is fixed to 0.3 and the grating height is varied from 0.1 to 0.9 μm (dashed line). We observe that peak rectification ratio is achieved when grating height is around 0.5 μm. Solid line in **Figure 9** shows variation

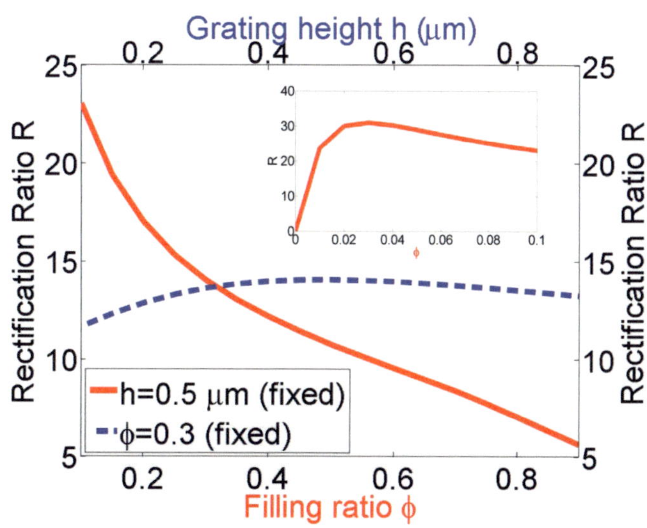

Figure 11. Effect of design parameters (filling ratio ϕ and grating height h) on the rectification ratio of thermal diode using rectangular gratings. Inset figure shows variation of rectification ratio near zero filling ratio. (Horizontal axis in main figure starts at 0.1).

in filling ratio for fixed grating height of 0.5 μm. Dependence of rectification on filling ratio is clearly stronger, as it directly affects the optical properties of grating structure therefore influencing the surface waves across the interfaces. For higher filling ratios, rectification values are lower, and understandably, are close to what is predicted for a thin film design (case II). Thermal rectification even higher than 20 can be predicted at smaller filling ratios. Rectification reaches maximum value when filling ratio is around 0.03, and it becomes zero when filling ratio is zero (**Figure 9** inset). However, filling ratio of 0.03 is impractical as grating period is 50 nm. Another parameter of interest would be grating period. Since our focus is on gap of 100 nm we have to limit ourselves to periods less than 100 nm. As the period is much smaller than the dominant thermal wavelength, it has virtually no effect on rectification.

4. Methods

To calculate heat flux for forward and reverse bias across near-field thermal diode, we can use the well-known expression of near-field radiative transfer obtained through dyadic Green's function formalism [55]. Radiative transfer between closely spaced objects can be calculated by

$$Q_{1\to 2}(T_1, T_2, L) = \int_0^\infty \frac{d\omega}{2\pi}[\Theta(\omega, T_1) - \Theta(\omega, T_2)]T_{1\to 2}(\omega, L) \tag{1}$$

where $\Theta(\omega, T) = (\hbar\omega/2)\coth(\hbar\omega/2k_BT)$ is the energy of harmonic oscillator at frequency ω and temperature T, $\hbar$ is the reduced Planck constant, and k_B is the Boltzmann constant. The function $T_{1\to 2}(\omega, L)$ is known as the spectral transmissivity in radiative transfer between media 1 and 2 separated of distance L [26, 32, 36, 55] and is expressed as

$$T_{1\to 2}(\omega) = \int_0^\infty \frac{k_\rho dk_\rho}{2\pi}\xi(\omega, k_\rho) \tag{2}$$

Here, k_ρ is the parallel component of wavevector and the integrand is known as energy transmission coefficient and can be defined as

$$\xi(\omega, k_\rho \le \omega/c) = \sum_{\mu=s,p} \frac{\left(1-\left|\tilde{R}_1^{(\mu)}\right|^2\right)\left(1-\left|\tilde{R}_2^{(\mu)}\right|^2\right)}{\left|1-\tilde{R}_1^{(\mu)}\tilde{R}_2^{(\mu)}e^{2jk_zL}\right|^2} \tag{3a}$$

$$\xi(\omega, k_\rho > \omega/c) = \sum_{\mu=s,p} \frac{4\Im\left(\tilde{R}_1^{(\mu)}\right)\Im\left(\tilde{R}_2^{(\mu)}\right)e^{-2|k_z|L}}{\left|1-\tilde{R}_1^{(\mu)}\tilde{R}_2^{(\mu)}e^{-2|k_z|L}\right|^2} \tag{3b}$$

where $\tilde{R}_1^{(\mu)}$ and $\tilde{R}_2^{(\mu)}$ are polarization dependent reflection coefficients of the two half spaces, $\mu = s$ (or p) refers to transverse electric (or magnetic) polarization, and k_z is the z-component of wavevector in vacuum. $k_\rho \le \omega/c$ and $k_\rho > \omega/c$ correspond to propagating and evanescent modes, respectively. For a structure having N-layer media having (N − 1) interfaces, by solving the boundary conditions at the interfaces, one can obtain the expression for the generalized reflection coefficient at the interface between regions i and i + 1 [56],

$$\tilde{R}_{i,i+1}^{(\mu)} = \frac{R_{i,i+1}^{(\mu)} + \tilde{R}_{i+1,i+2}^{(\mu)} e^{2jk_{i+1,z}(d_{i+1}-d_i)}}{1 + R_{i,i+1}^{(\mu)} \tilde{R}_{i+1,i+2}^{(\mu)} e^{2jk_{i+1,z}(d_{i+1}-d_i)}} \tag{4}$$

where $R_{i,i+1}^{(\mu)}$ is the Fresnel reflection coefficient at the interface between the layers i and i + 1, and $\tilde{R}_{i+1,i+2}^{(\mu)}$ is the generalized reflection coefficient at the interface between the layers i + 1 and i + 2, $z = -d_i$ is the location of the ith interface. $k_{i,z} = \sqrt{\varepsilon_i(\omega)\omega^2/c^2 - k_\rho^2}$ is the normal z-component of the wave vector in medium i, wherein $\varepsilon_i(\omega)$ is the relative permittivity of the medium i as a function of angular frequency ω, c is the speed of light in vacuum. With $\tilde{R}_{N,N+1}^{(\mu)} = 0$, the above equation provides a recursive relation to calculate the reflection coefficients $\tilde{R}_{i,i+1}^{(\mu)}$ in all regions. For the far-field configuration, $\xi(\omega, k_\rho > \omega/c)$ is ignored as separation between the two half spaces is much larger than the thermal wavelength ($L \gg \lambda_{th}$). The hemispherical emissivity of the active component can be expressed as [32]

$$e(\omega) = \frac{c^2}{\omega^2}\int_0^{\omega/c} dk_\rho k_\rho \sum_{\mu=s,p}\left(1 - \left|\tilde{R}_h^{(\mu)}\right|^2\right) \tag{5}$$

Note that the term for transmissivity has been omitted as a layer of gold makes the structure opaque.

Since the proposed designs involve 1-D grating structure of VO_2 in vacuum, we can use second order approximation of effective medium theory to obtain the dielectric properties given by the expressions [47, 27, 8]

$$\varepsilon_{TE,2} = \varepsilon_{TE,0}\left[1 + \frac{\pi^2}{3}\left(\frac{\Lambda}{\lambda}\right)^2 \phi^2(1-\phi)^2 \frac{(\varepsilon_A - \varepsilon_B)^2}{\varepsilon_{TE,0}}\right] \tag{6a}$$

$$\varepsilon_{TM,2} = \varepsilon_{TM,0}\left[1 + \frac{\pi^2}{3}\left(\frac{\Lambda}{\lambda}\right)^2 \phi^2(1-\phi)^2(\varepsilon_A - \varepsilon_B)^2 \varepsilon_{TE,0}\left(\frac{\varepsilon_{TM,0}}{\varepsilon_A\varepsilon_B}\right)^2\right] \tag{6b}$$

where ε_A and ε_B are dielectric functions of the two materials (VO_2 and vacuum) in surface gratings, λ is the wavelength, Λ is grating period and filling ratio $\phi = w/\Lambda$ where w is width of VO_2 segment. The expressions for zeroth order effective dielectric functions $\varepsilon_{TE,0}$ and $\varepsilon_{TM,0}$ are given by [57, 58]

$$\varepsilon_{TE,0} = \phi\varepsilon_A + (1-\phi)\varepsilon_B \tag{7a}$$

$$\varepsilon_{TM,0} = \left(\frac{\phi}{\varepsilon_A} + \frac{1-\phi}{\varepsilon_B}\right)^{-1} \tag{7b}$$

For triangular gratings as shown in **Figure 6(b)**, gratings can be treated as a composition of multiple layers of rectangular gratings each having decreasing filling ratio and period equal to that of parent grating [58]. It was observed that slicing the triangular structure into 100 layers is sufficient to achieve converging values of near-field heat flux.

Effective medium approximation (EMA) holds true when grating period is much less than wavelength of interest [59]. As this study deals with temperatures around 341 K, grating period (50 nm) is much less than the thermal wavelength ($\sim 8.5\ \mu m$). Thus the condition of EMA is satisfied. When dealing with near-field radiative transfer pertaining to periodic gratings, criterion for the validity of EMA is somewhat different and effective medium theory holds true as long as gap between the planar structures (100 nm in this case) is greater than the grating period [60] (see discussions about dashed lines in **Figure 7**).

VO_2 in insulator state (below 341 K) is anisotropic. In a plane perpendicular to optical axis ($x - y$ plane in our case) known as ordinary mode, its dielectric function is ε_O and it is ε_E along the optical axis (z axis, extraordinary mode). Both ε_O and ε_E can be calculated using classical oscillator formula $\varepsilon(\omega) = \varepsilon_\infty + \sum_{i=1}^{N} \frac{S_i \omega_i^2}{\omega_i^2 - j\gamma_i\omega - \omega^2}$. Experimental values of high-frequency constant ε_∞, phonon frequency ω_i, scattering rate γ_i and oscillator strength S_i can be found in Ref. [61]. Here, j is the imaginary unit. There exist eight phonon modes for ordinary and nine phonon modes for extraordinary dielectric function. In metallic state, VO_2 is isotropic and Drude model [61] is used to describe the dielectric function that is given by $\varepsilon(\omega) = \frac{-\omega_p^2 \varepsilon_\infty}{\omega^2 - j\omega\Gamma}$. Dielectric function for BN [62] is of the form $\varepsilon(\omega) = \varepsilon_\infty \frac{(\omega^2 - \omega_{LO}^2 + j\omega\gamma)}{(\omega^2 - \omega_{TO}^2 + j\omega\gamma)}$. Here ω_{TO} and ω_{LO} are transverse and longitudinal optical phonon frequencies and γ is the damping constant. The values of ε_∞, ω_{TO}, ω_{LO} and γ for BN are 4.46, 0.1309, 0.1616 and 6.55×10^{-4} eV respectively. Dielectric properties of gold can be found in Ref. [63].

5. Conclusion

We saw two configurations of radiative thermal diode that exploit metal-insulator transition of VO_2. The VO_2 based far-field radiative thermal diode structure presented here has a high rectification ratio of 11.3. The active component of the device has a tri-layer structure consisting thin films of VO_2, KBr and gold. As VO_2 undergoes phase change around 341 K, reflecting properties of the surface are dramatically changed in the spectral region that contributes to significant amount of thermal radiation. Facilitated by Fabry-Perot type of resonance around 5.3 μm, metallic VO_2 makes the structure behave like a wide-angle antireflection coating while insulating VO_2 makes it highly reflecting. As a result, high degree of asymmetry in radiative heat transfer is predicted across the tri-layer structure and a blackbody. Contrasting reflecting properties of the structure can be explained using constructive and destructive interference of partial reflections across the interfaces. Layer thicknesses are optimized to maximize rectification. Thermal rectification greater than 11 is predicted for temperature difference of 20 K and it is highest among far-field radiative diodes that have been

http://dx.doi.org/10.5772/intechopen.72805

studied. Possibility of attaining higher rectification could be investigated in future by using alternate transparent materials, thinner films of VO_2 and/or using more number of alternating VO_2/dielectric layers. Such devices can find numerous applications such as thermal logic devices and thermal management systems. In the near-field configuration, it has been demonstrated that an enhanced thermal rectification can be achieved using 1-D grating of phase change material VO_2. The calculations presented here indicate that a very high value of rectification ratio ($\sim$ 16) can be obtained at a gap of 100 nm. Rectification ratio can be optimized by tuning parameters such as grating height and especially, filling ratio. Materials and structures on passive side also play a significant role in the rectification. Improved rectification is attributed to reduced tunneling of surface waves across the interfaces for reverse bias. Rectification ratio can be further increased to much higher values for distance shorter than 100 nm and it is a viable candidate for future investigations.

Acknowledgements

This project was supported in part by a National Science Foundation through grant number 1655221, Institutional Development Award (IDeA) Network for Biomedical Research Excellence from the National Institute of General Medical Sciences of the National Institutes of Health under grant number P20GM103430, and Rhode Island Foundation Research Grant number 20164342.

Conflict of interest

The authors declare no competing interests.

Author details

Alok Ghanekar, Yanpei Tian and Yi Zheng*

*Address all correspondence to: zheng@uri.edu

Department of Mechanical, Industrial and Systems Engineering, University of Rhode Island, Kingston, RI, USA

References

[1] Li B, Wang L, Casati G. Thermal diode: Rectification of heat flux. Physical Review Letters. 2004;**93**(18):184301

[2] Li B, Wang L, Casati G. Negative differential thermal resistance and thermal transistor. Applied Physics Letters. 2006;**88**(14):143501

[3] Wang L, Li B. Thermal memory: A storage of phononic information. Physical Review Letters. 2008;**101**(26):267203

[4] Kobayashi W, Teraoka Y, Terasaki I. An oxide thermal rectifier. Applied Physics Letters. 2009;**95**(17):171905

[5] Sawaki D, Kobayashi W, Moritomo Y, Terasaki I. Thermal rectification in bulk materials with asymmetric shape. Applied Physics Letters. 2011;**1102**:4182

[6] Kobayashi W, Sawaki D, Omura T, Katsufuji T, Moritomo Y, Terasaki I. Thermal rectification in the vicinity of a structural phase transition. Applied Physics Express. 2012;**5**(2):027302

[7] Chang CW, Okawa D, Majumdar A, Zettl A. Solid-state thermal rectifier. Science. 2006;**314**(5802):1121-1124

[8] Martínez-Pérez MJ, Fornieri A, Giazotto F. Rectification of electronic heat current by a hybrid thermal diode. Nature Nanotechnology. 2015;**10**(4):303-307

[9] Ben-Abdallah P, Biehs S-A. Near-field thermal transistor. Physical Review Letters. 2014; **112**(4):044301

[10] Ben-Abdallah P, Biehs S-A. Phase-change radiative thermal diode. Applied Physics Letters. 2013;**103**(19):191907

[11] Otey CR, Lau WT, Fan S. Thermal rectification through vacuum. Physical Review Letters. 2010;**104**(15):154301

[12] Chen Z, Wong C, Lubner S, Yee S, Miller J, Jang W, Hardin C, Fong A, Garay JE, Dames C. A photon thermal diode. Nature Communications. 2014;**5**

[13] Wang L, Li B. Phononics gets hot. Physics World. 2008;**21**:27-29

[14] Wang L, Li B. Thermal logic gates: Computation with phonons. Physical Review Letters. 2007;**99**(17):177208

[15] Ben-Abdallah P, Biehs S-A. Towards boolean operations with thermal photons. Physical Review B. 2016;**94**(24):241401

[16] Li N, Ren J, Wang L, Zhang G, Hänggi P, Li B. Colloquium: Phononics: Manipulating heat flow with electronic analogs and beyond. Reviews of Modern Physics. 2012;**84**(3):1045

[17] Song B, Fiorino A, Meyhofer E, Reddy P. Near-field radiative thermal transport: From theory to experiment. AIP Advances. 2015;**5**(5):053503

[18] Yang Y, Basu S, Wang L. Radiation-based near-field thermal rectification with phase transition materials. Applied Physics Letters. 2013;**103**(16):163101

[19] Huang J, Li Q, Zheng Z, Xuan Y. Thermal rectification based on thermochromic materials. International Journal of Heat and Mass Transfer. 2013;**67**:575-580

[20] Nefzaoui E, Joulain K, Drevillon J, Ezzahri Y. Radiative thermal rectification using superconducting materials. Applied Physics Letters. 2014;**104**(10):103905

[21] Joulain K, Ezzahri Y, Drevillon J, Ben-Abdallah P. Modulation and amplification of radiative far field heat transfer: Towards a simple radiative thermal transistor. Applied Physics Letters. 2015;**106**(13):133505

[22] Van Zwol PJ, Joulain K, Abdallah PB, Greffet JJ, Chevrier J. Fast nanoscale heat-flux modulation with phase-change materials. Physical Review B. 2011;**83**(20):201404

[23] Van Zwol PJ, Ranno L, Chevrier J. Tuning near field radiative heat flux through surface excitations with a metal insulator transition. Physical Review Letters. 2012;**108**(23):234301

[24] Menges F, Dittberner M, Novotny L, Donata P, Parkin SSP, Spieser M, Riel H, Gotsmann B. Thermal radiative near field transport between vanadium dioxide and silicon oxide across the metal insulator transition. Applied Physics Letters. 2016;**108**(17):171904

[25] Ghanekar A, Ji J, Zheng Y. High-rectification near-field thermal diode using phase change periodic nanostructure. Applied Physics Letters. 2016;**109**(12):123106

[26] Yang Y, Basu S, Wang L. Vacuum thermal switch made of phase transition materials considering thin film and substrate effects. Journal of Quantitative Spectroscopy and Radiative Transfer. 2015;**158**:69-77

[27] Prod'homme H, Ordonez-Miranda J, Ezzahri Y, Drevillon J, Joulain K. Optimized thermal amplification in a radiative transistor. Journal of Applied Physics. 2016;**119**(19):194502

[28] Zhu L, Otey CR, Fan S. Negative differential thermal conductance through vacuum. Applied Physics Letters. 2012;**100**(4):044104

[29] Lee K-T, Ji C, Jay Guo L. Wide-angle, polarization-independent ultrathin broadband visible absorbers. Applied Physics Letters. 2016;**108**(3):031107

[30] Zheng Y, Ghanekar A. Radiative energy and momentum transfer for various spherical shapes: A single sphere, a bubble, a spherical shell, and a coated sphere. Journal of Applied Physics. 2015;**117**(6):064314

[31] Zhu L, Otey CR, Fan S. Ultrahigh-contrast and large-bandwidth thermal rectification in near-field electromagnetic thermal transfer between nanoparticles. Physical Review B. 2013;**88**(18):184301

[32] Ghanekar A, Lin L, Junwei S, Sun H, Zheng Y. Role of nanoparticles in wavelength selectivity of multilayered structures in the far-field and near-field regimes. Optics Express. 2015;**23**(19):A1129-A1139

[33] Ghanekar A, Lin L, Zheng Y. Novel and efficient mie-metamaterial thermal emitter for thermophotovoltaic systems. Optics Express. 2016;**24**(10):A868-A877

[34] Rinnerbauer V, Lenert A, Bierman DM, Yeng YX, Chan WR, Geil RD, Senkevich JJ, Joannopoulos JD, Wang EN, Soljačić M, et al. Metallic photonic crystal absorber-emitter for efficient spectral control in high- temperature solar thermophotovoltaics. Advanced Energy Materials. 2014;**4**(12)

[35] Chou JB, Yeng YX, Lenert A, Rinnerbauer V, Celanovic I, Soljačić M, Wang EN, Kim S-G. Design of wide-angle selective absorbers/emitters with dielectric filled metallic photonic crystals for energy applications. Optics Express. 2014;**22**(101):A144-A154

[36] Biehs S-A, Rosa FSS, Ben-Abdallah P. Modulation of near-field heat transfer between two gratings. Applied Physics Letters. 2011;**98**(24):243102

[37] Zheludev NI, Kivshar YS. From metamaterials to metadevices. Nature Materials. 2012; **11**(11):917-924

[38] Nanfang Y, Capasso F. Flat optics with designer metasurfaces. Nature Materials. 2014; **13**(2):139-150

[39] Selim Ünlü M, Strite S. Resonant cavity enhanced photonic devices. Journal of Applied Physics. 1995;**78**(2):607-639

[40] Kishino K, Unlu MS, Chyi JI, Reed J, Arsenault L, Morkoc H. Resonant cavity-enhanced (rce) photodetectors. IEEE Journal of Quantum Electronics. 1991;**27**(8):2025-2034

[41] Fante RL, Mccormack MT. Reflection properties of the salisbury screen. IEEE Transactions on Antennas and Propagation. 1988;**36**(10):1443-1454

[42] RaâĂŹDi Y, Simovski CR, Tretyakov SA. Thin perfect absorbers for electromagnetic waves: Theory, design, and realizations. Physical Review Applied. 2015;**3**(3):037001

[43] Bosman H, Lau YY, Gilgenbach RM. Microwave absorption on a thin film. Applied Physics Letters. 2003;**82**(9):1353-1355

[44] Hägglund C, Peter Apell S, Kasemo B. Maximized optical absorption in ultrathin films and its application to plasmon-based two-dimensional photovoltaics. Nano Letters. 2010;**10**(8): 3135-3141

[45] Mingbo P, Feng Q, Wang M, Hu C, Huang C, Ma X, Zhao Z, Wang C, Luo X. Ultrathin broadband nearly perfect absorber with symmetrical coherent illumination. Optics Express. 2012;**20**(3):2246-2254

[46] Nefzaoui E, Drevillon J, Ezzahri Y, Joulain K. Simple far-field radiative thermal rectifier using fabry–perot cavities based infrared selective emitters. Applied Optics. 2014;**53**(16): 3479-3485

[47] Kats MA, Sharma D, Lin J, Genevet P, Blanchard R, Yang Z, Mumtaz Qazilbash M, Basov DN, Ramanathan S, Capasso F. Ultra-thin perfect absorber employing a tunable phase change material. Applied Physics Letters. 2012;**101**(22):221101

[48] Taylor S, Yang Y, Wang L. Vanadium dioxide based fabry-perot emitter for dynamic radiative cooling applications. Journal of Quantitative Spectroscopy and Radiative Transfer; 2017

[49] Qazilbash MM, Brehm M, Andreev GO, Frenzel A, Ho P-C, Chae B-G, Kim B-J, Yun SJ, Kim H-T, Balatsky AV, Shpyrko OG. Infrared spectroscopy and nano-imaging of the insulator-to-metal transition in vanadium dioxide. Physical Review B. 2009;**79**(7):075107

[50] Frenzel A, Qazilbash MM, Brehm M, Chae B-G, Kim B-J, Kim H-T, Balatsky AV, Keilmann F, Basov DN. Inhomogeneous electronic state near the insulator-to-metal transition in the correlated oxide vo 2. Physical Review B. 2009;**80**(11):115115

[51] Ordonez-Miranda J, Ezzahri Y, Drevillon J, Joulain K. Transistorlike device for heating and cooling based on the thermal hysteresis of vo 2. Physical Review Applied. 2016;**6**(5): 054003

[52] Kats MA, Capasso F. Optical absorbers based on strong interference in ultra-thin films (laser photonics rev. 10 (5)/2016). Laser & Photonics Reviews. 2016;**10**(5):699-699

[53] Narayanaswamy A, Mayo J, Canetta C. Infrared selective emitters with thin films of polar materials. Applied Physics Letters. 2014;**104**(18):183107

[54] Liu B, Shen S. Broadband near-field radiative thermal emitter/absorber based on hyperbolic metamaterials: Direct numerical simulation by the wiener chaos expansion method. Physical Review B. 2013;**87**(11):115403

[55] Narayanaswamy A, Zheng Y. A Green's function formalism of energy and momentum transfer in fluctuational electrodynamics. Journal of Quantitative Spectroscopy and Radiative Transfer. 2014;**132**:12-21

[56] Chew WC. Waves and Fields in Inhomogeneous Media. New York, NY: IEEE Press; 1995

[57] Raguin DH, Michael Morris G. Antireflection structured surfaces for the infrared spectral region. Applied Optics. 1993;**32**(7):1154-1167

[58] Glytsis EN, Gaylord TK. High-spatial-frequency binary and multilevel stairstep gratings: Polarization-selective mirrors and broadband antireflection surfaces. Applied Optics. 1992;**31**(22):4459-4470

[59] Chen Y-B, Zhang ZM, Timans PJ. Radiative properties of patterned wafers with nanoscale linewidth. Journal of Heat Transfer. 2007;**129**(1):79-90

[60] Liu XL, Bright TJ, Zhang ZM. Application conditions of effective medium theory in near-field radiative heat transfer between multilayered metamaterials. Journal of Heat Transfer. 2014;**136**(9):092703

[61] Barker AS Jr, Verleur HW, Guggenheim HJ. Infrared optical properties of vanadium dioxide above and below the transition temperature. Physical Review Letters. 1966;**17**(26): 1286

[62] Palik ED. Handbook of Optical Constants of Solids. Vol. 3. Cambridge, Massachusetts: Academic Press; 1998

[63] Johnson PB, Christy R-W. Optical constants of the noble metals. Physical Review B. 1972; **6**(12):4370

Section 3

Enhancement of Heat Transfer

Chapter 9

Heat Transfer Analysis without and with Forward Facing Spike Attached to a Blunt Body at High Speed Flow

Rakhab Chandra Mehta

Additional information is available at the end of the chapter

http://dx.doi.org/10.5772/intechopen.74522

Abstract

The present chapter deals with heat transfer analysis around unspiked and spiked bodies at high speeds. A spike attached to a blunt-nosed body drastically alters its flowfield and influences the aerodynamic heating in a high speed flow. The effect of spike length, shape and spike-nose configuration is numerically studied at zero angle of incidence. The numerical analysis describes overall flowfield features over without and with forward facing spike attached to a blunt body at high speed flow. The shock stand-off distance, sonic line, stagnation point velocity gradient and stagnation point heat flux are analyzed and compared with different aerodisk configurations. It is found that the hemispherical aerodisk experiences high wall heat flux as compared to the flat-faced aerodisk. Numerical and experimental studies reveal that the wall heat flux levels are decreased in the presence of the spikes and aerospike as compared to without attached spiked to the blunt-nose basic configuration.

Keywords: aerospike, blunt body, CFD, compressible flow, convective heat transfer

1. Introduction

The shock wave dominates aerodynamic drag and aerodynamic heating at high speeds of a blunt body. The analytical [1] and experimental [2] investigations have shown that for the blunt body noses with a fixed length, a pointed geometry with a blunt nose tip is most beneficial to minimize the wave drag at high speeds. If aerodynamic heating is considered, a large blunt nose radius R_N is preferred since the wall heat flux, q_w, is reciprocally proportional to the square root of R_N [3]. A blunt nose [4] is beneficial in increasing the volumetric space to accommodate more payload and avionic components in a space vehicle.

There are several alternative concepts to the aerospike developed such as changing catalyst properties [5], pulse heating in front of the blunt body [6–8], DC arc discharging [9], electrically heated wire pointing upstream [10], injection of jets of air [11–13], or plasma technology [14–16], forward facing cavity [17], self-aligning aerodisk [18], opposite jet injection [19], focused energy deposition [20–22], non-ablative thermal protection [23] and multiaerodisk attached to the blunt body [24]. Many review articles have appeared [25–27] to summarize the flowfield characteristics in front of the blunt body to the aerospike.

The features of the high speed flowfield can be delineated through these experimental [28] and numerical [29, 30] studies. Based on these investigations a schematic of the flowfield around the blunt body, the conical, the hemispherical and the flat-face spiked blunt body at zero angle of incidence is delineated in **Figure 1**. A hemi-spherical portion of the blunt body is accomplished by a bow shock wave as depicted in **Figure 1(a)**. The total pressure loss over the shock leads to a high wave drag. A well-known concept for reducing the impact of the bow shock wave on a blunt body, while keeping a blunt nose, is the aerospike. The simplest aerospike design is a thin rod mounted on the tip of a blunt body as in **Figure 1(b)**. For the aerospike in ideal case, the boundary layer on the rod separates along the whole rod surface due to the pressure rise over the bow shock wave [31]. The separated boundary layer forms a shear layer that is reattached on the blunt nose. Due to the shear layer, the outer supersonic bow is detected and a weaker conical shock is formed instead of the initial bow shock. The conical shock unites with the reattachment shock further downstream. A recirculation zone forms inside the shear surface and shows significantly lower pressure levels compared with the blunt body without an aerospike. In the model of the flat-face and the hemispherical aerodisk, we observe a formation of a bow shock wave ahead of the body as delineated in **Figure 1(c)** and **(d)**. The flow separation zone is noticed around the root of the spike up to the reattachment point of the flow at the corner of the blunt body. Due to the recirculating region, the pressure at the stagnation region of the blunt body will reduce.

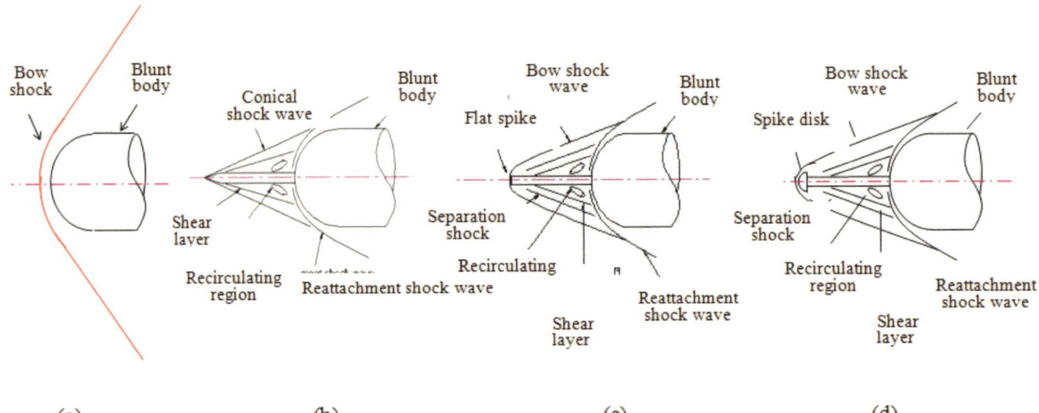

Figure 1. Flowfield features around unspiked and spiked bodies at high-speeds: (a) blunt body (b) conical spike (c) flat-face spike and (d) hemispherical disc spike.

However, because of the reattachment of the shear layer on the corner of the blunt body, the pressure near the reattachment point becomes large. However, the reattachment point can depend on the geometrical parameters of the spike or the blunt body configuration. The spike is characterized by a free shear layer, which is formed as a result of the flow separating from spike leading edge and reattaching to the blunt body, essentially bridging the spike. The separating shear layer from the spike leading edge of the aerospike attaches to the blunt body, after entering through an expansion fan at the leading edge corner and a recompression shock at attachment point. The attached shear layer then separates near the trailing edge and generates a separation shock before reattaching point at the trailing edge prior to undergoing flow expansion. The separated boundary layer forms a shear layer that reattaches on the blunt nose.

The above descriptions of the flowfield features show that the flowfield past a spiked blunt body appears to be very complicated and complex and having a number of interesting flow phenomena and characteristics, which have been further studied in order to compute the aerodynamic heating at high speeds.

2. Governing equations

The time-dependent axisymmetric compressible fluid dynamics equations were written in integral form, and the system of equations was augmented by the ideal gas law for numerical simulation [31]. The coefficient of molecular viscosity was calculated using Sutherland's law. A laminar flow is considered in the numerical simulation which also agrees with Bogdonoff and Vas [32], Fujita and Kubota [33], Yamauchi et al. [34] and Ahmed and Qin [25].

3. Numerical algorithm

The flowfield solver uses a finite-volume discretization employing the method of lines. The spatial computational region was divided into a number of finite non-overlapping quadrilateral grids. Thus, the discretized solution to the Navier-Stokes equations results in a set of volume-averaged state variables of mass, momentum and energy, which are in balance with their area-averaged fluxes (inviscid and viscous) across the grid faces [35]. The finite-volume flow solver algorithm written in this way reduces to a central difference scheme and is second-order accurate in space provided that the grid is generated in an orderly manner and is smooth enough. The cell-centered spatial discretization method is non-dissipative [36]; therefore, artificial dissipation terms are added as a blend of a Laplacian and biharmonic operator in an analogous to the second and fourth difference. The artificial dissipation term was added in the algorithm explicitly to prevent numerical oscillations near flow discontinuity to dampen high-frequency undamped modes. Temporal integration was performed using a multistage time-stepping scheme of Jameson et al. [36] and numerical integration used on the Runge–Kutta method. The artificial dissipation is evaluated only at the first stage.

3.1. Initial and boundary conditions

An initial condition corresponding to freestream conditions is considered. All flow quantities were extrapolated at the outer-boundary, and the no-slip condition was imposed on the wall. An isotherm wall condition was used for the wall of the model, that is, a surface temperature of 300 K. The symmetric condition is imposed on the centerline.

3.2. Computational grid

One of the controlling factors for the numerical simulation is the proper grid arrangement. The grid points are generated by a homotopy scheme [37]. Mesh-independence tests were performed, taking into consideration the influence of the computational region, the stretching factor to control the mesh intensity in the vicinity of body surface and the number of mesh points in the axial and normal directions. The outer surface of the computational zone is varied about 5–8 times the maximum blunt body diameter *D*. the mesh stretching factor in the radial direction is varied in order to resolve boundary layer. The convergence criterion is based on the difference in density value at any grid point between two successive iterations, that is, $|\rho^{n+1} - \rho^{n}| \leq 10^{-5}$, where n is the iteration index.

4. Geometrical details of model

4.1. Axisymmetric blunt body

The spherical blunted-cone/flare configuration without spike is illustrated in **Figure 2(a)**. The spherical forebody has $R_N = 0.51$ m, D = 2.03 m, L = 1.67 m, and cone angle of 20°. The flare has a semi-cone angle 15° and is terminated with a right circular cylinder. A close-up view of the computational mesh is shown in **Figure 2(b)**.

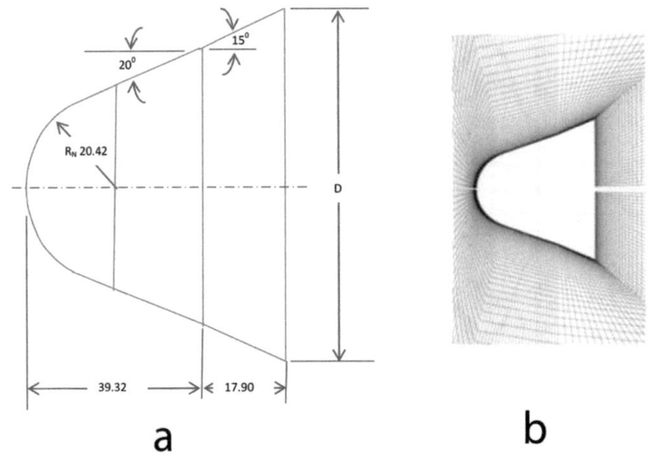

Figure 2. (a) Geometrical parameters of reentry capsule and (b) enlarged view of computational grid.

http://dx.doi.org/10.5772/intechopen.74522

4.2. Heat shield with forward facing spike

The fore and afterbody diameters of the heat shield [38–39] are 43.26×10^{-3} and 35×10^{-3} m, respectively, as shown in **Figure 3(a)**. The semi-cone angle of the heat shield is 20°. The hemispherical nose of the forward facing spike has radius of 0.55×10^{-3} m and length of the spike is 10×10^{-3} m. The other end of spike has diameter of 0.98×10^{-3} m and is attached to the blunt spherical cap of the heat shield of radius 8.75×10^{-3} m. The computational grid is displayed in **Figure 3(b)**.

4.3. Conical spike attached to blunt body

The model is axisymmetric, the main body has a hemisphere-cylinder nose and diameter $D = 7.62 \times 10^{-2}$ m. The spike's stem has a conical and cylindrical shape. The angle of the spike's cone is 10° and diameter of the cylinder of the spike is $0.1D$. Length to diameter ratio of spike are considered as 0.5, 1.0 and 2.0 in the flowfield and heat transfer analysis. **Figure 4(a)** depicts the geometrical dimensions and enlarged view of mesh over the model in **Figure 4(b)**.

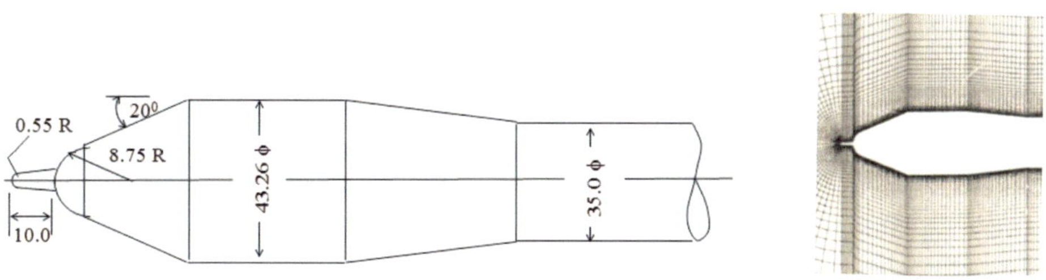

(a) (b)

Figure 3. (a) Dimensions of the heat shield with spike and (b) enlarged view of the grid.

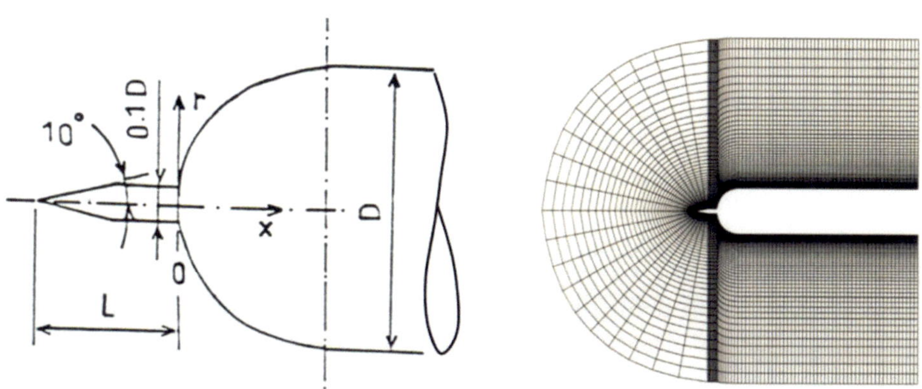

(a) (b)

Figure 4. (a) Dimensions of the spiked blunt body and (b) computational grid.

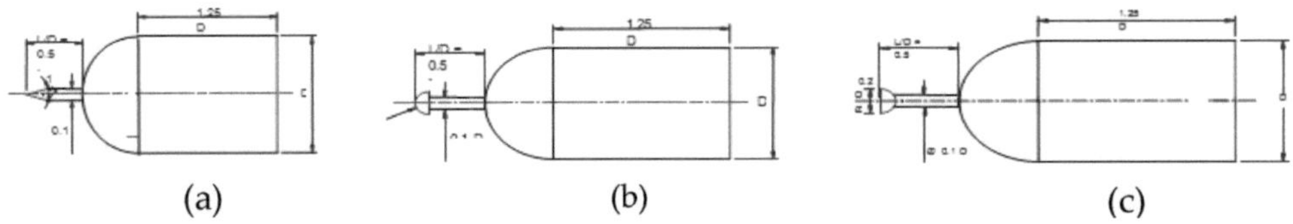

Figure 5. Dimensions of the aerospike blunt bodies: (a) conical spike (b) hemispherical spike and (c) flat-face disc.

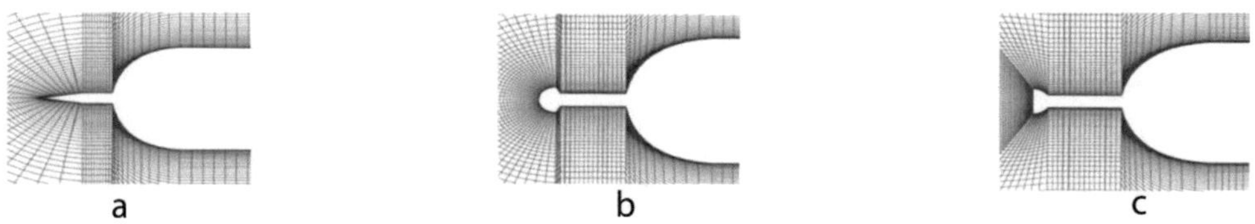

Figure 6. Close-up view of computational grid over the aerospike blunt bodies: (a) conical spike (b) hemispherical spike and (c) flat-face disc.

4.4. Conical, disk and flat spiked body

The dimensions of the spiked blunt body are depicted in **Figure 5**. The basic body has a hemispherical-cylinder nose and diameter $D = 4.0 \times 10^{-2}$ m. The conical spike is as shown in **Figure 5(a)**. The semi cone is 15°, the spike length $L = 0.5\ D$ and stem diameter of 0.1 D. The spike having a hemispherical cap of diameter D_s is 0.2 D, attached to a stem of diameter of 0.1D. The aerodisk type configuration utilizes a hemispherical disk on its nose of diameter D_s is 0.2D as depicted in **Figure 5(b)**. The flat-faced aerodisk with spike length $L = 0.5\ D$ is shown in **Figure 5(c)**. A close-up view of the computational grid over the conical, the hemispherical and the flat-faced aerospike is depicted in **Figure 6**. The structured grid generation and the mono-block are suitable to accommodate aerospike shape.

5. Results and discussions

Characteristic features of the flowfield around the hemispherical and the flat-disk aerospike attached to the blunt body at high speeds were investigated with the help of velocity vector plots, density, pressure and Mach contours diagram.

5.1. Flow and heat transfer analysis of axisymmetric blunt body

Computed Mach and temperature contours around the unspiked reentry module are shown in **Figure 7** for M_∞ = 2.0–6.0. The Mach contours exhibit the vortices formation at the shoulder region of the module. Behaviors of the flowfield around the blunt body at supersonic speeds shows the formation of the bow shock wave ahead of the blunt body; the wake, and the recompression shock waves coming out from the neck point are observed in the Mach contours. The flow expands at the base region and is followed by the recompression shock downstream of the base. Then, the flow progress in the wake region. The flow ground of the capsule is divided

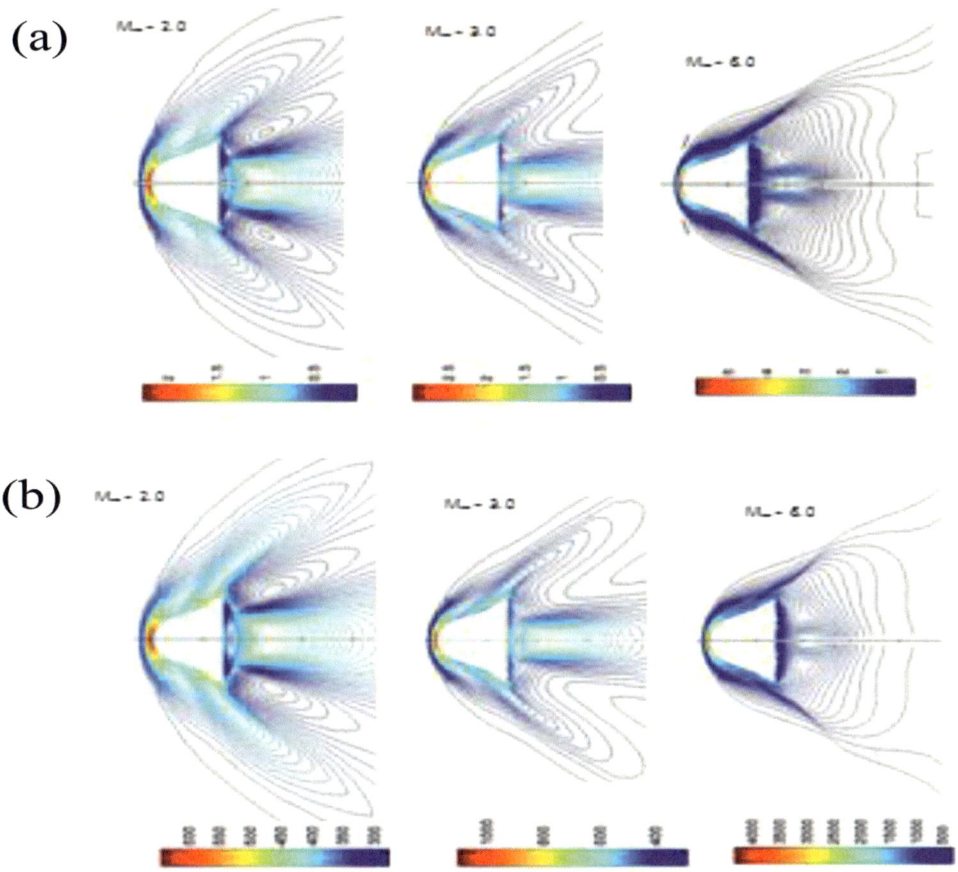

Figure 7. (a) Mach and (b) temperature contours over the blunted-cone/flare module.

into regions inside and outside of the flow recirculation zone, and two flow zones are separated by the shear layer. From the temperature contours, a rapid raise of temperature with the increase of M_∞ can be observed.

Figure 8 shows the pressure coefficient $[Cp = 2\{(p/p_\infty)-1\}/\gamma M_\infty^2]$ and wall heat flux q_w variation along the surface of the reentry module for $M_\infty = 2.0–6.0$. The $s/D = 0$ location is the stagnation point, where s is the distance taken along the surface of the blunt body from the stagnation point and D is the maximum diameter of the module. The pressure coefficient on the spherical cap of the capsule decreases gradually for a given M_∞. It can be visualized from **Figure 8(a)** that the Cp falls on the sphere-cone junction and remains constant over the cone. A sudden drop in Cp is observed on the shoulder of the module followed by a negative Cp variation in the base flow region. A low pressure is occurred immediately downstream of the base of the module which is characterized by a low speed recirculating flow region. In the base region, Cp is decreases with increasing M_∞. The variation of wall heat flux along the body is shown in **Figure 8(b)**. The wall heat flux q_w distribution along the wall shows similar characteristics as the pressure coefficient variation and q_w increases with increasing M_∞.

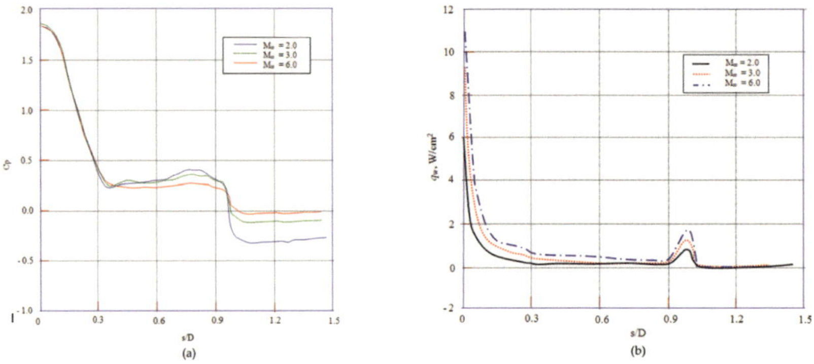

Figure 8. Variation of (a) pressure coefficient and (b) wall heat flux over reentry module.

5.2. Heat shield with a forward-facing spike

Figure 9 depicts the close-up view of the velocity vector plots and Mach contours unspiked and spiked over the heat shield [39]. We can visualize from the vector plot in **Figure 9(a)** the interaction between bow shock wave and reattachment shock. The recirculation region behaves as if it has a spike boundary. A significant flowfield is found around the unspiked and spiked heat shield as depicted in **Figure 9(b)**. **Figure 10** depicts the surface pressure coefficient and wall heat flux variation along blunt body surface for M_∞ = 2.0. A very high-pressure peak can be seen in the *Cp* variation. The wall heat flux variation along the model exhibits similar characteristics as the *Cp* variation. The peak in the heat flux is found at about the same location as in the *Cp* distribution.

5.3. Conical spike attached to the blunt body

Interaction between the conical oblique shock wave starting from the tip of the spike and the reattachment shock wave of blunt body is observed in the pressure, density and Mach contours in **Figure 11**. The reflected reattachment wave and shear layer from the interaction are shown behind the reattachment shock wave. A large separated region is found in front of the blunt body and the shear layer, the boundary of the separated region is

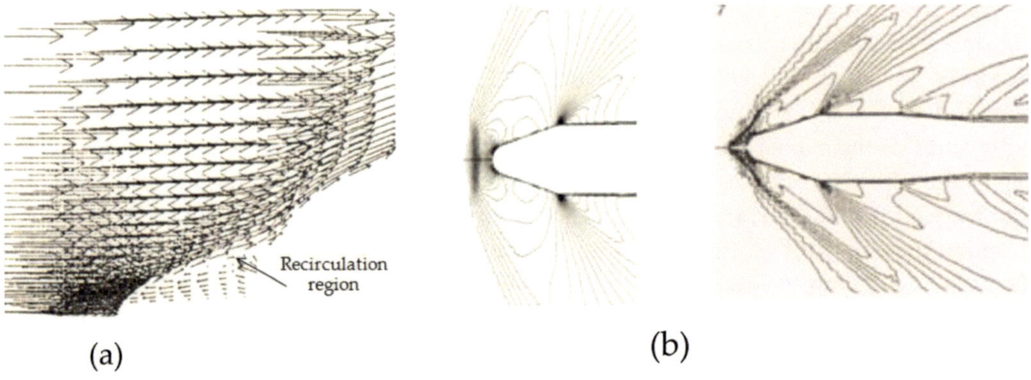

Figure 9. (a) Velocity vector and (b) Mach contours over the heat shield at M_∞ = 2.0.

http://dx.doi.org/10.5772/intechopen.74522

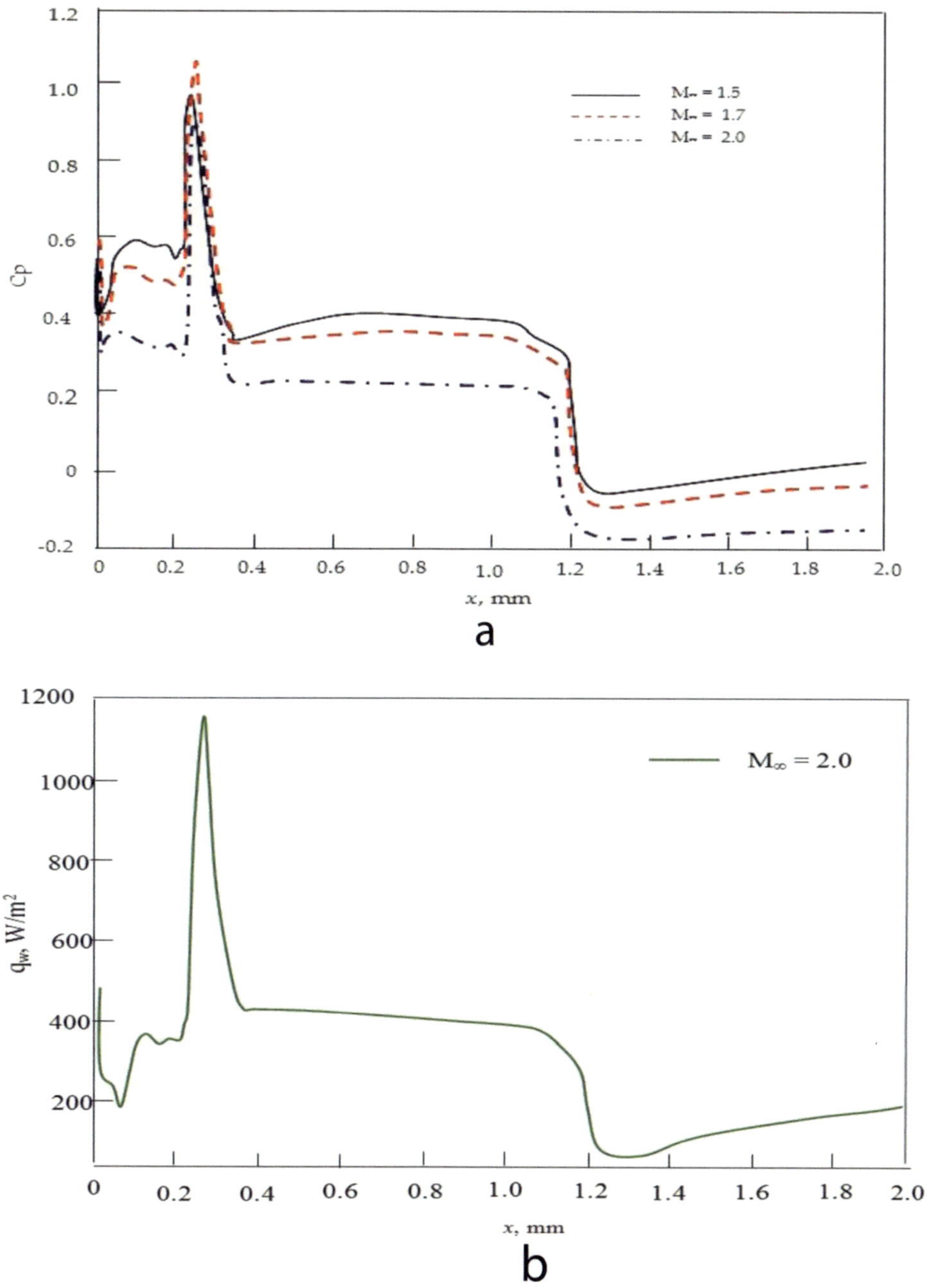

Figure 10. Variation of (a) pressure coefficient and (b) wall heat flux over the spiked heat shield.

visible. The variation of p/p_∞ and q_w along the spike blunt body is shown in **Figure 12(a)** and **(b)**, respectively. A sharp and sudden rise of heat flux is noticed very close to the spike tip, which is attributed to flow stagnation. The secondary peak heat flux is seen at about the reattachment point.

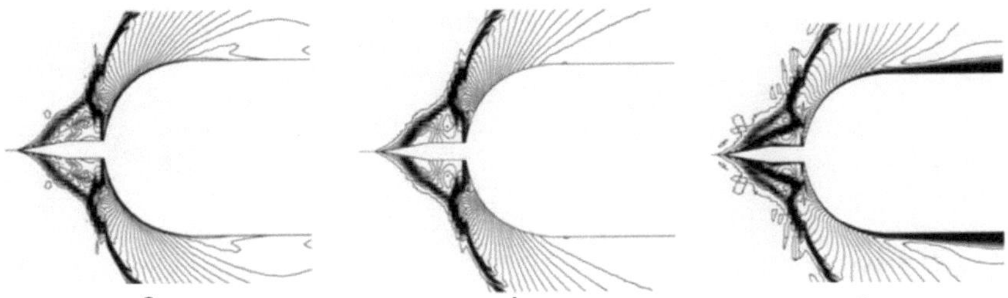

Figure 11. Contour plots over the spiked blunt body: (a) pressure (b) density and (c) Mach number.

5.4. Conical, disk and flat attached to the blunt body

5.4.1. Shock stand-off distance

The computed density contour plots over the conical, the hemispherical and the flat-faced aerospiked configurations are shown in **Figure 13(a), (b)** and **(c)**, respectively. The separated shear layer and the recompression shock from the reattachment point on the corner of the unspiked blunt body are observed in the flowfield region of contour diagram. The bow shock wave in front of the aerospike disk will decrease the aerodynamic drag as compared to the case the unspiked body as observed in the contour plots. In the fore-region of the aerodisk, the flow velocity decreases after the bow shock wave. At the corner of the aerodisk, the flow turns and expands rapidly, the boundary layer separates, gives a free-shear layer that separates the recirculating flow zone after the base region to the outer region. For the case of a button type spike flying at hypersonic speeds, a detached bow shock wave is observed in front of the spike which is appeared normal at to the body axis [29]. Since the flow behind the normal shock is subsonic, simple continuity considerations may show that the shock-detachment distance and stagnation-velocity gradient are essentially a function of ρ_2/ρ_∞ across the normal shock. The flow behind the bow shock wave is subsonic; the flow is no longer independent of the far-downstream influences. An alteration of the spike geometry in the subsonic region alters the complete flowfield up to the bow shock wave. **Figure 14(a)** shows schematic bow shock stand-off distance and the position of the sonic line. A schematic sketch of the bow shock stand-off distance of the flat-faced and the hemispherical spike attached to the blunt body is shown in **Figure 14(b)** and **(c)**. The bow shock-detachment distance becomes smaller with increasing ρ_2/ρ_∞. Probstein [40] gives an expression for the shock detachment distance Δ_F (**Figure 14(b)**) with diameter of the flat-disk D_S radio as:

$$\frac{\Delta_F}{D_S} = 2.8\sqrt{\frac{\rho_\infty}{\rho_0}} \tag{1}$$

The gas is considered thermally and calorically perfect. The ratio of the flow properties across the normal shock wave [41, 42] can be calculated as a function of M_∞ and γ, the relations are

$$\frac{\rho_2}{\rho_\infty} = \frac{(\gamma+1)M_\infty^2}{(\gamma-1)M_\infty^2+2} \tag{2a}$$

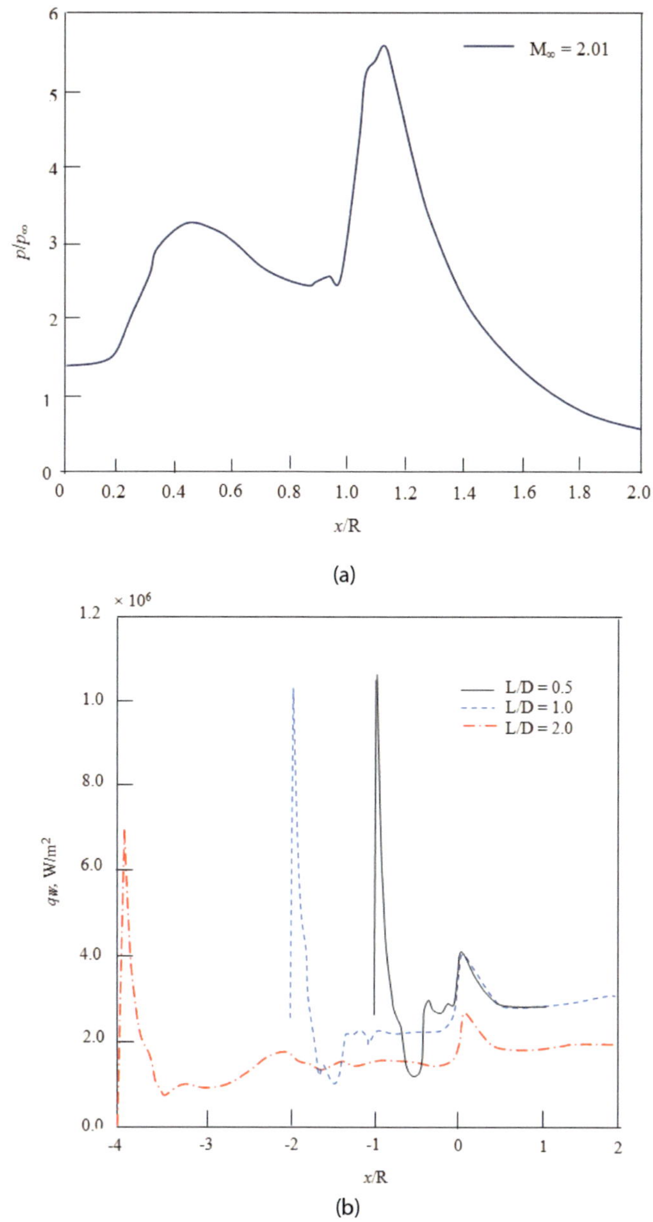

Figure 12. Variation of (a) pressure ratio and (b) wall heat flux over the spiked blunt body.

$$\frac{\rho_e}{\rho_\infty} = \frac{(\gamma+1)M_\infty^2}{(\gamma-1)M_\infty^2+2}\left[1+\frac{\gamma-1}{2}\frac{(\gamma-1)M_\infty^2+2}{2\gamma M_\infty^2-\gamma+1}\right]^{1/(\gamma-1)} \tag{2b}$$

$$\frac{p_e}{p_\infty} = \left[\frac{(\gamma+1)M_\infty^2}{2}\right]^{\gamma/(\lambda-1)}\left[\frac{\gamma+1}{2\gamma M_\infty^2+(\gamma-1)}\right]^{1/(\lambda-1)} \tag{2c}$$

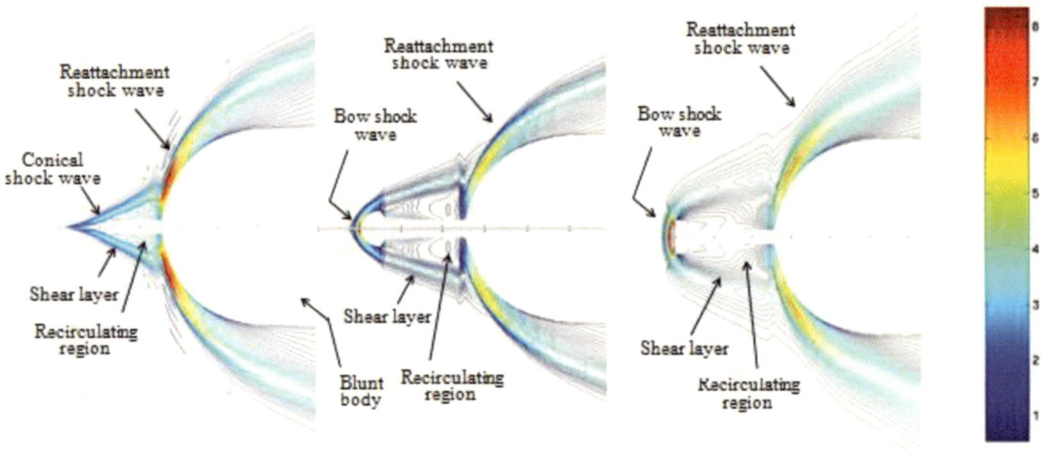

Figure 13. Mach contours over (a) the conical spike (b) the hemisphere and (c) flat-faced aerospike attached to the blunt-body.

The effects of the subsonic flow on the hemispherical and the flat-face disk bodies have been investigated by Truitt [43]. The nomenclature is illustrated in **Figure 14(a)**. The freestream flow passes through the normal portion of the shock wave reaching state 2 then decelerates isentropically to state *e*, which constitutes the edge condition for the thermal boundary layer at the stagnation point. The ratio of shock stand-off distance Δ_S with hemispherical spike of diameter, D_S (**Figure 14(c)**) is

$$\frac{\Delta_S}{D_S} = \frac{2\left(\frac{\rho_\infty}{\rho_0}\right)}{1+\sqrt{\frac{8\left(\frac{\rho_\infty}{\rho_0}\right)}{3}}} \tag{3}$$

The values of Δ_F/D_S and Δ_S/D_S are found to be 0.1898 and 0.1109, respectively. The spherical spike shows the greatest change in velocity gradient as compared to the flat-face disk. The flow is compressed to subsonic condition, which appears as inviscid and incompressible [44]. The shock wave stands in front of the blunt body and forms a region of subsonic flow around the stagnation region. The bow shock wave stand-off distance is calculated employing the following asymptotic formula of Frank and Zierep [45]:

$$\frac{\Delta}{R_N} = \frac{2(b)^{2/3}}{\left[\frac{(M_\infty^2-1)}{(\gamma+1)M_\infty^2}\right]^{2/3}} - 1 \tag{4}$$

where the value of *b* is taken as 0.14 [45]. The bow shock stand-off distance Δ/R_N is 0.17. The present numerical data is in agreement with the experimental data [46], asymptotic formula of Frank and Zierep [45] and with the theoretical results of Van Dyke and Gordon [47].

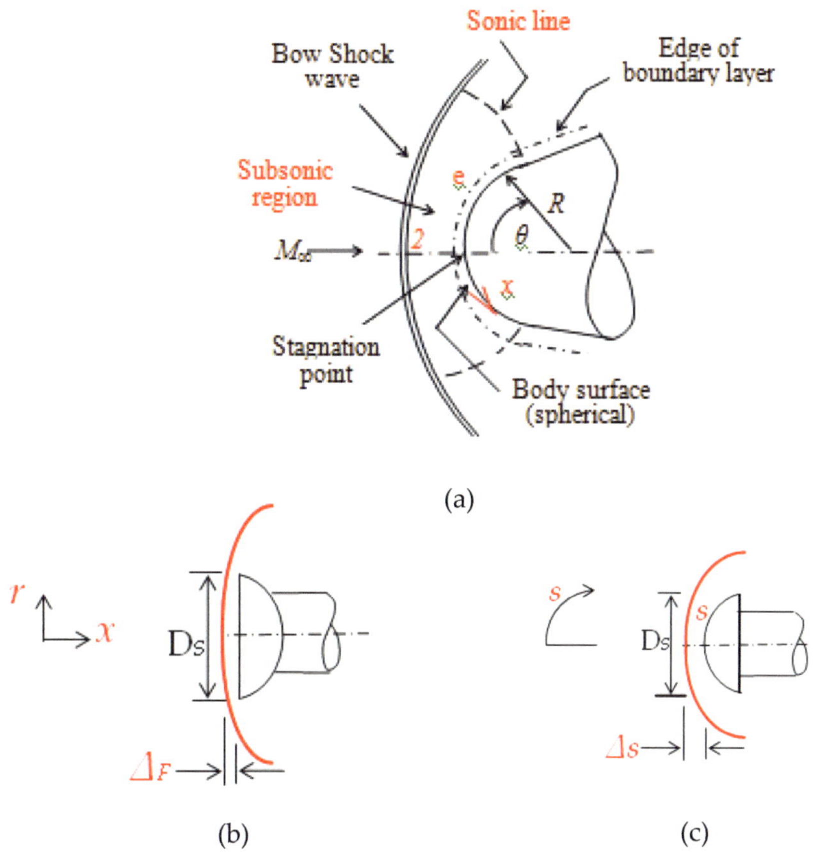

Figure 14. (a) Schematic sketch of flowfield over the blunt body, (b) the hemisphere and (c) the flat-face disc spiked.

5.4.2. Surface pressure variations

The pressure coefficient distribution on the spiked blunt-body with other aerospikes configurations is given in **Figure 15**. The $x/R = 0$ is the location of the spike tip, where R is radius of the blunt body. The maximum pressure on the surface of the spiked blunt body is found at an angle of about $40°$. This is the point of flow reattachment. It is interesting to notice that the maximum pressure is obtained on the same location on the hemisphere body. The low-pressure region in front of the body attributes reason for the drag reduction.

Tables 1 and **2** depict the variation of pressure coefficient, non-dimensional pressure, skin friction coefficient and wall heat flux over the spike surface facing the flow direction along the spike. The $s/D_S = 0$ is measured from the stagnation point. The s is location along the surface of spike and D_S is diameter of the spike as shown in **Figure 14(b)** and **(c)**. The p/p_∞ on the stagnation point is 38.23 and 48.84 for the flat-disk aerospike and the hemispherical disk aerospike, respectively. The pressure ratio across the normal shock is 41.83. It shows the percentage p/p_∞ difference of the order of −8.59% and 16.77% for the flat-disk and the hemispherical disk aerospike, respectively. The difference is attributed to the finite compressibility in the shock and the spike surface.

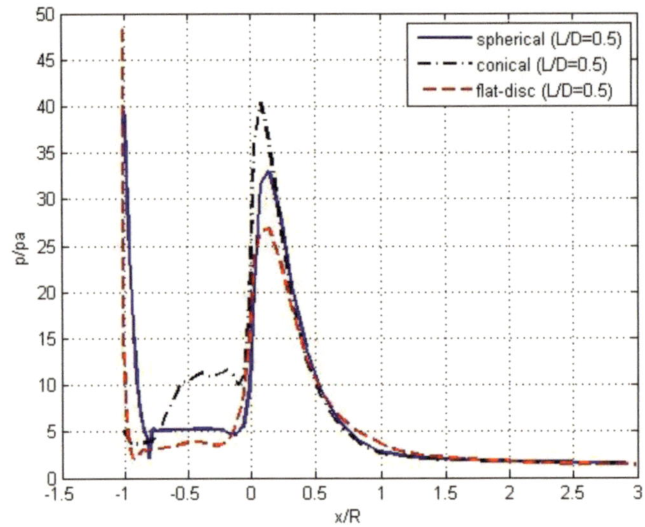

(a)

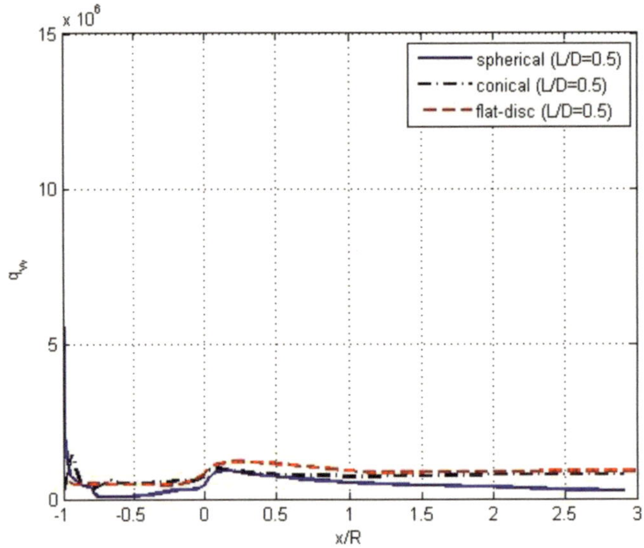

(b)

Figure 15. Variations of (a) pressure coefficient and (b) wall heat flux over the spiked blunt body.

5.4.3. Stagnation point heating and wall heat flux

The nose tip of a high-speed vehicle usually is hemi-spherically in shape. Consequently, a normal detached shock is formed in front of the stagnation point as depicted in **Figure 14(a)** which extends around the body as a curved oblique shock. The shock wave stands in front of the blunt body and forms a region of subsonic flow around the stagnation region and sonic line. The flow in the shock is at low subsonic speeds in the stagnation region and accelerates to sonic speeds in the shoulder region. For flat-nosed body, the detached bow shock wave of the

r/D_S	Cp	p/p_∞	cf	q_w, W/m^2
.00000E + 00	.14776E + 01	.38234E + 02	.42341E-05	.21158E + 06
.21500E-02	.15140E + 01	.39153E + 02	.11539E-03	.55592E + 07
.51000E-02	.15278E + 01	.39502E + 02	.68686E-04	.30266E + 07
.88500E-02	.15021E + 01	.38854E + 02	.55081E-04	.21678E + 07
.13500E-01	.14458E + 01	.37433E + 02	.49153E-04	.17160E + 07
.18950E-01	.13656E + 01	.35414E + 02	.45983E-04	.14250E + 07
.25200E-01	.12628E + 01	.32823E + 02	.43713E-04	.12121E + 07
.32300E-01	.11468E + 01	.29898E + 02	.40917E-04	.10449E + 07
.40250E-01	.10324E + 01	.27016E + 02	.36909E-04	.90977E + 06
.49000E-01	.92245E + 00	.24246E + 02	.33641E-04	.79750E + 06
.58600E-01	.82295E + 00	.21738E + 02	.32589E-04	.71001E + 06
.69000E-01	.71613E + 00	.19046E + 02	.33443E-04	.64654E + 06
.80250E-01	.59834E + 00	.16078E + 02	.35029E-04	.59998E + 06
.92300E-01	.48029E + 00	.13103E + 02	.35604E-04	.55673E + 06

Table 1. Variations of Cp, p/p_∞, cf and q_w over flat-face disk aerospike.

nose is particularly normal to the body axis. A compressible subsonic region is formed between the body and the shock which is a function of density ratio across the normal shock. The shock-detachment distance Δ and stagnation-velocity gradient K are essentially functions of density ratio across the normal shock [41, 42]. One of the areas of concern is the stagnation point heating of a spiked and unspiked body, when the incoming high-velocity flow is come to stagnation on the wall by a normal shock and adiabatically compression process. The problem now becomes finding out the heat flux in the vicinity of the stagnation point. It requires a solution of the entire flowfield from shock to the body.

The inviscid flowfield in the vicinity of the stagnation point is described in a fluid dynamics sense as the conversion of a unidirectional high-velocity stream by a normal shock wave into a high temperature subsonic layer, which is taken to be inviscid and incompressible [43]. The heat transfer rate is directly proportional to the enthalpy gradient on the body surface and square root of the velocity gradient, ($\beta = du_e/dx)_s$. The inviscid flowfield in the vicinity of the stagnation point is described as the conversion of the unidirectional, high-velocity flow by a normal shock wave into a high temperature subsonic flow. The enthalpy gradient is depended on the shape of the velocity profile in the boundary layer and by the variation of the air properties with temperature. Wall heat flux at the stagnation point of the blunt body can be calculated using the following expression of Fay and Riddell [48].

$$q_w = 0.763\mathrm{Pr}^{-0.6}\sqrt{(\rho_e\mu_e K)}\sqrt{\left(\frac{\rho_w\mu_w}{\rho_e\mu_e}\right)}(h_e - h_w) \qquad (5)$$

The stagnation point velocity gradient can be written as non-dimension parameter as ($K = \beta D/V_\infty$), where D is the diameter of the blunt body. Newtonian flow and similarity

r/D_S	Cp	p/p_∞	Cf	q_w, W/m^2
0	1.8987	48.846	1.3293e-06	326,130
0.005	1.8974	48.814	5.7879e-05	1.4549e + 07
0.01	1.8906	48.643	2.7381e-05	7.2722e + 06
0.015	1.8798	48.372	1.6742e-05	4.8444e + 06
0.02	1.8669	48.046	1.1241e-05	3.6302e + 06
0.025	1.8529	47.694	7.9517e-06	2.9018e + 06
0.03	1.8386	47.333	5.8737e-06	2.4167e + 06
0.035	1.8244	46.974	4.5396e-06	2.0707e + 06
0.04	1.8102	46.616	3.6880e-06	1.8115e + 06
0.045	1.7957	46.253	3.1615e-06	1.6101e + 06
0.05	1.7807	45.873	2.8655e-06	1.4489e + 06
0.055	1.7642	45.458	2.7391e-06	1.3167e + 06
0.06	1.7449	44.972	2.7632e-06	1.2057e + 06
0.065	1.7221	44.396	3.0004e-06	1.1111e + 06
0.07	1.6957	43.731	3.5721e-06	1.0305e + 06
0.075	1.6573	42.765	4.8788e-06	960,610
0.08	1.5942	41.173	7.8607e-06	897,760
0.085	1.5726	40.629	1.0038e-05	853,270
0.09	1.3785	35.737	1.1276e-05	826,300
0.095	1.2663	32.910	1.0595e-05	855,090
0.1	0.5229	14.177	1.2575e-05	708,300

Table 2. Variations of Cp, p/p_∞, cf and q_w over hemispherical disk aerospike.

method [41, 43] methods do not take into consideration the finite compressibility that exists between the shock wave and the spike surface. Experiments were conducted by Boison et al. [49] in the supersonic wind tunnel to obtain the values of the stagnation point velocity gradient. The numerical analysis is able to take into consideration the compressibility effects in the subsonic region. The stagnation point stream-wise velocity gradient is given by the Newtonian impact theory, viz.,

$$\left(\frac{du_e}{dx}\right)_s = \frac{1}{R_N}\sqrt{\frac{2(\rho_2 - \rho_\infty)}{\rho_2}} \tag{6}$$

It is important to mention here that the sphere shows the greatest change in velocity gradient as compared to the flat disk. The magnitude of the stagnation-velocity gradient indicates the maximum heat transfer rate. The disk or flat plate with free streamlines will experience the lowest stagnation wall temperature of blunt body. The value K is considered as 0.3 [50]. The hemispherical aerospike exhibits the significant changes in the magnitude of K as compared to the flat disk aerospike. The magnitude of the K also indicates the maximum heat transfer rate over the hemispherical disk. The computed values of the stagnation point heat flux $q_{w,s}$ are 0. 556 × 10^7 W/m^2 and 1.45 × 10^7 W/m^2 for the flat-disk and the hemispherical disk aerospike, respectively. The $q_{w,s}$ calculated using Eq. (5) is 0. 831 × 10^7 W/m^2 and 1.55 × 10^7 W/m^2 for the flat-disk and the hemispherical disk spike, respectively. The discrepancy is due to the value of the K. The value of K is difficult to calculate analytically and experimentally for the flat-face disk spike [50]. The expression for the stagnation point heat transfer [26] for a sphere is

$$\dot{q}_{w,s} = 1.83 \times 10^{-4} \left(\frac{\rho_w}{R_n}\right)^{0.5} V_\infty^3 \left[1 - \frac{H_w}{H_s}\right] \tag{7}$$

where R_N is nose radius in m, V_∞ is freestream velocity, H_w is the wall enthalpy and H_S the stagnation enthalpy. Experiments were carried out to evaluate the stagnation point velocity gradient.

5.5. Heat transfer measurements

Experiments [51] were carried out in shock tunnel to measure the heat flux on the blunt body attached with the hemispherical aerospike of L/D = 1.5 and 2.0 at zero angle of attack. **Figure 16** depicts the locations of heat transfer measurement. The shock tunnel consists of a shock tube divided into driver and driven sections, separated by a metallic diaphragm. The shock tube is connected to the wind tunnel portion by a convergent-divergent conical nozzle, test-section and dump tank separated by a diaphragm to obtain M_∞ = 6. The platinum sensors were fabricated by depositing a thin coating of platinum on the surface of Macor kept on the surface of the spiked model. The stagnation wall heat flux for the unspiked blunt body configuration was evaluated for the shock-tunnel conditions, using Fay and Riddell [48] equations. Surface heat flux on the basic body was obtained employing Lees [52] equations as

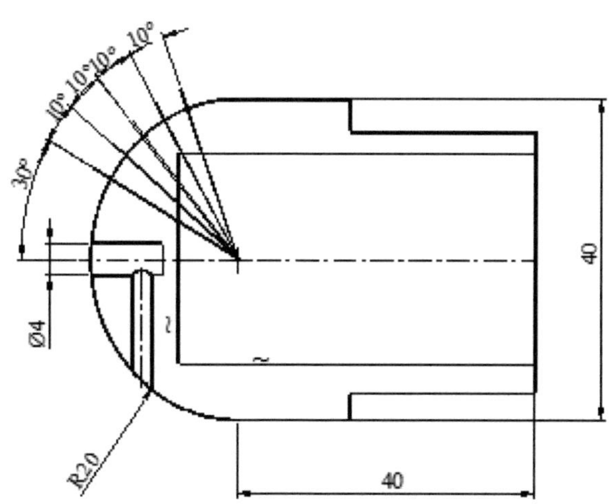

Figure 16. Locations of the heat transfer probes.

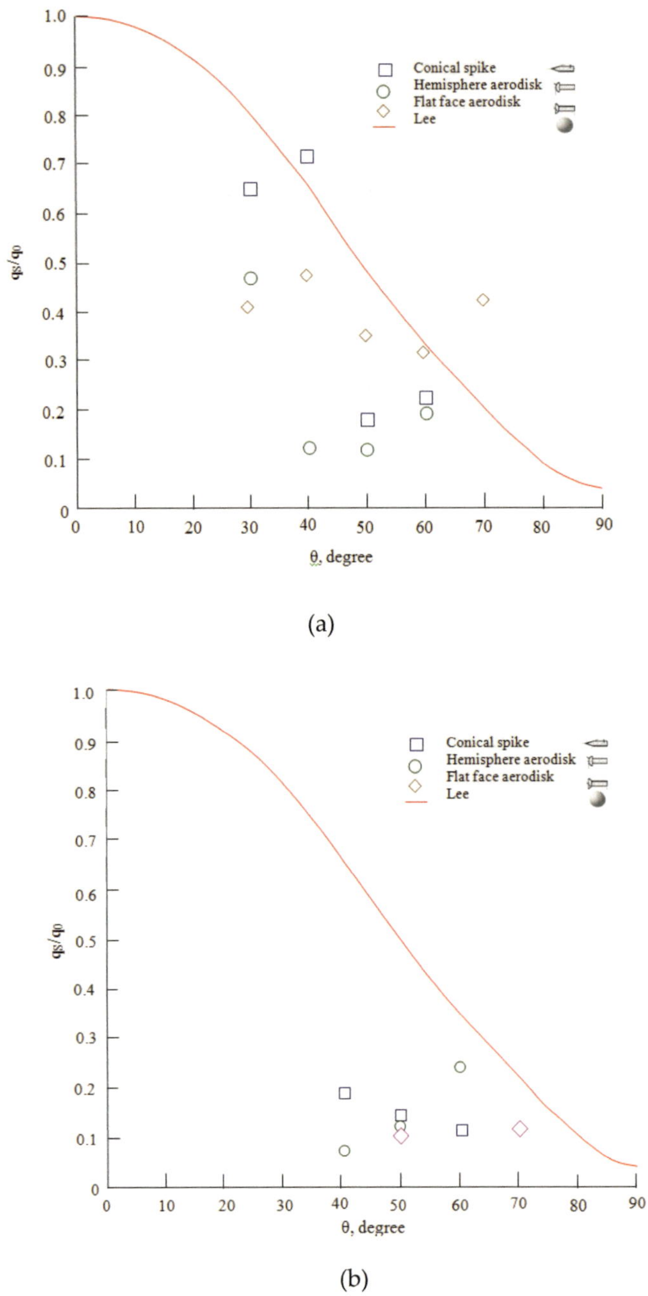

Figure 17. Variations of heat flux over the various spiked blunt body (a) $L/D = 1.5$ and (b) $L/D = 2.0$ at $M_\infty = 6.0$.

$$\frac{q_w}{q_S} = \frac{2\theta \sin\theta \left[\left\{ 1 - \frac{1}{\gamma M_\infty^2} \left(\cos^2\theta + 1 \right) \right\} \right]}{\sqrt{\left(1 - \frac{1}{\gamma M_\infty^2}\right)\left(\theta^2 - \frac{\theta \sin 4\theta}{2} + \frac{1 - \cos 4\theta}{8}\right) + \frac{4}{\gamma M_\infty^2}\left(\theta^2 - \theta \sin 2\theta + \frac{1 - \cos 2\theta}{2}\right)}} \tag{8}$$

The wall heat flux values are non-dimensionlised with the stagnation heat flux value. The temperature variations on the model surface that can be recorded using platinum thin film gauge can be used to find the heat transfer rate to the model by using the following expression [53]:

$$q(t) = \frac{\sqrt{k\rho C_p}}{2\sqrt{\pi}}\left[\frac{2T(t)}{\sqrt{t}} + \int_0^t \frac{T(t) - T(\tau)}{(t-\tau)^{3/2}} d\tau\right] \tag{9}$$

Evaluate above equation by applying approximate numerical schemes. Numerical methods involve division of the time interval (0, t) into a finite number of increments and the evaluation of the integrand at each of the division points. The heat flux over the basic configuration with and without spiked for a fixed L/D of 1.5 and 2.0 is shown in **Figure 17**. It is seen that for the basic body with spike the heat flux values is lower than the basic body without spike. Except the conical aerospike, all other spike configurations result in lower heat flux values than the basic body alone. For this length also, the spikes could reduce the heat flux on the basic body substantially, as seen from these results.

6. Conclusions

The flowfield around a forward facing a hemispherical and a flat disk spike attached to blunt-nosed body has been numerically simulated at high speeds at zero angle of attack. The flow visualizations were done using the velocity vector and contour plots in order to analyze the influence of the shape of the spike on the drag reduction and wall heat flux. The formation of the bow shock wave is observed over the unspiked and spiked blunt body. Different flow separation zones depend on the shape of the spike attached to the blunt body. The stand-off distances of the bow shock wave for the hemispherical and the flat aerospike are compared with the analytical solutions and are seen in good agreement. The variations of surface pressure, the skin friction coefficient and the wall heat flux along the surface of the spike facing the flow direction is significantly influenced by the geometrical shape of the spike. The density and pressure ratio and the heat flux at the stagnation point are computed and compared with the analytical results. The numerical analysis delineates complete flowfield information over the unspiked and the spiked blunt-body surface including the bow shock, shock stand-off distance shock, sonic line and stagnation point velocity gradient.

Nomenclature

C_f	skin friction coefficient
C_p	pressure coefficient
D	cylinder diameter

L	length of the spike
M	Mach number
q	heat flux
p	pressure
R_N	radius of the spherical body
s	distance along the surface of the spike
T	temperature
t	time
x, r	coordinate direction
γ	ratio of specific heats
ρ	density
τ	dummy line variable
Δ	shock stand-off distance

Subscripts.

e	edge of the boundary layer
o	stagnation
w	wall
∞	freestream condition

Author details

Rakhab Chandra Mehta

Address all correspondence to: drrakhab.mehta@gmail.com

Department of Aeronautical Engineering, Noorul Islam Center for Higher Education, Noorul Islam University, Kumaracoil, India

References

[1] Takovitskii SA. Analytical solution in the problem of constructing axisymmetric noses with minimum wave drag. Fluid Dynamics. 2005;**41**(2):308-312

[2] Eggers AJ, Meyer M, Resniko D, Dennis H. Bodies of Revolution Having Minimum Drag at High Supersonic Airspeeds. Moffett Field, USA, NACA TR-1306: Ames Research Laboratory; 1957

[3] Allen HJ, Eggers AJ. A Study of the Motion and Aerodynamic Heating of Ballistic Missiles Entering the Earth's Atmosphere at High Supersonic Speeds. Moffett Field, USA, NACA-TR 1381: Ames Research Laboratory; 1958

[4] Yamamoto Y, Yoshioka M. CFD and FEM coupling analysis of OREX aero-thermodynamic flight data. 29th AIAA Thermophysics Conference, June 19–22, 1995, San Diego, California, USA, AIAA 95–2087

[5] Kovalev V, Yakuchikov A. Simulation of hydrogen adsorption in carbon nanotube arrays. Acta Astronatica. 2011;**68**:681-685

[6] Tret'yakov PK, Garanin AF, Grachev GN, Krainev VL, Ponomarenko AG, Tishchenko VN, Yakolec VI. Control of supersonic flow around bodies by means of high-power recurrent optical breakdown. Physics-Doklady. 1996;**41**(11):566-567

[7] Kim J-H, Matsuda A, Sakai T, Sasoh A. Wave drag reduction with acting spike induced by laser-pulse energy deposition. AIAA Journal. 2011;**49**:2076-2078

[8] Schülein E, Zheltovodov AA, Pimonov EA, Loginov MS. Experimental and numerical modeling of the bow shock interaction with pulse-heated air bubbles. International Journal Aerospace Innovation. 2010;**2**(3):183-205

[9] Schülein E, Zheltovodov A. Effects of steady flow heating by arc diskharge upstream of non-slender bodies. Shock Waves. 2011;**24**(4):383-390

[10] Artem'ev VI, Bergel'son VI, Nemchinoy TI, Orlova VA, Smirnov VM, Khazins VM. Change of regime in supersonic flow past an obstacle preceded by a thin channel of reduced density. Fluid Dynamics. 1989;**24**(5):779-784

[11] Meyer B, Nelson HF, Riggins DW. Hypersonic drag and heat-transfer reduction using a forward-facing jet. Journal of Aircraft. 2001;**38**(4):680-686

[12] Eghlima Z, mansour K, Fardipour K. Heat transfer reduction using combination of spike and counter flow jet on blunt body at high Mach number flow. Acta Astronautica. 2018; **143**:92-104

[13] Finley PJ. The flow of a jet from a body opposing a supersonic freestream. Journal of Fluid Mechanics. 1966;**26**(2):337-368

[14] Gordeev VP, Krasilnikov AV, Lagutin VI, Otmennikov VN. Experimental study of the possibility of reducing supersonic drag by employing plasma technology. Fluid Dynamics. 1996;**31**(2):313-317

[15] Shang JS. Plasma injection for hypersonic blunt-body drag reduction. AIAA Journal. 2002;**40**(6):1178-1186

[16] Fomin VM, Maslov AA, Malmuth ND, Fomichev VP, Shashkin AP, Korotaeva TA, Shiplyuk AN, Pozdnyakov GA. Influence of a counter flow plasma jet on supersonic blunt-body pressures. AIAA Journal. 2002;**40**(6):1170-1177

[17] Ladoon DW, Schneider SP, Schmisseur JD. Physics of a supersonic forward-facing cavity. Journal of Spacecraft and Rockets. 1998;**35**(5):626-632

[18] Schnepf C, Wysocki O, Schülein E. Wave drag reduction due to self-aligning aerodisk. Progress in Flight Physics. 2015;**7**:475-488

[19] Hayashi K, Aso S, Tani, Y: Numerical study of thermal protection system by opposing jet. 43rd AIAA Aerospace Sciences Meeting and Exhibit, Reno, Nevada, USA, 2005; AIAA 2005–188

[20] Georgievskii PY, Levin VA. Control of the flow past bodies using localized energy addition the supersonic oncoming flow. Fluid Dynamics. 2003;**38**(5):794-805

[21] Knight D, Kolesnichenko YF, Brovkin V, Khmara D. High speed flow control using microwave energy deposition. 16th Australasian Fluid Mechanics Conference, Gold Coast, Australia; Dec. 2007

[22] Reggins D, Nelson HF, Johnson E. Blunt-body wave drag reduction using focused energy deposition. AIAA Journal. 1999;**47**(4):460-467

[23] Liu Y, Jiang Z. Concept of non-ablative thermal protection system for hypersonic vehicles. AIAA Journal. 2013;**51**(3):584-590

[24] Motoyama N, Mihara K, Miyajima R, Watanuki T, Kubota H. Thermal protection and drag reduction with use of spike in hypersonic flow, AIAA/NAL-NASDA-ISAS. AIAA 2001–1828; 2001

[25] Ahmed MYM, Qin N. Recent advances in the aerothermodynamics of spiked hypersonic vehicles. Progress in Aerospace Science. 2011;**47**(6):425-449

[26] Tauber, ME: A review of high speed, convective, heat transfer computational methods. Ames Research Laboratory, Moffett Field, USA, NASA TP-2914, July 1989

[27] Wang Z-G, Sun X-W, Huang W, Li S-B, Yan L. Experimental investigation on drag and heat flux reduction in supersonic/hypersonic flows: A review. Acta Astronautica. 2016; **129**:95-110

[28] Kalimuthu R, Mehta RC, Rathakrishnan E. Experimental investigation on spiked body at hypersonic flow. The Aeronautical Journal. 2008;**112**(1136):593-598

[29] Mehta RC. Numerical simulation of the flowfield over conical, disk and flat spiked body at Mach 6. The Aeronautical Journal. 2010;**114**(1154):225-236

[30] Mehta RC. Numerical heat transfer study around a spiked blunt-nose body at Mach 6. Heat and Mass Transfer. 2013;**49**:485-496

[31] Mehta RC. Numerical heat transfer study over spiked blunt bodies at Mach 6.8. Journal of Spacecraft and Rockets. 2000;**37**(5):700-703

[32] Bogdonov SM, Vas IE. Preliminary investigations of spiked bodies at hypersonic speeds. Journal of Aerospace Sciences. 1959;**26**(2):65-74

[33] Fujita M, Kubota H. Numerical simulation of flowfield over a spiked blunt nose. Computational Fluid Dynamics Journal. 1992;**1**(2):187-195

[34] Yamauchi M, Fujjii K, Tamura Y, Higashino F. Numerical investigation of hypersonic flow around a spiked blunt body. AIAA Paper 93-0887; 1993

[35] Blazek J. Computational Fluid Dynamics: Principles and Applications. 1st ed. Oxford, UK: Elsevier Science Ltd; 2001

[36] Jameson A, Schmidt W, Turkel E. Numerical simulation of Euler equations by finite volume methods using Runge-Kutta time-stepping schemes. AIAA paper 81–1259; 1981

[37] Mehta RC. Isoperimetric finite element method to generate structured grid for numerical flow simulation. Journal of Aerospace Sciences and Technologies. 2017;**69**(4):588-596

[38] Mehta RC, Jayachandran T. Navier-stokes solution for a heat shield with and without a forward-facing spike. Computers and Fluids. 1997;**26**(7):741-754

[39] Mehta RC. Heat transfer study of high speed flow over a spiked blunt body. International Journal of Numerical Methods for Heat and Fluid Flow. 2000;**10**(7):750-769

[40] Probstein RF. Inviscid flow in the stagnation region of very blunt-nosed bodies at hypersonic flight speeds. WADC–TN 56-395; Sept. 1956

[41] Ames Research Staff. Equations, Tables and Charts for Compressible Flow. Moffett Field, USA, NACA report 1135: Ames Research Laboratory. p. 1953

[42] Liepmann HW, Roshko A. Elements of Gas Dynamics. 1st South Asian ed. New Delhi: Dover Publications Inc; 2007

[43] Truitt RW. Hypersonic Aerodynamic. New York, USA: Ronald Press Co.; 1959

[44] Hayer WD, Probstein RF. Hypersonic Flow Theory. New York: Academic Press; 1959

[45] Frank W, Zierep J. Schallnahe überschallströmung um rotationssymmetrische körper. Acta Mech. 1974;**19**:277-287

[46] Stilp, A: Strömungsuntersuchungen an Kgeln mit tranonischen und supersonischen Geschwidigkeiten in Luft und Frigen-Luftgemischen Bericht Nr. 10.65, Fckerstrasse 4, Germany; 1965

[47] Van Dyke MD, Gordon HD. Supersonic flow past a family of blunt axisymmetric bodies. NASA TR-R 1; 1959

[48] Fay JA, Riddell FR. Theory of stagnation point heat transfer in dissociated air. Journal of Aeronautical Sciences. 1958;**25**:73-85

[49] Boison JC, Curtiss HA. An experimental investigation of blunt body stagnation point velocity gradient. ARS Journal. 1959;**29**(2):130-135

[50] White FM. Viscous Fluid Flow. 2nd ed. Singapore: McGraw Hill International Edition; 1991

[51] Kalimuthu R. Experimental Investigation of Hemispherical Nosed Cylinder with and without Spike in a Hypersonic Flow. Ph. D. Thesis. Kanpur, India: Department of Aerospace Engineering, Indian Institute of Technology; April 2009

[52] Lees L. Laminar heat transfer over blunt-nosed bodies at hypersonic flight speeds. Jet Propulsion. 1956;**26**(4):259-269

[53] Cook WJ, Felderman EJ. Reduction of data from thin-film heat-transfer gages: A concise numerical technique. AIAA Journal. 1966;**4**(4):561-562

A Review of Heat Transfer Enhancement Methods Using Coiled Wire and Twisted Tape Inserts

Orhan Keklikcioglu and Veysel Ozceyhan

Additional information is available at the end of the chapter

http://dx.doi.org/10.5772/intechopen.74516

Abstract

Heat transfer enhancement is categorized into passive and active methods. Active methods need external power to input the process; in contrast, passive methods do not require any additional energy to improve the thermohydraulic performance of the system. Passive methods are widely used in both experimental and numerical applications when investigating heat transfer enhancement and friction losses to save energy and costs. The many passive methods for increasing heat transfer rate include various components located in the fluid flow path, such as twisted tapes, coiled or tangled wires, and nozzle turbulators. The present paper represents a comprehensive review that focused on heat transfer enhancement methods with coiled wire and twisted tape inserts since the installation of inserts is easier and more economical. The thermodynamic performance of heat exchange components is also affected by the flow conditions such as laminar or turbulence. The present review comprises investigations on the enhancement of heat transfer using twisted tape and coiled wire inserts in laminar and turbulent flow region.

Keywords: heat transfer enhancement, coiled wire, twisted tape, heat transfer, friction factor

1. Introduction

As a result of the global energy crisis, which is one of the most crucial problems due to the large and continuous increase in the consumption and the increment shortage of energy resources as well as the high cost, many researchers have performed to increase the efficiency of thermal systems and reduction of the size and thus energy consumption rates.

Heat transfer enhancement is a process of increasing the heat transfer rate and thermohydraulic performance of a system using various methods. The methods of heat transfer enhancement are employed for developing the heat transfer without affecting the overall realization of the systems significantly, and it covers a wide range of areas where heat exchangers are used for such functions as air-conditioning, refrigeration, central heating systems, cooling automotive components, and many uses in the chemical industry.

Heat transfer enhancement methods exist on three general classifications which are passive, active, and compound techniques. Active methods require external power to input the process; in contrast, passive methods do not require any additional energy to improve the thermohydraulic performance of the system. Also, two or more passive and active techniques can be used together and that is called compound technique, which is employed to produce a higher augmentation than using one passive or active technique independently.

1.1. Passive techniques

In the passive techniques, any external power is not required; rather, geometry or surface of the flow channel is modified to increase the thermohydraulic performance of the systems. The inserts, ribs, and rough surfaces are utilized to promote fluid mixing and the turbulence in the flow, which results in an increment of the overall heat transfer rate. Passive techniques have also some advantages in relation to the other heat transfer enhancement techniques such as low cost, easy production, and installation.

1.2. Active techniques

Active techniques are more complex than the passive techniques in the expression of design and application because of the necessity of external energy to adjust the flow of fluid so as to obtain an improvement in thermal efficiency. Providing external energy in most applications is not easy; for this reason, the use of active techniques in scientific fields is limited.

1.3. Compound techniques

A compound technique consists of the combination of more than one heat transfer enhancement method (active and/or passive) to increase the thermohydraulic performance of heat exchangers. It can be employed simultaneously to generate an augmentation that promotes the performance of the system either of the techniques operating independently. Preliminary studies on compound passive augmentation technique of this kind are quite encouraging.

2. Simple definitions used to evaluate heat transfer enhancement

The basics of performance evaluation criteria (PEC) were determined for the fixed geometry criteria (FG), which was related with heat transfer and friction factor characteristics of various augmentation techniques in Ref. [1]. The PEC define the performance advantages of a heat exchanger having enhanced surfaces, relative to a reference exchanger, for example, one having smooth surfaces [1].

The fixed geometry criteria are used for many passive techniques which include various types of inner ribs or inserts in the tube. The investigations of heat transfer enhancement with passive techniques are generally conducted at the same pumping power in accordance with the fixed geometry criteria. Thus, comparison of results for the tube with inserts and without inserts can be carried out in ease. The Nusselt number can be written in below as to calculate the thermal performance of the system:

$$Nu = \frac{hD}{k} \tag{1}$$

And, the Reynolds number for fluid is

$$\mathrm{Re} = \frac{UD}{\nu} \tag{2}$$

The friction factor, f, is calculated as follows, for the fully developed isothermal flow:

$$f = \frac{\Delta P}{\frac{1}{2}\rho.U^2\frac{L}{D}} \tag{3}$$

where U represents the mean fluid velocity in the tube.

The heat transfer rate and friction factor of the smooth tube and smooth tube fitted with inserts are evaluated under the same pumping power as below [2]:

$$\left(\dot{V}\,\Delta P\right)_s = \left(\dot{V}\,\Delta P\right)_a \tag{4}$$

$$\left(f\mathrm{Re}^3\right)_s = \left(f\mathrm{Re}^3\right)_a \tag{5}$$

$$\mathrm{Re}_s = \mathrm{Re}_a\left(f_a/f_s\right)^{1/3} \tag{6}$$

The overall enhancement ratio (OER) is the parameter that is usually used in heat transfer augmentation to determine the performance of different variations of heat exchangers. The parameter can be written as in Eq. (7) for the same pumping power based on the fixed geometry criteria [2]:

$$\eta = \left.\frac{h_a}{h_s}\right|_{pp} = \left.\frac{Nu_a}{Nu_s}\right|_{pp} = \left(\frac{Nu_a}{Nu_s}\right)\left(\frac{f_s}{f_a}\right)^{1/3} \tag{7}$$

3. Heat transfer enhancement with twisted tape and coiled wire inserts

Many researchers have performed the reviewing on passive or active heat transfer enhancement methods [3–10]. This paper focuses on the reviewing of the recent investigations about the heat transfer enhancement with coiled and twisted tape inserts which are widely used as a passive technique in tubular flow. The coiled wire inserts intensify the disturbance of viscous sublayer and promote redevelopment of the thermal and hydrodynamic boundary layers in the tube flow effectively [11]. Also, twisted tape inserts are used commonly on heat transfer process. Due to the simple design and easy installation, twisted tape inserts are widely preferred to generate swirl flow and increase the turbulence rate in the flow. Additionally, they are used extensively over decades in scientific research as well as industrial applications.

3.1. Twisted tape inserts

Twisted tape inserts are one of the most used enhancement methods of heat transfer. Twisted tape inserts increase both convective heat transfer and fluid friction in the flow region. They induce the turbulence and promote the swirl flow. Moreover, geometric configurations of twisted tape inserts can disturb the boundary layer; with this way, better heat transfer rate can be obtained. However, increment of the fluid friction can negatively affect the overall enhancement ratio for a heat exchanger tube. The performance of a heat exchanger with twisted tape inserts depends on pitch and twist ratios. In recent investigations, a lot of researchers have conducted both experimental and numerical studies to determine the optimal configuration in accordance with the ratios of pitch and twist.

Man et al. [12] carried out an experimental investigation on heat transfer and friction characteristics of dual-pipe heat exchanger for single-phase forced convective flow with alternate clockwise and counterclockwise twisted tape (ACCT tape) and typically twisted tape (TT tape) for the Reynolds number ranging from 3000 to 9000. They reported that the maximum values of performance evaluation criteria (PEC) with the full-length ACCT tape insert reached 1.42 in experimental flowing conditions. Suri et al. [13] experimentally investigated the augmentation in heat transfer and friction in a flow through heat exchanger tube with multiple square perforated twisted tape inserts. The experiments were conducted with the Reynolds number between 5000 and 27,000, perforation width ratio a/W_T from 0.083 to 0.333, and twist ratio $T_L = W_T$ from 2.0 to 3.5. The maximum enhancement is observed at a/W_T of 0.250 and $T_L = W_T$ of 2.5. Sundar et al. [14] investigated the effectiveness of solar flat-plate collectors with and without twisted tapes in Al_2O_3/water nanofluid flow region. The experimental results indicated that the heat transfer rate enhanced 49.75% with the twist ratio of 5 at the Reynolds number 13,000. The maximum friction penalty of 1.25 times was observed for 0.3% nanofluid with twist ratio of 5 with the comparison of water in a smooth tube. Saysroy and Eiamsa-ard [15] conducted a numerical study to determine the thermohydraulic performance of a multichannel twisted tape inserts in laminar and turbulent flow regions. The numerical results showed that for the laminar flow, the maximum thermal performance factor of 7.28 was obtained by using the tube with the multichannel twisted tape with $N = 2$ and $y/w = 2.5$ at the Reynolds number of 2000. Heat transfer and pressure drop of CuO/water nanofluid with twisted tape inserts were explored by Wongcharee and Eiamsa-ard [16]. The results demonstrated that using CuO nanofluid with twisted alternate axis (TA) obtain a higher Nusselt number and thermal performance, and the twisted tape in alternate axis was about 89% more effective than typical twisted tape. Various types of twisted tape such as hollow, double, perforated, and dimpled configurations based on the physical properties, e.g., hollow widths and hole diameters, were investigated in many studies [17–21]. Li et al. [22] carried out a numerical study on her transfer enhancement in a tube with centrally hollow narrow twisted tape under laminar flow conditions. They reported that the tube with cross hollow twisted tape inserts has the best overall heat transfer performance for different hollow widths of the tape. Eiamsa-ard et al. [23] investigated the effect of helically twisted tapes on the thermal performance. Three different twist and helical pitch ratios were used under the turbulent flow region (**Tables 1–3**).

The results were compared to conventional helical tape, and it was reported that the helically twisted tape showed higher thermal performance than the conventional helical tape. Piriyarungrod et al. [40] studied the effects of inserted tapered twisted tapes, their taper angle and twist ratio on heat transfer rate, pressure drop, and thermal performance factor characteristics. They reported that thermal performance factor tended to increase with increasing taper angle and decreasing tape twist ratio. Promvonge et al. [41] investigated turbulent convective heat transfer characteristics in a helical-ribbed tube fitted with twin twisted tapes. The results obtained from the ribbed tube and the twin twisted tape insert were compared with those from the smooth tube and the ribbed tube acting alone. Prasad and Gupta [42] studied to enhance the rate of heat transfer of Al_2O_3 nanofluid in a U-tube heat exchanger with twisted tape insert for the Reynolds number ranging from 3000 to 30,000. The Nusselt number of the entire pipes for 0.03% concentrations of nanofluid with twisted tape inserts was enhanced by 31.28% compared to water. Pal and Saha [43] presented an experimental study to investigate the friction factor and Nusselt number in a circular duct having spiral rib roughness and fitted with twisted tapes with oblique teeth. The experiments were conducted for laminar flow region using viscous oil. The thermo-hydraulic performance was also evaluated, and the twisted tapes with oblique teeth in combination with integral spiral rib roughness performed significantly better than the individual enhancement technique. Eiamsa-ard and Kiatkittipong [44] investigated the enhancement of thermal performance in a heat exchanger with multiple twisted tapes. TiO_2/water nanofluid was used in experiment as working fluid. They reported that the tube inserted with multiple twisted tapes showed superior thermal performance factor when compared with the plain tube or the tube inserted with a single twisted tape.

3.2. Coiled wire inserts

Transverse or helical ribs, for example, coiled wire inserts, are an attractive method to create the surface roughness [45]. The coiled wire inserts intensify the disturbance of laminar boundary layer and promote redevelopment of the thermal and hydrodynamic boundary layers in the tube flow effectively [11]. Moreover, helically coiled wire can be used to generate secondary flow which helps to enhance the heat transfer rate with the increment of vorticity in the tubular flow. They have also some advantages in relation to the other passive methods such as easy manufacturing and installation; lower manufacturing cost; better fluid mixing and disturbance of laminar boundary layer; possibility to use different fluid types, e.g., nanofluid, water, viscous oil, and air; and possibility to install with various passive techniques together.

The enhancement of heat transfer rate by using coiled wire inserts has been widely studied both experimentally and numerically by many researchers. Garcia et al. [46] analyzed the thermohydraulic performance of three types of passive heat transfer enhancement based on

Authors	Fluid	Type	Flow region	OER	Sample of twisted tape inserts
Jaisankar et al. [24]	Water	Helically twisted tape	3000 ≤ Re ≤ 23,000 turbulence	1.2	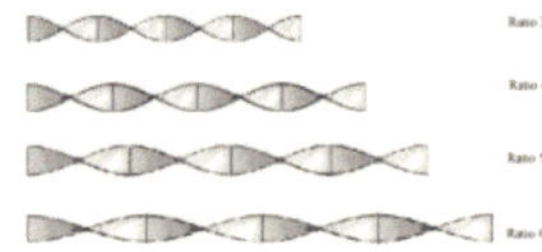
Jaisankar et al. [25]	Water	Full-length left-right twisted and fitted with rod	700 ≤ Re ≤ 1600 laminar	1.1–1.9	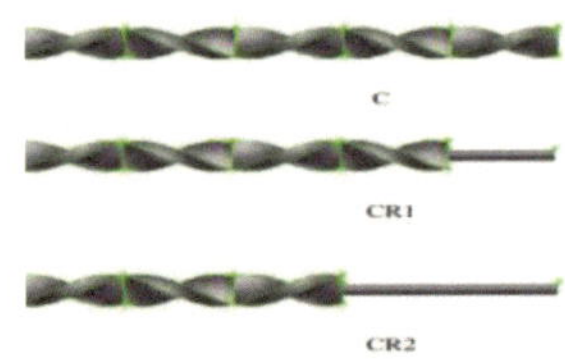
Wongcharee and Eiamsa-ard [26]	Water	Twisted tapes with wing shape including triangle, rectangle, and trapezoid	5500 ≤ Re ≤ 20,200 turbulence	1.42	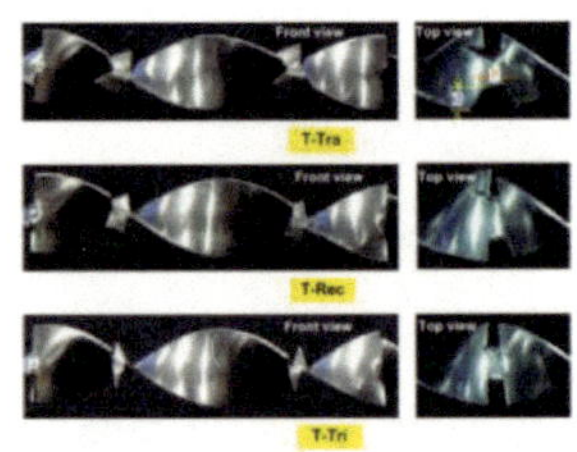
Eiamsa-ard et al. [27]	Water	Peripherally cut twisted tape	1000 ≤ Re ≤ 20,000 laminar-turbulence	1.29–4.88	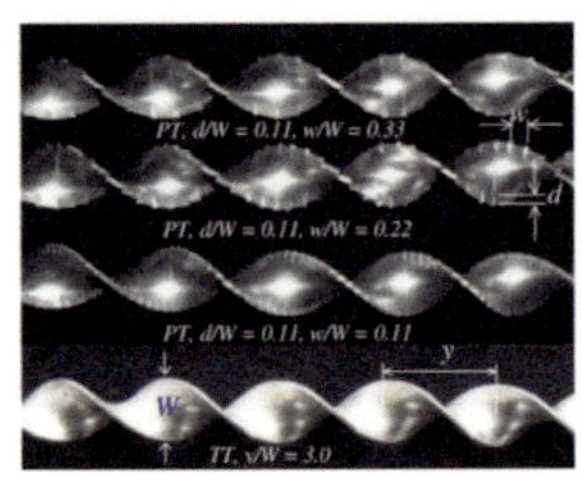
Eiamsa-ard et al. [28]	Air	Full- and short-length twisted tape	4000 ≤ Re ≤ 20,000 turbulence	0.95–1.04	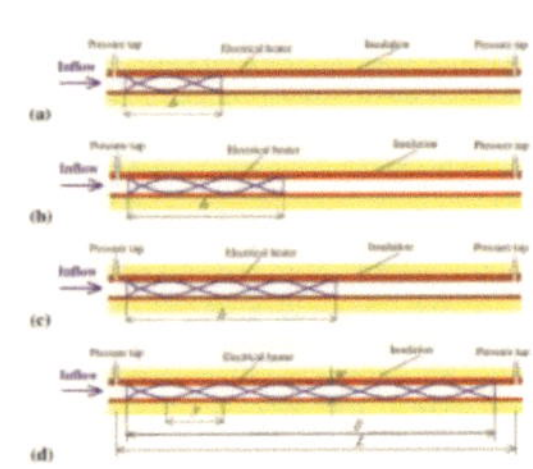
Bas and Ozceyhan [29]	Air	Twisted tape	5132 ≤ Re ≤ 24,989 turbulence	1.2–1.74	

http://dx.doi.org/10.5772/intechopen.74516

Authors	Fluid	Type	Flow region	OER	Sample of twisted tape inserts
Salam et al. [30]	Water	Rectangular cut	$10,000 \le Re \le 19,000$ turbulence		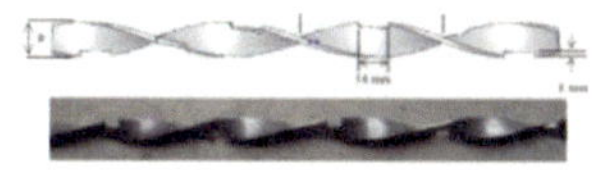
Sarada et al. [31]	Air	Twisted tape	$6000 \le Re \le 13,500$ turbulence	1.36–1.48	
Maddah et al. [32]	Al_2O_3/ water nanofluid	Modified twisted tape	$5000 \le Re \le 21,000$ turbulence	1.03–1.6	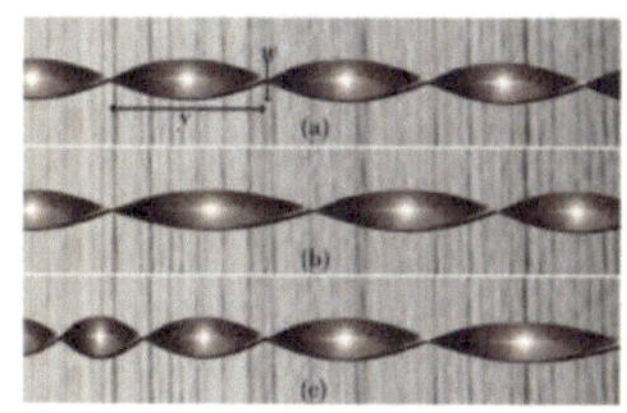
Zheng et al. [20]	Al_2O_3/ water nanofluid	Dimpled twisted tape	$1000 \le Re \le 10,000$ Laminar-turbulence	1.25	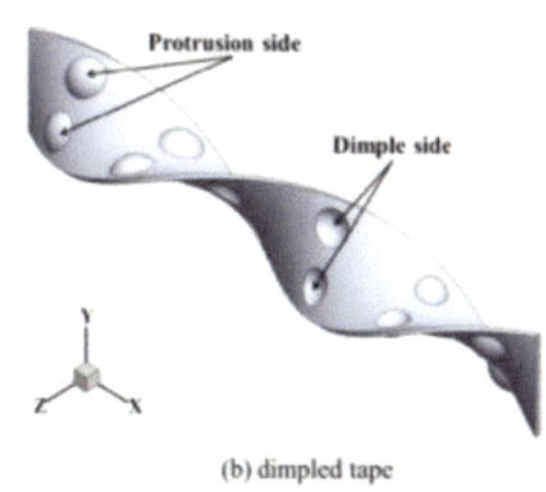
Murugesan et al. [33]	Water	Twisted tape with wire nails	$2000 \le Re \le 12,000$ Laminar-turbulence	1.05–1.75	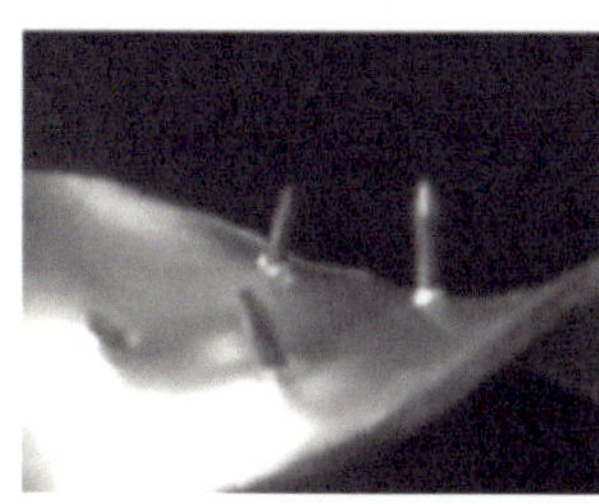
Wongcharee and Eiamsa-ard [16]	CuO/ water nanofluid	Twisted tape with alternate axis	$830 \le Re \le 1990$ turbulence	5.53	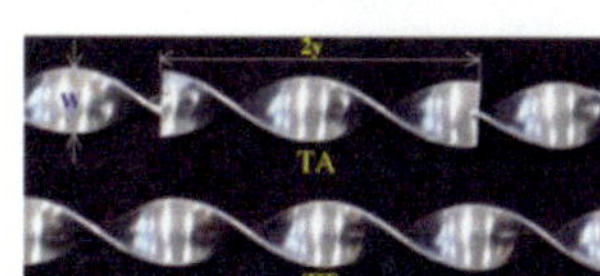

Authors	Fluid	Type	Flow region	OER	Sample of twisted tape inserts
Nanan et al. [34]	Air	Twisted tape	$6000 \leq Re \leq 20{,}000$ turbulence	1.28	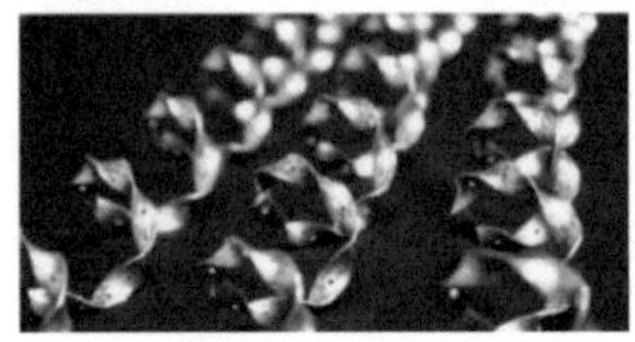

Table 1. Various examples of twisted tape inserts for heat transfer enhancement.

Authors	Fluid	Flow region	Nusselt number	Friction factor
Jaisankar et al. [24]	Water	$3000 \leq Re \leq 23{,}000$ Turbulence	$Nu = 0.000115\ Re^{1.169}\ Pr^{2.424}\ Y^{-0.511}$	$f = 271.1\ Re^{-0.947}\ Y^{-0.584}$
Ibrahim [35]	Water	$570 \leq Re \leq 1310$ Laminar	$Nu = 6.11\ Re^{0.199}\ (1 + x)^{-0.064}\ Y^{-0.318}$	$f = 54.41\ Re^{-0.87}\ (1 + x)^{-0.045}\ Y^{-0.146}$
Sivashanmugam and Suresh [36]	Water	Laminar	$Nu = 0.017\ Re^{0.996}\ Pr\ Y^{-0.5437}$	$f = 10.7564\ Re^{-0.387}\ Y^{-1.054}$
He et al. [17]	Air	$5600 \leq Re \leq 18{,}000$ Turbulence	$Nu = 0.3415\ Re^{0.5911}\ Pr^{0.32}$ $(0.9058c^3 + 0.5439c^2 - 1.345c + 1.271)$	$f = 9.348$ $Re^{-0.3959}(5.53c^3 + 2.578c^2 - 7.307c + 3.499)$
Paisarn Naphon [37]	Water	$7000 \leq Re \leq 23{,}000$ Turbulence	$Nu = 0.648\ Re^{0.36}\ (1 + D/H)^{2.475}\ Pr^{1/3}$	$f = 3.517\ Re^{-0.414}\ (1 + D/H)^{1.045}$
Tamna et al. [18]	Air	$5300 \leq Re \leq 24{,}000$ Turbulence	$Nu = 0.1687\ Re^{0.701}\ Pr^{0.4}\ B_R^{0.172}$	$f = 5.494\ Re^{-0.263}\ B_R^{0.729}$
Eiamsa-ard et al. [27]	Water	$1000 \leq Re \leq 20{,}000$ Turbulence	$Nu = 0.244\ Re^{0.625}\ Pr^{0.4}\ (d/W)^{0.168}(w/W)^{-0.112}$	$f = 39.46\ Re^{-0.591}\ (d/W)^{0.195}(w/W)^{-0.201}$
Eiamsa-ard et al. [38]	Water	$2000 \leq Re \leq 12{,}000$ Turbulence	$Nu = 0.01014\ Re^{0.929} Pr^{1/3}\ (1 + S)^{-0.266}$	$f = 4.143\ Re^{-0.398}\ (1 + S)^{-0.376}$
Seemawute and Eiamsa-ard [39]	Water	$5000 \leq Re \leq 20{,}000$ Turbulence	$Nu = 0.076\ Re^{0.718}\ Pr^{0.4}$	$f = 6.42\ Re^{-0.428}$
Jaisankar et al. [25]	Water	Laminar	*Phase 1* $Nu = 0.00395\ Re^{1.067} Pr^{0.757} Y^{0.033}(1 + S/D)^{-0.0304}$ *Phase 2* $Nu = 0.00363\ Re^{1.433} Pr^{0.266} Y^{0.154}(1 + S/D)^{-0.024}$	*Phase 1* $f = 1.30\ Re^{-0.310}\ Y^{-0.124}(1 + S/D)^{-0.063}$ *Phase 2* $f = 3.527\ Re^{-0.436}\ Y^{-0.145}(1 + S/D)^{-0.066}$

Table 2. Correlations for the Nusselt number and friction factor for twisted tape inserts.

artificial roughness. According to the results, the use of coiled wire at a lower Reynolds number was more advantageous than corrugated and dimpled tubes. Ozceyhan [47] numerically analyzed conjugate heat transfer and thermal stress in a tube with coiled wire inserts.

Gunes et al. [58] investigated the effects of coiled wire on heat transfer enhancement and pressure drop in a tube. It was found that the best overall enhancement efficiency was achieved for the configuration with P/D = 1. In another study Gunes et al. [59] reported on the characteristics of heat transfer and pressure drop in a tube with coiled wire inserts separated from the tube wall by two different distances. The results showed that the Nusselt number and pressure drop increase with decreasing distance between the coiled wire and the tube wall.

Authors	Fluid	Type	Flow region	OER	Sample of twisted tape inserts
Keklikcioglu and Ozceyhan [48]	Air	Triangle cross-sectional coiled wire	$2851 \leq Re \leq 27{,}732$ turbulence	1.67	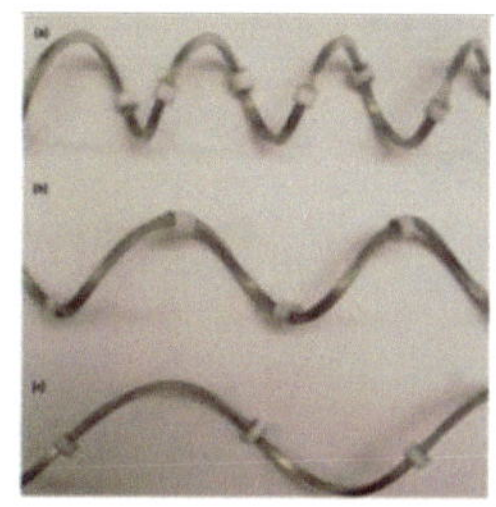
Reddy and Rao [49]	Ethylene glycol water-based TiO_2	Helically coiled wire	$4000 \leq Re \leq 15{,}000$ Turbulence	1.06–1.38	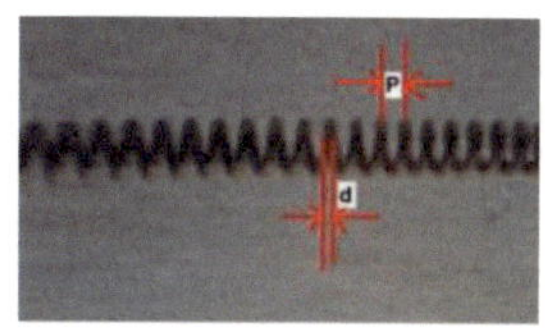
Keklikcioglu and Ozceyhan [50]	Air	Triangle cross-sectional coiled wire	$3429 \leq Re \leq 26{,}663$ Turbulence	1.82	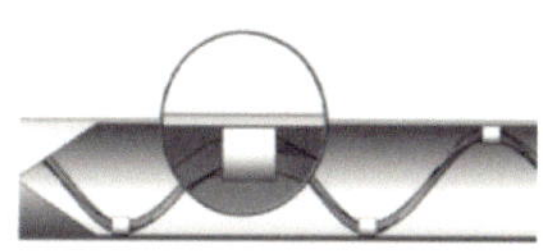
Chandrasekar et al. [51]	Al_2O_3/water nanofluid	Coiled wire	$600 \leq Re \leq 2275$ laminar	–	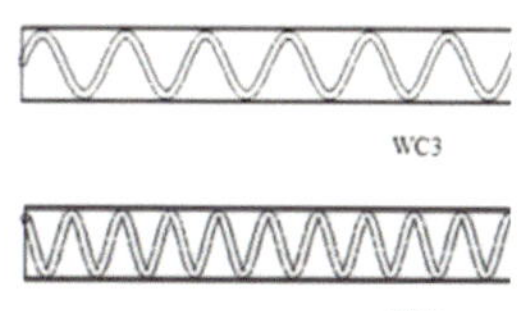
Roy and Saha [52]	Servotherm medium oil	Coiled wire	Laminar	–	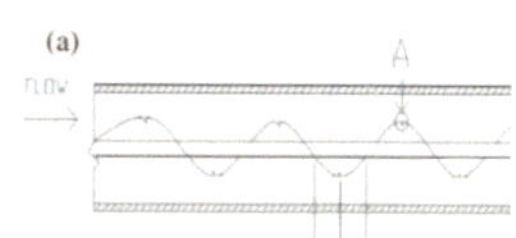
Panahi and Zamzamian [53]	Water and air	Coiled wire	$4000 \leq Re \leq 18{,}000$ turbulence	1.26–1.52	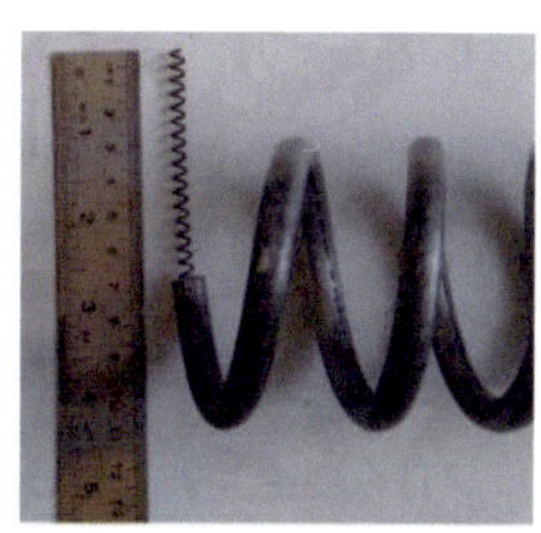

Authors	Fluid	Type	Flow region	OER	Sample of twisted tape inserts
Goudarzi and Jamali [54]	Al_2O_3-EG nanofluid	Coiled wire	$18{,}500 \leq Re \leq 22{,}700$ turbulence		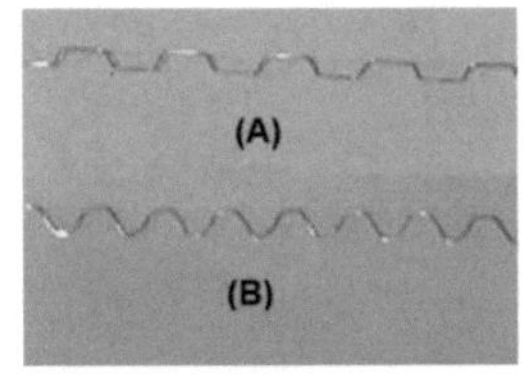
Goodarzi et al. [55]	Water	Square cross-sectional coiled wire	$5800 \leq Re \leq 18{,}500$ turbulence	0.93–0.97	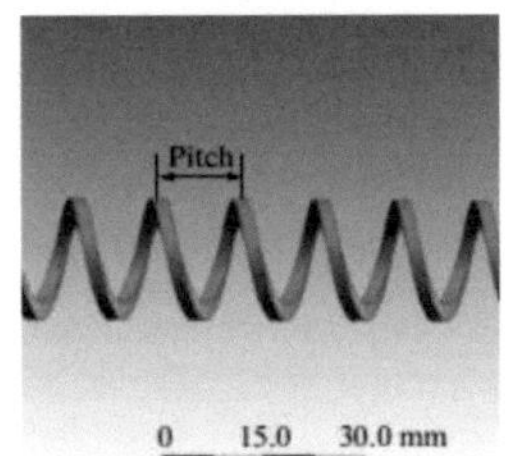
Sharifi et al. [56]	Engine oil	Helically coiled wire	$100 \leq Re \leq 1200$ Laminar	1.77	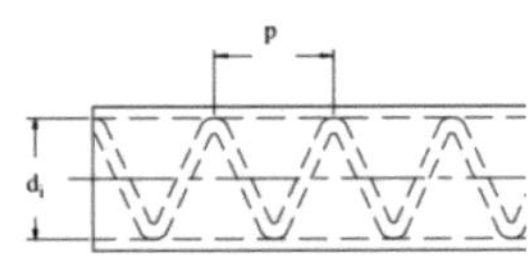
Zhang et al. [57]	Air	Double spiral coiled wire	$6000 \leq Re \leq 31{,}000$ Turbulence	1.5	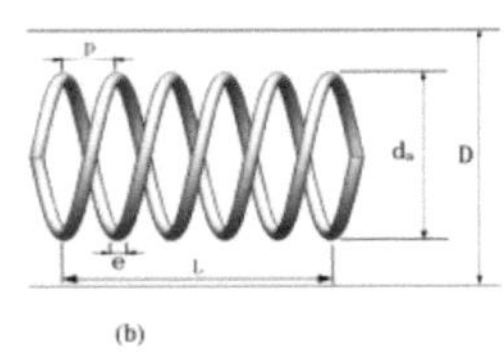

Table 3. Various examples of coiled inserts for heat transfer enhancement.

Promvonge [60] presented the experimental results for the heat transfer and flow friction characteristics in a circular tube in which coiled wire with a square cross section was inserted. The thermal performance of helically twisted tapes was experimentally investigated by Eiamsa-ard et al. [61]. They reported that the heat transfer rate and the friction factor increase as the twist ratio and helical pitch ratio increase. In another study, Eiamsa-ard et al. [62] studied the thermal performance of a tube fitted with both twisted tape and a coiled wire. Heat transfer and friction factor analyses of coiled wire inserted tube using various types of nanofluid as working fluid were performed by many researchers [63–73]. Promvonge [74] conducted another investigation of thermohydraulic performance in a tube with coiled wire insert and a snail entry. The heat transfer and the friction factor characteristics of the laminar flow of oil in square and rectangular ducts with transverse ribs and coiled wire were studied by Saha [75]. Feng et al. [76] carried out a numerical study of the laminar liquid flow and coupled heat transfer performance in rectangular microchannel heat sink (MCHS) equipped with coiled wire inserts. In the study the effects of the

http://dx.doi.org/10.5772/intechopen.74516

length and arrangement of coiled wire inserts on heat transfer enhancement were analyzed. Chougulea et al. [77] investigated heat transfer and friction factor characteristics of MWCNT/water nanofluid flowing through a uniformly heated horizontal tube with and without coiled wire. Solano et al. [78] carried out a numerical study of the flow pattern and heat transfer enhancement in oscillatory-baffled reactors with helical coil inserts. The heat transfer enhancement rate was discussed, considering the compound effect of oscillatory motion and helical coil inserts in the paper. Garcia et al. [79] investigated the thermal performance in a round tube with a coiled wire insert for laminar and transitional flow fields, while Arici and Asan [80] studied enhancement of heat transfer for turbulent flows in a tube with coiled wire inserts. Zohir et al. [81] studied the effect of pulsation on the heat transfer rate in a pipe with coiled wire inserts. Apart from the pulsation, the thermohydraulic performances of the process were solely studied for a coiled wire insert with different pitch values. Vahidifar and Kahrom [82] studied the characteristics of the heat transfer and the pressure drop of a heat exchanger with an inserted coiled wire and rings. San et al. [83] measured the heat transfer and pressure drop data for air flow and water flow in a tube fitted with a coiled wire.

4. Conclusions

The present review comprises both recent and important investigations on the enhancement of heat transfer using twisted tape and coiled wire inserts in laminar and turbulent flow region. Most of the studies for both twisted tape and coiled wire inserts emphasize the increased heat transfer rate and fluid friction or pressure drop. The main objective to design a heat exchanger is to enhance the heat transfer without causing more pressure drop. For this purpose many articles have been evaluated in terms of overall enhancement ratio (OER) in this review study. In conclusion, the following statements can be generalized as a result of this review study:

- Both twisted tape and coiled wire inserts can be used together with nanofluid flow to eliminate the pressure drop penalty on OER.
- Most of the researchers oriented to find better configuration of internal ribs to determine the minimum pressure drop and the maximum heat transfer.
- Twisted tape inserts generally show better performance in laminar flow region.
- Twisted tape inserts cause high-pressure drop penalty in turbulent region so they are not good at enhancing thermohydraulic performance.
- Coiled wire shows better performance to create swirl flow inside the tube.
- Coiled wire inserts have the ability to destruct the laminar boundary layer.
- When the pressure drop penalty in considered, coiled wire inserts show better performance than twisted tape inserts.
- The configuration and physical properties of inserts are the most important parameters to enhance the overall thermohydraulic performance of a heat exchanger.

Acknowledgements

The authors would like to thank both Erciyes University and Erciyes University Faculty of Engineering for funding and supporting the book chapter.

Appendices and nomenclature

D inner diameter of the tube (m)

f friction factor

h convective heat transfer coefficient (W/m^2 K)

k thermal conductivity (W/m K)

L length of the test tube (m)

Nu Nusselt number (hD/k)

ΔP pressure drop (N/m^2)

Re Reynolds number (UD/v)

T steady-state temperature (K)

U mean fluid velocity (m/s)

$\dot{V}$ volumetric flow rate (m^3/s)

Greek letters

ρ fluid density (kg/m^3)

U overall enhancement efficiency

υ kinematic viscosity (m^2/s)

Subscripts

a augmented tube

pp. pumping power

s smooth tube

Author details

Orhan Keklikcioglu and Veysel Ozceyhan*

*Address all correspondence to: ozceyhan@erciyes.edu.tr

Department of Mechanical Engineering, Erciyes University, Kayseri, Turkey

References

[1] Webb RL. Performance evaluation criteria for use of enhanced heat transfer surfaces in heat exchanger design. International Journal of Heat and Mass Transfer. 1981;**715**:726-724. DOI: 10.1016/0017-9310(81)90015-6

[2] Eiamsa-ard S, Promvonge P. Thermal characteristics in round tube fitted with serrated twisted tape. Applied Thermal Engineering. 2010;**30**:1673-1682. DOI: 10.1016/j.applthermaleng.2010.03.026

[3] Kumar B, Srivastava GP, Kumar M, Patil AK. A review of heat transfer and fluid flow mechanism in heat exchanger tube with inserts. Chemical Engineering & Processing: Process Intensification. 2018;**123**:126-137. DOI: 10.1016/j.cep.2017.11.007

[4] Hasanpour A, Farhadi M, Sedighi K. A review study on twisted tape inserts on turbulent flow heat exchangers: The overall enhancement ratio criteria. International Communication in Heat and Mass Transfer. 2014;**55**:53-62. DOI: 10.1016/j.icheatmasstransfer.2014.04.008

[5] Kareem ZS, Mohd Jaafar MN, Lazim TM, Abdullah S, Abdulwahid AF. Passive heat transfer enhancement review in corrugation. Experiment Thermal and Fluid Science. 2015;**68**:22-38. DOI: 10.1016/j.expthermflusci.2015.04.012

[6] Liu S, Sakr M. A comprehensive review on passive heat transfer enhancements in pipe exchangers. Renewable and Sustainable Energy Reviews. 2013;**19**:64-81. DOI: 10.1016/j.rser.2012.11.021

[7] Sheikholeslami M, Gorji-Bandpy M, Ganji DD. Review of heat transfer enhancement methods: Focus on passive methods using swirl flow devices. Renewable and Sustainable Energy Reviews. 2015;**49**:444-469. DOI: 10.1016/j.rser.2015.04.113

[8] Varun MOG, Nautiyal H, Khurana S, Shukla MK. Heat transfer augmentation using twisted tape inserts: A review. Renewable and Sustainable Energy Reviews. 2016;**63**:193-225. DOI: 10.1016/j.rser.2016.04.051

[9] Dewan A, Mahanta P, Sumithra Raju K, Suresh Kumar P. Review of passive heat transfer augmentation techniques. Proceedings of Institution of Mechancial Engineers. 2004;**218**:509-527

[10] Alam T, Kim M-H. A comprehensive review on single phase heat transfer enhancement techniques in heat exchanger applications. Renewable and Sustainable Energy Reviews. 2018;**81**:813-839. DOI: 10.1016/j.rser.2017.08.060

[11] Feng Z, Luo X, Guo F, Li H, Zhang J. Numerical investigation on laminar flow and heat transfer in rectangular microchannel heat sink with coiled wire inserts. Applied Thermal Engineering. 2017;**116**:597-609. DOI: 10.1016/j.applthermaleng.2017.01.091

[12] Changzhong Man XL, Jingwei H, Sun P, Tang Y. Experimental study on effect of heat transfer enhancement for single-phase forced convective flow with twisted tape inserts. International Journal of Heat and Mass Transfer. 2017;**106**:877-883. DOI: 10.1016/j.ijheatmasstransfer.2016.10.026

[13] Suria ARS, Kumara A, Maithani R. Heat transfer enhancement of heat exchanger tube with multiple square perforated twisted tape inserts: Experimental investigation and correlation development. Chemical Engineering and Processing:Process Intensification. 2017;**116**:76-96. DOI: 10.1016/j.cep.2017.02.014

[14] Syam Sundar L, Singh MK, Punnaiah V, Sousa ACM. Experimental investigation of Al_2O_3/water nanofluids on the effectiveness of solar flat-plate collectors with and without twisted tape inserts. Renewable Energy. 2017:1-14. DOI: 0.1016/j.renene.2017.10.056

[15] Saysroy A, Eiamsa-ard S. Enhancing convective heat transfer in laminar and turbulent flow regions using multi-channel twisted tape inserts. International Journal of Thermal Sciences. 2017;**121**:55-74. DOI: 10.1016/j.ijthermalsci.2017.07.002

[16] Wongcharee K, Eiamsa-ard S. Enhancement of heat transfer using Cuo/water Nanofluid and twisted tape with alternate Axis. International Communication of Heat and Mass Transfer. 2011;**38**:742-748. DOI: 10.1016/j.icheatmasstransfer.2011.03.011

[17] He Y, Li L, Li P, Ma L. Experimental study on heat transfer enhancement characteristics of tube with cross hollow twisted tape inserts. Applied Thermal Engineering. 2018;**131**:743-749. DOI: 10.1016/j.applthermaleng.2017.12.029

[18] Tamna S, Kaewkohkiat Y, Skullong S, Promvonge P. Heat transfer enhancement in tubular heat exchanger with double V-ribbed twisted-tapes. Case Studies in Thermal Engineering. 2016;**7**:14-24. DOI: 10.1016/j.csite.2016.01.002

[19] Mashoofi N, Pourahmad S, Pesteei SM. Study the effect of axially perforated twisted tapes on the thermal performance enhancement factor of a double tube heat exchanger. Case Studies in Thermal Engineering. 2017;**10**:161-168. DOI: 10.1016/j.csite.2017.06.001

[20] Zheng L, Xie Y, Di Z. Numerical investigation on heat transfer performance and flow characteristics in circular tubes with dimpled twisted tapes using Al_2O_3-water nanofluid. International Journal of Heat and Mass Transfer. 2017;**111**:962-981. DOI: 10.1016/j.ijheatmasstransfer.2017.04.062

[21] Bhuiya MMK, Chowdhury MSU, Saha M, Islam MT. Heat transfer and friction factor characteristics in turbulent flow through a tube fitted with perforated twisted tape inserts. International Communication of Heat and Mass Transfer. 2013;**46**:49-57. DOI: 10.1016/j.icheatmasstransfer.2013.05.012

[22] Li P, Liu Z, Liu W, Chen G. Numerical study on heat transfer enhancement characteristics of tube inserted with centrally hollow narrow twisted tapes. International Journal of Heat and Mass Transfer. 2015;**88**:481-491. DOI: 10.1016/j.ijheatmasstransfer.2015.04.103

[23] Eiamsa-ard S, Yongsiri K, Nanan K, Thianpong C. Heat transfer augmentation by helically twisted tapes as swirl and turbulence promoters. Chemical Engineering and Processing. 2012;**60**:42-48. DOI: 10.1016/j.cep.2012.06.001

[24] Jaisankar S, Radhakrishnan TK, Sheeba KN. Experimental studies on heat transfer and friction factor characteristics of forced circulation solar water heater system fitted with helical twisted tapes. Solar Energy. 2009;**83**:1943-1952. DOI: 10.1016/j.solener.2009.07.006

[25] Jaisankar S, Radhakrishnan TK, Sheeba KN, Suresh S. Experimental investigation of heat transfer and friction factor characteristics of thermosyphon solar water heater system fitted with spacer at the trailing edge of left–right twisted tapes. Energy Conversion and Management. 2009;**50**:2638-2649. DOI: 10.1016/j.enconman.2009.06.019

[26] Wongcharee K, Eiamsa-ard S. Heat transfer enhancement by twisted tapes with alternate-axes and triangular, rectangular and trapezoidal wings. Chemical Engineering and Processing. 2011;**50**:211-219. DOI: 10.1016/j.cep.2010.11.012

[27] Eiamsa-ard S, Seemawute P, Wongcharee K. Influences of peripherally-cut twisted tape insert on heat transfer and thermal performance characteristics in laminar and turbulent tube flows. Experimental Thermal and Fluid Science. 2010;**34**:711-719. DOI: 10.1016/j.expthermflusci.2009.12.013

[28] Eiamsa-ard S, Thianpong C, Eiamsa-ard P, Promvonge P. Convective heat transfer in a circular tube with short-length twisted tape insert. International Communication of Heat and Mass Transfer. 2009;**36**:365-371. DOI: 10.1016/j.icheatmasstransfer.2009.01.006

[29] Bas H, Ozceyhan V. Heat transfer enhancement in a tube with twisted tape inserts placed separately from the tube wall. Experimental Thermal and Fluid Science. 2012;**41**:51-58. DOI: 10.1016/j.expthermflusci.2012.03.008

[30] Salam B, Biswas S, Saha S, Bhuiya MMK. Heat transfer enhancement in a tube using rectangular-cut twisted tape insert. Procedia Engineering. 2013;**56**:96-103. DOI: 10.1016/j.proeng.2013.03.094

[31] Naga Sarada S, Sita Rama Raju AV, Kalyani Radha K, Shyam Sunder L. Enhancement of heat transfer using varying width twisted tape inserts. International Journal of Engineering, Science and Technology. 2010;**2**:107-118. DOI: 10.4314/ijest.v2i6.63702

[32] Maddah H, Alizadeh M, Ghasemi N, Alwi SRW. Experimental study of Al_2O_3/water nanofluid turbulent heat transfer enhancement in the horizontal double pipes fitted with modified twisted tapes. International Journal of Heat and Mass Transfer. 2014;**78**:1042-1054. DOI: 10.1016/j.ijheatmasstransfer.2014.07.059

[33] Murugesan P, Mayilsamy K, Suresh S. Heat transfer and friction factor studies in a circular tube fitted with twisted tape consisting of wire-nails. Chinese Journal of Chemical Engineering. 2010;**18**:1038-1042. DOI: 10.1016/S1004-9541(09)60166-X

[34] Nanan K, Thianpong C, Promvonge P, Eiamsa-ard S. Investigation of heat transfer enhancement by perforated helical twisted-tapes. International Communication Heat and Mass Transfer. 2014;**52**:106:112. DOI: 10.1016/j.icheatmasstransfer.2014.01.018

[35] Ibrahim EZ. Augmentation of laminar flow and heat transfer in flat tubes by means of helical screw-tape inserts. Energy Conversion and Management. 2011;**52**:250-257. DOI: 10.1016/j.enconman.2010.06.065

[36] Sivashanmugam P, Suresh S. Experimental studies on heat transfer and friction factor characteristics of laminar flow through a circular tube fitted with helical screw tape inserts. Applied Thermal Engineering. 2006;**31**:301-308. DOI: 10.1016/j.expthermflusci.2006.05.005

[37] Naphon P. Experimental studies on heat transfer and friction factor characteristics of laminar flow through a circular tube fitted with regularly spaced helical screw-tape inserts. International Communication of Heat and Mass Transfer. 2006;**33**:166-175. DOI: 10.1016/j.icheatmasstransfer.2005.09.007

[38] Eiamsa-ard S, Thianpong C, Promvonge P. Experimental investigation of heat transfer and flow friction in a circular tube fitted with regularly spaced twisted tape elements. International Communication of Heat and Mass Transfer. 2006;**33**:1225-1233. DOI: 10.1016/j.icheatmasstransfer.2006.08.002

[39] Seemawute P, Eiamsa-ard S. Thermohydraulics of turbulent flow through a round tube by a peripherally-cut twisted tape with an alternate axis. International Communication of Heat and Mass Transfer. 2010;**37**:652-659. DOI: 10.1016/j.icheatmasstransfer.2010.03.005

[40] Piriyarungrod N, Eiamsa-ard S, Thianpong C, Pimsarn M, Nanan K. Heat transfer enhancement by tapered twisted tape inserts. Chemical Engineering and Processing. 2015;**96**:62-71. DOI: 10.1016/j.cep.2015.08.002

[41] Promvonge P, Pethkool S, Pimsarn M, Thianpong C. Heat transfer augmentation in a helical-ribbed tube with double twisted tape inserts. International Communication of Heat and Mass Transfer. 2012;**39**:953-959. DOI: 10.1016/j.icheatmasstransfer.2012.05.015

[42] Durga Prasad PV, Gupta AVSSKS. Experimental investigation on enhancement of heat transfer using Al_2O_3/water nanofluid in a u-tube with twisted tape inserts. International Communication of Heat and Mass Transfer. 2016;**75**:154-161. DOI: 10.1016/j.icheatmasstransfer.2016.03.019

[43] Pal S, Saha SK. Laminar fluid flow and heat transfer through a circular tube having spiral ribs and twisted tapes. Experimental Thermal and Fluid Science. 2015;**60**:173-181. DOI: 10.1016/j.expthermflusci.2014.09.005

[44] Eiamsa-ard S, Kiatkittipong K. Heat transfer enhancement by multiple twisted tape inserts and TiO_2/water nanofluid. Applied Thermal Engineering. 2014;**70**:896-924. DOI: 10.1016/j.applthermaleng.2014.05.062

[45] Ravigururajan T, Bergles A. Development and verification of general correlations for pressure drop and heat transfer in single-phase turbulent flow in enhanced tubes. Experimental Thermal and Fluid Science. 1996;**13**:55-70. DOI: 10.1016/0894-1777(96)00014-3

[46] García A et al. The influence of artificial roughness shape on heat transfer enhancement: Corrugated tubes, dimpled tubes and coiled wires. Applied Thermal Engineering. 2012; **35**:196-201. DOI: 10.1016/j.applthermaleng.2011.10.030

[47] Ozceyhan V. Conjugate heat transfer and thermal stress analysis of coiled wire inserted tubes that are heated externally with uniform heat flux. Energy Conversion and Management. 2005;**46**:1543-1559. DOI: 10.1016/j.enconman.2004.08.003

[48] Keklikcioglu O, Ozceyhan V. Experimental investigation on heat transfer enhancement in a circular tube with equilateral triangle cross sectioned coiled-wire inserts. Applied Thermal Engineering. 2018;**131**:686-695. DOI: 10.1016/j.applthermaleng.2017.12.051

[49] Chandra Sekhara Reddy M, Rao VV. Experimental investigation of heat transfer coefficient and friction factor of ethylene glycol water based TiO_2 nanofluid in double pipe heat exchanger with and without helical coil inserts. International Communication of Heat and Mass Transfer. 2014;**50**:68:76. DOI: 10.1016/j.icheatmasstransfer.2013.11.002

[50] Keklikcioglu O, Ozceyhan V. Experimental investigation on heat transfer enhancement of a tube with coiled-wire inserts installed with a separation from the tube wall. International Communication of Heat and Mass Transfer. 2016;**78**:88-94. DOI: 10.1016/j.icheatmasstransfer.2016.08.024

[51] Chandrasekar M, Suresh S, Chandra Bose A. Experimental studies on heat transfer and friction factor characteristics of Al_2O_3/water nanofluid in a circular pipe under laminar flow with coiled wire inserts. Experimental Thermal and Fluid Science. 2010;**34**:122-130. DOI: 10.1016/j.expthermflusci.2009.10.001

[52] Roy S, Saha SK. Thermal and friction characteristics of laminar flow through a circular duct having helical screw-tape with oblique teeth inserts and coiled wire inserts. Experimental Thermal and Fluid Science. 2015;**68**:733-743. DOI: 10.1016/j.expthermflusci.2015.07.007

[53] Panahi D, Zamzamian K. Heat transfer enhancement of shell-and-coiled tube heat exchanger utilizing helical wire turbulator. Applied Thermal Engineering. 2017;**115**:607-615. DOI: 10.1016/j.applthermaleng.2016.12.128

[54] Goudarzi K, Jamali H. Heat transfer enhancement of Al_2O_3-EG nanofluid in a car radiator with coiled wire inserts. Applied Thermal Engineering. 2017;**118**:510-517. DOI: 10.1016/j.applthermaleng.2017.03.016

[55] Goodarzi K, Goudarzi SY, Zendehbudi G. Investigation of the effect of using tube inserts for the intensification of heat transfer. Heat and Mass Transfer and Properties of Working Fluids and Materials. 2015;**62**:68-75. DOI: 10.1134/S004060151501005X

[56] Sharifi K, Sabeti M, Rafiei M, Mohammadi AH, Shirazi L. Computational fluid dynamics (CFD) technique to study the effects of helical wire inserts on heat transfer and pressure drop in a double pipe heat exchanger. Applied Thermal Engineering. 2018;**128**:898-910. DOI: 10.1016/j.applthermaleng.2017.08.146

[57] Zhang C, Wang D, Zhu Y, Han Y, Jinxing W, Peng X. Numerical study on heat transfer and flow characteristics of a tube fitted with double spiral spring. International Journal of Thermal Sciences. 2015;**94**:18-27. DOI: 10.1016/j.ijthermalsci.2015.02.001

[58] Gunes S, Ozceyhan V, Buyukalaca O. Heat transfer enhancement in a tube with equilateral triangle cross sectioned coiled wire inserts. Experimental Thermal and Fluid Science. 2010;**34**:684-691. DOI: 10.1016/j.expthermflusci.2009.12.010

[59] Gunes S, Ozceyhan V, Buyukalaca O. The experimental investigation of heat transfer and pressure drop in a tube with coiled-wire inserts placed separately from the tube wall. Applied Thermal Engineering. 2010;**30**:1719-1725. DOI: 10.1016/j.applthermaleng.2010.04.001

[60] Promvonge P. Thermal performance in circular tube fitted with coiled square wires. Energy Conversion and Management. 2008;**49**:980-987. DOI: 10.1016/j.enconman.2007.10.005

[61] Eiamsa-ard S, Yongsiri K, Nanan K, Thianpong C. Heat transfer augmentation by helically twisted tapes as swirl and turbulence promoters. Chemical Engineering and Processing. 2012;**60**:42-48. DOI: 10.1016/j.cep.2012.06.001

[62] Eiamsa-ard S, Nivesrangsan P, Chokphoemphun S, Promvonge P. Influence of combined non-uniform coiled wire and twisted tape inserts on thermal performance characteristics. International Communication of Heat and Mass Transfer. 2010;**37**:850-856. DOI: 10.1016/j.icheatmasstransfer.2010.05.012

[63] Syam Sundar L, Bhramara P, Ravi Kumar NT, Singh MK, Sousa ACM. Experimental heat transfer, friction factor and effectiveness analysis of Fe_3O_4 nanofluid flow in a horizontal plain tube with return bend and coiled wire inserts. International Journal of Heat and Mass Transfer. 2017;**109**:440-453. DOI: 10.1016/j.ijheatmasstransfer.2017.02.022

[64] Yang San J, Chieh Huang W, An Chen C. Experimental investigation on heat transfer and fluid friction correlations for circular tubes with coiled-wire inserts. International Communication of Heat and Mass Transfer. 2015;**65**:8-14. DOI: 10.1016/j.icheatmasstransfer.2015.04.008

[65] Akhavan-Behabadi MA, Aligoodarz MSMR. An experimental study on heat transfer and pressure drop of MWCNT–water nano-fluid inside horizontal coiled-wire inserted tube. International Communication of Heat and Mass Transfer. 2015;**63**:62-72. DOI: 10.1016/j.icheatmasstransfer.2015.02.013

[66] Saeedinia M, Akhavan-Behabadi MA, Nasr M. Experimental study on heat transfer and pressure drop of nanofluid flow in a horizontal coiled wire inserted tube under constant heat flux. Experimental Thermal and Fluid Science. 2012;**36**:158-168. DOI: 10.1016/j.expthermflusci.2011.09.009

[67] Safikhani H, Abbasi F. Numerical study of nanofluid in flat tubes fitted with multiple twisted tapes. Advances in Powder Technology. 2015;**26**:1609-1617. DOI: 10.1016/j.apt.2015.09.002

[68] Safikhani H, Zare Mehrjardi AR, Safari M. Effect of inserting coiled wires in tubes on the fluid flow and heat transfer performance of nanofluids. Transparent Phenomena in Nano Micro Scales. 2016;**4**:9-16. DOI: 7508/tpnms.2016.02.002

[69] Akhavan-Behabadi MA, Shahidi M, Aligoodarz MR, Ghazvini M. Experimental investigation on thermo-physical properties and overall performance of MWCNT–water nanofluid flow inside horizontal coiled wire inserted tubes. Heat and Mass Transfer. 2017;**53**:291-304. DOI: 10.1007/s00231-016-1814-5

[70] Naik MT, Fahad SS, Syam Sundar L, Singh MK. Comparative study on thermal performance of twisted tape and coiled wire inserts in turbulent flow using CuO/water nanofluid. Experimental Thermal and Fluid Science. 2014;**57**:65-76. DOI: 10.1016/j.expthermflusci.2014.04.006

[71] Chandrasekar M, Suresh S, Chandra Bose A. Experimental studies on heat transfer and friction factor characteristics of Al_2O_3/water Nanofluid in a circular pipe under transition flow with coiled wire inserts. Heat Transfer Engineering. 2011;**32**:485-496. DOI: 10.1080/01457632.2010.506358

[72] Mirzaei M, Azimi A. Heat transfer and pressure drop characteristics of Graphene oxide/water Nanofluid in a circular tube fitted with coiled wire insert. Experimental Heat Transfer. 2016;**29**:173-187. DOI: 10.1080/08916152.2014.973975

[73] Fallahiyekta M, Jafari Nasr MR, Rashidi A, Arjmand M. Convective heat transfer enhancement of CNT-water Nanofluids in plain tube fitted with coiled wire inserts. Iranian Journal of Chemical Engineering. 2014;**11**:43-55

[74] Promvonge P. Thermal enhancement in a round tube with snail entry and coiled-wire inserts. International Communication of Heat and Mass Transfer. 2008;**35**:623-629. DOI: 10.1016/j.icheatmasstransfer.2007.11.003

[75] Saha SK. Thermal and friction characteristics of laminar flow through rectangular and square ducts with transverse ribs and coiled wire inserts. Experimental Thermal and Fluid Science. 2010;**34**:63-72. DOI: 10.1016/j.expthermflusci.2009.09.003

[76] Feng Z, Luo X, Guo F, Li H, Zhang J. Numerical investigation on laminar flow and heat transfer in rectangular microchannel heat sink with coiled wire inserts. Applied Thermal Engineering. 2017;**116**:597-609. DOI: 10.1016/j.applthermaleng.2017.01.09

[77] Chougulea SS, Nirgudea VV, Ghargeb PD, Modaka M, Sahu SK. Heat transfer enhancements of low volume concentration CNT/water Nanofluid and coiled wire inserts in a circular tube. Energy Procedia. 2016;**90**:552-558. DOI: 10.1016/j.egypro.2016.11.223

[78] Solano JP, Herrero R, Espín S, Phan AN, Harvey AP. Numerical study of the flow pattern and heat transfer enhancement in oscillatory baffled reactors with helical coil inserts. Chemical Engineering Research and Design. 2012;**90**:732-742. DOI: 10.1016/j.cherd.2012.03.017

[79] Garcia A, Solano JP, Vicente PG, Viedma A. Enhancement of laminar and transitional flow heat transfer in tubes by means of coiled wire inserts. International Journal of Heat and Mass Transfer. 2007;**50**:3176-3189. DOI: 10.1016/j.ijheatmasstransfer.2007.01.015

[80] Arici ME, Asan H. Enhancement of turbulent flow heat transfer in tubes by means of coiled wire inserts. ASME, PD Advance in Heat Transfer. 1994;**113**:117-164

[81] Zohir AE, Aziz AAA, Habib MA. Heat transfer characteristics and pressure drop of the concentric tube equipped with coiled wires for pulsating turbulent flow. Experimental Thermal and Fluid Science. 2015;**65**:41-51. DOI: 10.1016/j.expthermflusci.2015.03.003

[82] Vahidifar S, Kahrom M. Experimental study of heat transfer enhancement in a heated tube caused by wire-coil and rings. Journal of Applied Fluid Mechanics. 2015;**8**:885-892. DOI: 10.18869/acadpub.jafm.73.238.23359

[83] Yang San J, Chieh Huang W, An Chen C. Experimental investigation on heat transfer and fluid friction correlations for circular tubes with coiled-wire inserts. International Communication of Heat and Mass Transfer. 2015;**65**:8-14. DOI: 10.1016/j.icheatmasstransfer.2015.04.008

Experimental Study of Concave Rectangular Winglet Vortex Generators Effect on Thermal-Hydrodynamic Performances of Airflow inside a Channel

Syaiful

Additional information is available at the end of the chapter

http://dx.doi.org/10.5772/intechopen.74518

Abstract

The present study focuses on the effect of CDW VG on the rate of heat transfer and flow pressure drop. The present experimental study was conducted by observing the effect of attack angles of VGs varied from 15 to 45°. In addition, the effect of number of pairs of VGs on convection heat transfer coefficients and flow pressure drop was also investigated. In order to observe the longitudinal vortex (LV) formed, flow visualization was performed in this work. The velocity of the airflow varied from 0.4 to 2.0 m/s with an interval of 0.2 m/s. In this work, the authors compared the thermal and hydrodynamic performance between cases using CRW and RW VGs. The results showed that the use of CRW VGs at an attack angle 45° increased convection heat transfer coefficient of up to 188% against the baseline (without using VG), whereas the use of RW VGs at the same attack angle increased the convection heat transfer coefficient by up to 100% against the baseline. The value of convection heat transfer coefficient increased with increasing angle of attack of VGs. However, the increase of attack angle has an impact on increasing pressure drop.

Keywords: concave rectangular winglet VG, convection heat transfer coefficient, pressure drop

1. Introduction

The fin and tube heat exchanger is mostly found in power plants, HVACR (heating, ventilating, air conditioning and refrigeration), chemical industry, and automotive. Improved performance of the fin and tube heat exchanger is a necessary thing to do to achieve high efficiency.

The thermal performance of heat exchanger can be indicated by increasing the value of its heat transfer coefficient. The improvement of heat transfer can be done in various ways; one effective way is to use a vortex generator (VG). Fiebig et al. used wing-type VG to obtain heat transfer enhancement [1]. They observed the effect of using delta winglet VG with in-line and staggered arrangements on the heat transfer enhancement. They concluded that the use of VG can reduce the size and mass of heat exchangers at the same heat load. Zhu et al. investigated the incorporation of rib-roughness with rectangular winglet VG in enhanced heat transfer [2]. They found that the use of rib-roughness with a rectangular winglet VG increases Nusselt number (Nu) by 405%.

The improvement of convection heat transfer on the surface using longitudinal vortex generator (LVG) has been expressed by Jacobi and Shah [3]. They said that the use of LVG to enhance heat transfer is promising. In order to achieve this goal, however, a deep understanding of the interaction between the flow structure and heat transfer is required. Biswas et al. studied numerically and experimentally the interaction between the flow structure and heat transfer in the presence of LVG [4]. Their work was able to identify the existence of main vortex, corner vortex and induced vortex in complex flow structures. The combined effect of these vortices was capable of distorting the temperature field within the channel resulting in an increase in heat transfer between the main fluid and the fluid near the surface. Gentry and Jacobi investigated in more detail the interaction between LV and the boundary layer experimentally [5]. They installed a delta wing VG to enhance the thermal performance of the fluid flow over the flat plate. Their experiments indicated that there was an increase in heat and mass transfer from the baseline (not using VG). Guo et al. proposed the concept of heat transfer improvement by estimating the angle between the flow velocity and the temperature gradient [6]. They found that the improvement of heat transfer may be indicated by increasing angle between fluid velocity and temperature gradient. They claimed that this novel method can estimate the heat transfer enhancement more effectively.

Lee et al. studied numerically the effect of LV on the characteristics of heat transfer within the turbulent boundary layer [7]. They found that Reynolds Stress Model (RSM) was able to predict anisotropy more accurately than the standard k-ε model. Their work also showed that disturbance in the boundary layer caused the rise of Stanton number with the flow direction to the wall, but the mixing of fluid inside the vortex core was observed to be weak. One of various ways of VG installation was the punched surface to the wall as has been performed by Chen et al. [8]. Their result showed that the staggered arrangement of punched delta winglet VG was greater in heat transfer improvement compared to in-line arrangement. In order to observe the complexity of the flow structure within the channel in the presence of VG, the flow visualization needs to be performed as done by Wang et al. [9]. They presented the flow visualization on a fin-and-tube heat exchanger with/without VGs (vortex generators). The results of this study showed that LVs (longitudinal vortices) formed behind the tube in the presence of annular VG. Tiwari et al. studied numerically heat transfer improvement in cross-flow heat exchanger with oval tubes and multiple delta winglets [10]. They revealed that the combination of oval tubes with delta winglet pairs VGs enhanced heat transfer significantly.

Leu et al. analyzed numerically and experimentally heat transfer and fluid flow in plate fin and tube heat exchanger by mounting VG pair of block shape [11]. Their study results indicated that the pair of block shapes can generate LVs and improve heat transfer performance in the wake region. The use of delta wing VG to enhance heat transfer in refrigerator evaporator applications has been proposed by Sommers and Jacobi [12]. Their results found that the use of delta wing VG decreased thermal resistance by up to 42%. Ferrouillat et al. studied experimentally and numerically thermal performance improvement and fluid mixing capabilities in the presence of delta winglet VG [13]. Their goal was to investigate a multifunctional heat exchanger as a flow mixer. The main objective of their research was the determination of turbulent flow in various geometries by using computational fluid dynamic. Hiravennavar et al. showed heat transfer enhancement within a channel by using a pair of winglet delta VGs that generate counter rotating LV [14]. They observed that heat transfer was doubled up by using a pair of VGs compared with using a VG. They also found that delta winglet VG with a certain thickness was superior compared with that without thickness.

Wu and Tao verified heat transfer improvement by using the field synergy principle [15]. Their numerical analysis revealed that the improvement of heat transfer by using rectangular winglet VG can be explained well by using the field synergy principle. The concept of a field synergy principle is to observe the intersection angle between the flow velocity and the temperature gradient. It can be revealed that the smaller the intersection angle the greater the heat transfer. Wu and Tao also studied some parameters that affect heat transfer improvement [16]. These parameters included the location of VG in the channel, the geometry size and the shape of the VG. A three-dimensional numerical analysis of the fin and oval tube heat exchanger by installing a punched delta winglet has been carried out by Chu et al. [17]. They analyzed based on the field synergy principle to provide a fundamental understanding of the correlation of flow structures and enhancement of heat transfer. They found that the placement of VG in the downstream, 30° attack angle and the minimum number of rows gave the highest increase of heat transfer. Min et al. presented fluid flow and heat transfer characteristics within the rectangular channel in the presence of modified rectangular wing VG [18]. Their experimental results showed that modified rectangular wings have better fluid flow and heat transfer characteristics than that of rectangular wing VG.

Wu and Tao presented the impact of delta winglet VG on the performance of heat transfer and pressure drop with different tube diameter [19]. Their numerical simulation showed that the fin and tube heat exchanger with the diameter of the first row tube smaller than that of the second row tube indicated better heat transfer with a smaller pressure drop than the traditional fin and tube. Wu and Tao have analyzed experimentally and numerically to know the effect of VGs on the heat transfer in a rectangular channel [20]. They varied the angle of attack to determine its effect on heat transfer characteristics. They found that the Nusselt number at an attack angle of 60° was slightly higher than at 45°, but the pressure drop also increased with increasing angle of attack. He et al. numerically analyzed heat transfer and pressure drop in a fin and tube heat exchanger with the use of rectangular winglet VGs [21]. They tried to determine the effect of the attack angle, the number of pairs of VGs, and the placement of VGs. Their results showed that longitudinal vortices (LVs) generated by rectangular winglet VGs were the main reason for heat transfer improvement. The staggered arrangement of VGs produced a smaller pressure drop than the in-line arrangement. Zhou and Feng investigated experimentally the effect of curved winglet VG with punched hole on heat transfer enhancement [22]. Their experiment proved that the curved winglet VGs produced better thermal-hydraulic performance than the plane winglet VGs. Punched holes on VG improved heat transfer and decreased pressure drop. Syaiful et al. evaluated numerically and experimentally the effect of concave delta winglet VG on heat transfer augmentation in a rectangular channel [23]. From the results of their study, it might be concluded that the concave delta winglet (CDW) VG provided greater heat transfer augmentation than delta winglet (DW) VG at the same fluid velocity. Syaiful et al. studied numerically the attack angle effect of CDW VG on the improvement of heat transfer in the fin and tub heat exchanger for the EGR cooler application [24].

Oneissi et al. investigated the novel design of the delta winglet VG for heat transfer enhancement [25]. They examined the use of inclined projected winglet (IPW) VG against the increase of heat transfer and its effect on pressure drop. IPW VG exhibited the same heat transfer augmentation with that of DW VG but with a lower pressure drop. Song et al. studied experimentally the effect of geometric size of curved DW VGs and tube pitch on heat transfer augmentation in a fin-and-tube heat exchanger [26]. They found that small-size VGs resulted in improved heat transfer at lower Reynolds numbers, while large VGs resulted in heat transfer augmentation at high Reynolds numbers. Syaiful et al. analyzed the effect of the concave delta winglet VGs on the thermo-hydrodynamic performance of the fluid flow in a rectangular channel [27]. They observed that CDW VG resulted in a higher heat transfer enhancement compared to DW VG. However, this increase in heat transfer was followed by an increase in pressure drop.

This chapter discusses the experimental results of concave rectangular winglet vortex generator (CRW VG) application to enhance heat transfer in the flow. The negative impact of using CRW VG on pressure drop is also informed in this chapter. The angle of attack and the number of rows VG become parameters that are considered at present. The flow visualization was performed to observe the longitudinal vortex (LV) structure formed behind the VGs.

2. Materials and experimental set up

This work aims to obtain better performance on heat exchanger with the installation of a rectangular and concave rectangular vortex generator in it. VG producing LV increases fluid mixing resulting in a rise in the rate of heat transfer.

2.1. Test equipment

This test was performed inside a glass air duct with a thickness of 10 mm as can be shown in **Figure 1**. This air channel has a rectangular cross section with a length of 370 cm, a width of 8 cm, and a height of 18 cm. The air was sucked by a blower coming into the channel through the straightener which consists of a number of tubes with the diameter of 5 mm arranged stacked. The air from the straightener then passed through a wire mesh to obtain uniform flow within the channel entry region. A hot wire anemometer (Lutron AM4204, resolution of 0.1 m/s) was placed 26 cm from the wire mesh to measure the velocity of the airflow inside the channel. A number

http://dx.doi.org/10.5772/intechopen.74518

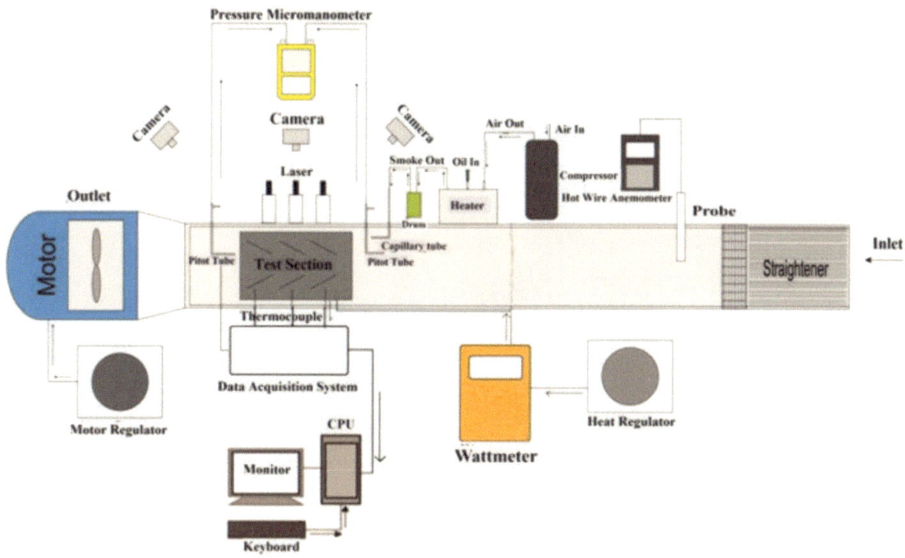

Figure 1. Schematic diagram of experimental set up.

of thermocouples (K-type with an accuracy of 0.75%) were placed on the test plate to measure the wall temperature of the plate. Several thermocouples were also set on the upstream and downstream of the test plate to measure inlet and outlet air temperatures, respectively. These thermocouples were connected to the acquisition data connected to the CPU. Data acquisition converts analog signals into digital signals and then stored in the CPU. A micro-manometer (Fluke 922 pitot tube with an accuracy of 1%) with two pitot tubes was set on the upstream and downstream of the test section to measure pressure drop. The airflow velocity inside the channel was varied from 0.4 to 2 m/s with an interval of 0.2 m/s using a motor regulator. A heater was placed in the test section to induce a constant heat rate of 35 W which was controlled by using a heating regulator. The volatile liquid was used to produce smoke for flow visualization purposes. A compressor was used to pump smoke into the smoke injector. Smoke was captured by a cross section of the green laser field that was fired from three laser pens. A camera was placed in three locations alternately to obtain the best flow visualization image.

2.2. Test specimen (VG)

The test specimen used in this experiment was a hot plate attached by RW VGs expressed by **Figure 2** and CRW VGs shown in **Figure 3**. Both VGs shown in **Figures 2** and **3** had the same aspect ratio. The height of the RW and CRW VGs was 27 mm with an attack angle of 30°. VGs were mounted on a 500 mm × 155 mm heating plate. VGs were arranged in-line with spacing between pairs of 20 mm and the distance of one pair of VGs with the other pair was 125 mm. Several parameters investigated in this experiment are shown in **Table 1**. Baseline means that the specimen test is a heating plate without VGs. The attack angle for CRW VG is the angle between the flow direction and the tangent at the front end of VG. **Figure 4** shows the test plate mounted with RW and CRW VGs.

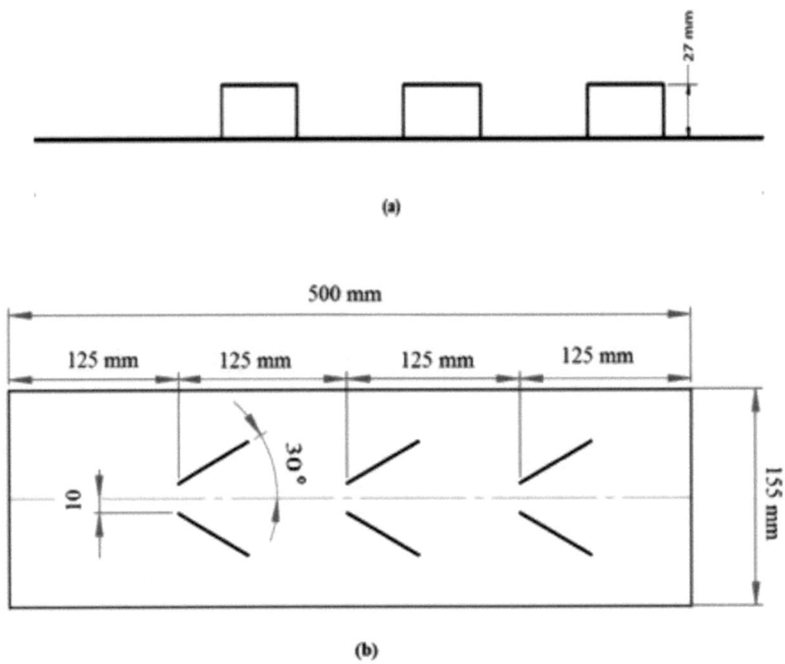

Figure 2. Three rows of rectangular winglet pairs (RWPs) of VGs: (a) side view and (b) top view.

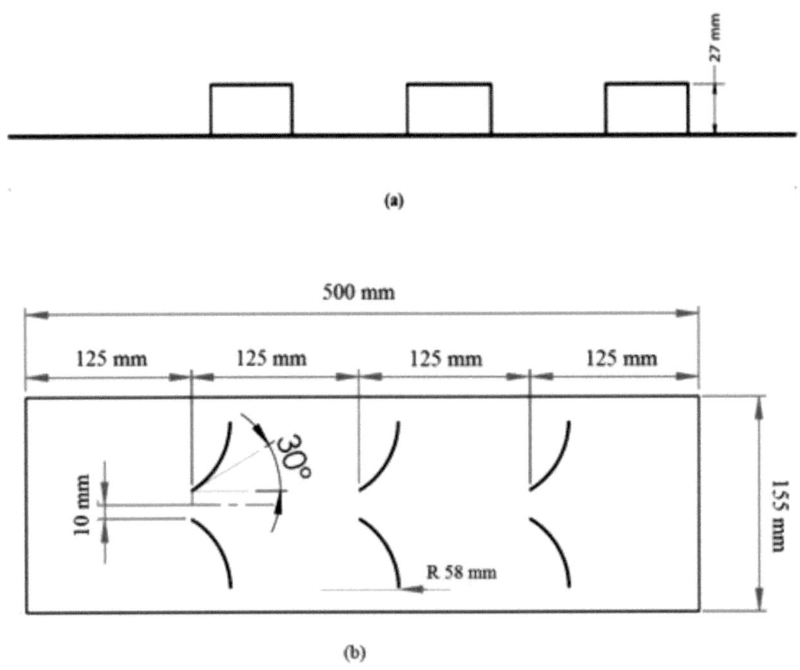

Figure 3. Three rows of concave rectangular winglet pairs (RWPs) of VGs: (a) side view and (b) top view.

No.	Test parameters
1.	Baseline, RW VG and CRW VG
2.	Attack angle: 15, 30 and 45°
3.	A number of row: one, two and three

Table 1. Some parameters in the experiment.

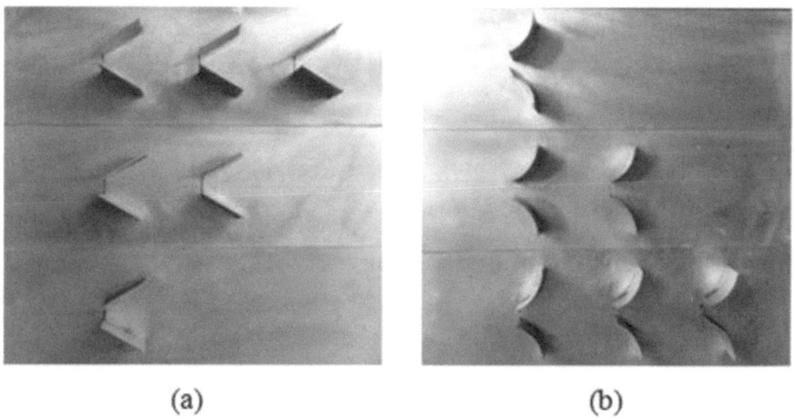

Figure 4. Test specimens. (a) RWPs one, two and three rows, (b) CRWPs one, two and three rows.

2.3. Testing procedures

Checking the temperature values of the thermocouple before the heat rate on the test plate was induced. All temperature values should be equal to ambient temperatures. The heat of 35 W was induced into the test plate by monitoring it through a watt meter. The heating of this plate was carried out in the absence of airflow until the surface temperature of the plate reaches a steady state of 54–55°C. Then the airflow speed was set from 0.4 to 2 m/s with 0.2 m/s interval using motor regulator by changing the frequency. The volatile liquid was flowed into the heater, and then the formed smoke was compressed into the capillary tube using a compressor and then injected into the airflow at a velocity of 0.4 m/s.

3. Results and discussion

3.1. Effect of attack angle on heat transfer

Figure 5 shows the effect of the attack angle on the convection heat transfer coefficient at various air velocities. In general, the convection heat transfer coefficient increases with increasing airflow velocity for all cases. This can be explained easily because the convection heat transfer coefficient is directly proportional to the flow velocity [28]. Higher velocity values decrease the thickness of the boundary layer resulting in higher heat transfer [21]. By installing a RW of three pairs of VG

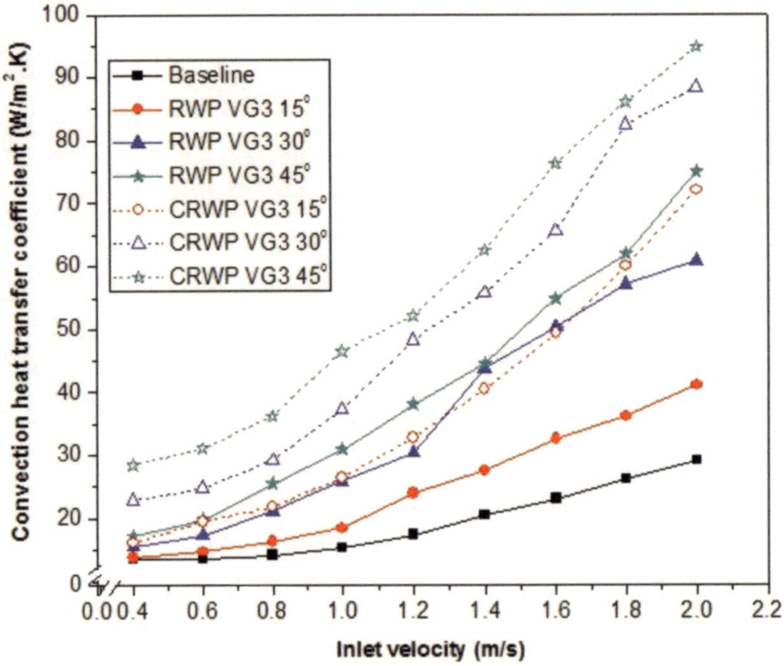

Figure 5. Convection heat transfer values with variations of inlet velocity at different angle attack for RW and CRW VGs.

(RWP VG3 15), the convection heat transfer coefficient increased 41.6% against the baseline at a flow rate of 2 m/s. This increase in heat transfer is due to the appearance of longitudinal vortex (LV) in the wake area of VG resulting in a good mixing of hot fluids near the wall with cold fluids in the main flow [27]. **Figure 5** also shows that the greater the attack angle of the flow to VG, the value of the convection heat transfer coefficient tends to increase at the same flow velocity. This is because the rectangular winglet with a larger angle of attack produces higher vortex circulation [4, 20]. LV promotes better fluid mixing resulting in a decrease in angle between the velocity field and the temperature gradient field. This results in an increase in the rate of heat transfer [6]. LV attenuates the boundary layer in the down wash region resulting in an increase in heat transfer [8].

For the case of RW VGs, an increase in the angle of attack from 15 to 45° at a velocity of 2 m/s results in 83% increase in heat transfer rate. In the case of CRW VG, the installation of VGs with 15, 30 and 45° attack angles resulted in a 158.1% increase in heat transfer, respectively, to the baseline at the flow velocity of 2 m/s. Increases in attack angles from 15 to 45° for CRW VGs cases resulted in 31.5% increase in heat transfer at a flow rate of 2 m/s. Just as the preceding reason, VGs with a larger angle of attack produce LV with higher strength which results in better heat transfer [10]. By installing CRW VGs, heat transfer increased 75.6, 45.4 and 26.3% against RW VGs at the same attack angles of 15, 30 and 45°, respectively, at the flow velocity of 2 m/s. This is due to the instability of the centrifugal force on the concave surface resulting in a larger radius of LV with higher strength of vortex circulation than that of RW VGs [29]. The formed LV when the flow passing through a concave surface is called the Görtler vortex in which the growth rate is 10 times of vortex generated by flat plate [30]. Disturbances caused by the presence of Görtler vortex increase local heat transfer [31].

3.2. Effect of a number of VG rows on heat transfer

Figure 6 shows the effect of the number of rows of VGs on convective heat transfer coefficient values at various flow velocities with 30° attack angle for RW and CRW VGs cases. Two and three rows of VG pairs were arranged in-line in this case. By mounting VGs, in general, improvement of heat transfer was found against the baseline. In the case of RW VGs, the use of two and three pairs of VGs increased the heat transfer by 5.5 and 26.4% of a pair of VGs, respectively. This increase was triggered by the acceleration of formed LV by LV produced by a previous pair of VGs [11]. In the case of CRW VGs, the increase in heat transfer rates at two and three pairs of VGs increased from 19.4 to 37.9% of a single pair of VGs. The use of CRW VGs in heat transfer improvement was better than RW VGs on the same number of rows of VGs pairs. As mentioned earlier, CRW VGs generate LV with a larger vortex radius and stronger vortex circulation than that of LV produced by RW VGs due to the instability of centrifugal force. In the baseline case, an increase in the flow rate from 0.4 to 2 m/s resulted in an increase of 114.7% heat transfer rate. **Table 2** presents the percentage increase in the rate of heat transfer to the increase of flow velocity inside the channel.

3.3. Effect of attack angle and number of pairs of VG on pressure loss penalty

Figure 7 expresses the effect of the number of pairs of VGs lines on the pressure loss penalty (pressure drop) for various flow velocities at the attack angle 15°. RWP VG1, RWP VG2 and RWP VG3 reveal one, two and three rows of pair of RW VG, respectively. The same is true for

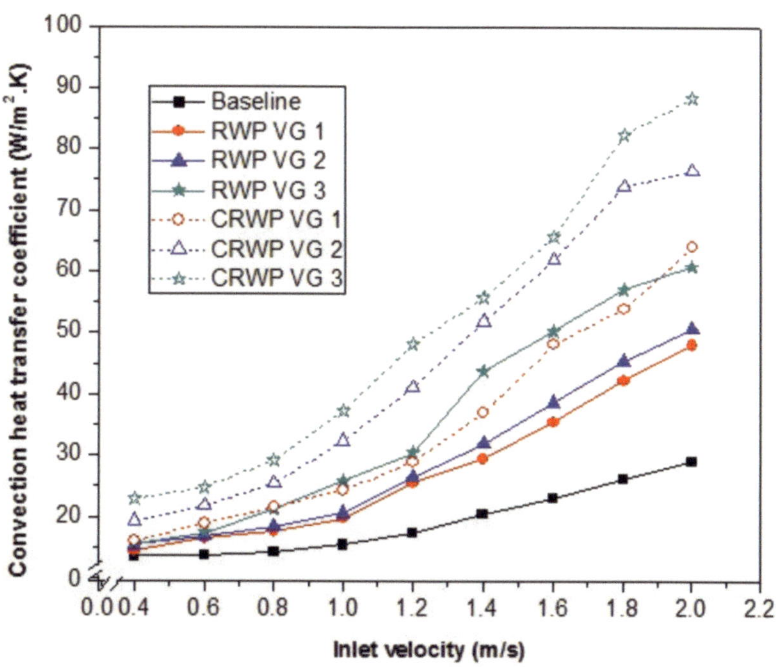

Figure 6. Convection heat transfer values with variations of inlet velocity in different a number of VG rows at 30° attack angle for RW and CRW VGs.

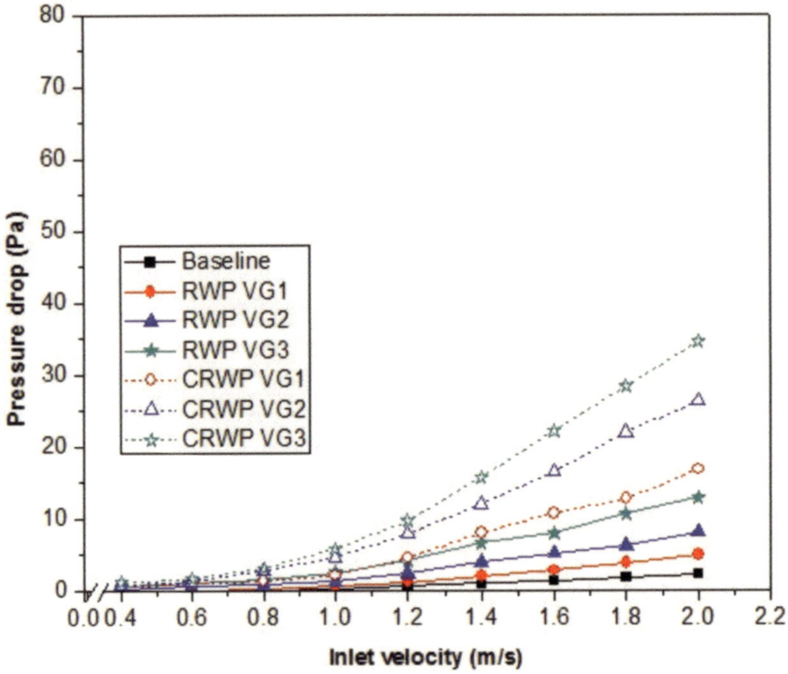

Figure 7. Pressure drop values with variations of inlet velocity in different a number of VG rows at 15° attack angle for RW and CRW VGs.

Baseline	RWP VG1	RWP VG2	RWP VG3	CRWP VG1	CRWP VG2	CRWP VG3
114.7	236.2	231.4	290.9	299.4	296.9	286.8

Table 2. Heat transfer augmentation for increasing in flow velocity (%).

CRW VGs. In general, the pressure drop increases with increasing flow velocity for all cases. This is because the pressure drop is directly proportional to the flow velocity [32]. The value of the pressure drop increases with the increase of the number of rows of VGs. The increase in pressure drop is mainly caused by the addition of formed drag from VGs [21]. **Table 3** shows the increase in pressure drop to flow velocity for each case. These values describe what times the pressure drop increases when the flow velocity inside the channel was increased from 0.4 to 2.0 m/s for each case. For baseline and RWP VG1 cases, the pressure drop values might not be compared because of the limited accuracy of micro-manometers especially for velocities below 1 m/s. From **Figure 6**, it is found that the use of CRW VGs increases higher pressure drop than the use of RW VGs due to the greater area of the obstruction on CRW VGs [33].

Figure 8 illustrates the pressure drop generated from the variation in the number of pairs of RW and CRW VGs as well as the baseline at the attack angle of 30°. The experimental results

http://dx.doi.org/10.5772/intechopen.74518

Baseline	RWP VG1	RWP VG2	RWP VG3	CRWP VG1	CRWP VG2	CRWP VG3
-	-	31.2	24.4	30.4	34.4	29.8

Table 3. Pressure loss penalty at attack angle of 15° for increasing in flow velocity (times).

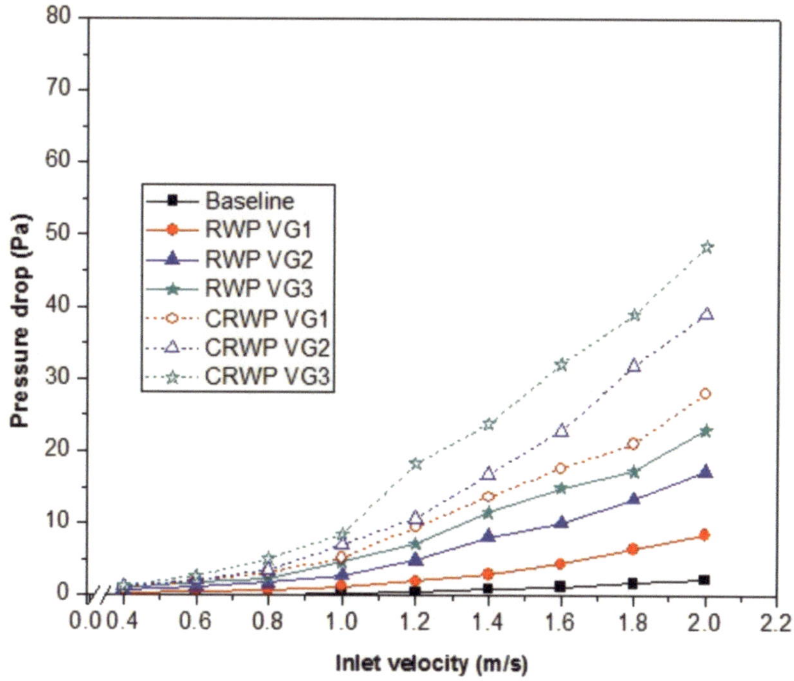

Figure 8. Pressure drop values with variations of inlet velocity in different a number of VG rows at 30° attack angle for RW and CRW VGs.

found that the pressure drop increased with increasing flow velocity. The increase in pressure drop to the increase in flow velocity in the channel for each case can be expressed in **Table 4**. For the baseline case, the increase in pressure drop might not be determined because the pressure drop value was not detected in the measurement at the flow velocity of 0.4 m/s.

This was triggered by the increase in formed drag by increasing flow velocity. At a flow rate of 2 m/s, the pressure drop increased almost three times from the baseline when a pair of RW VGs was mounted on the plate. This value increased to nine times when three pairs of RW VGs were installed. By installing three pairs of CRW VGs, the pressure drop value increased 20 times compared to the baseline at a flow rate of 2 m/s. This indicates that the use of CRW VGs increases the pressure drop greater than the use of RW VGs. This is the disadvantage of using CRW VGs. For the case of three pairs of CRW VGs, the pressure drop increased significantly at a flow velocity above 1.0 m/s (**Table 5**).

Baseline	RWP VG1	RWP VG2	RWP VG3	CRWP VG1	CRWP VG2	CRWP VG3
-	32.8	22.3	22.2	27.3	32.1	37.4

Table 4. Pressure loss penalty at attack angle of 30° for increasing in flow velocity (times).

Baseline	RWP VG1	RWP VG2	RWP VG3	CRWP VG1	CRWP VG2	CRWP VG3
-	67.3	34.1	38.8	33	37	30.8

Table 5. Pressure loss penalty at attack angle of 30° for increasing in flow velocity (times).

Figure 9 shows the pressure drop for various flow velocities and VG types at an angle of attack 45°. By comparing **Figure 9** with **Figures** 7 and **8**, the pressure drop increases very high against the baseline. The formed drag is large at large angle of attack which is the main reason for the significant increase in pressure drop. The area of the flow obstruction increases with the larger angle of attack [33]. A strong LV also contributed to the increase in pressure drop

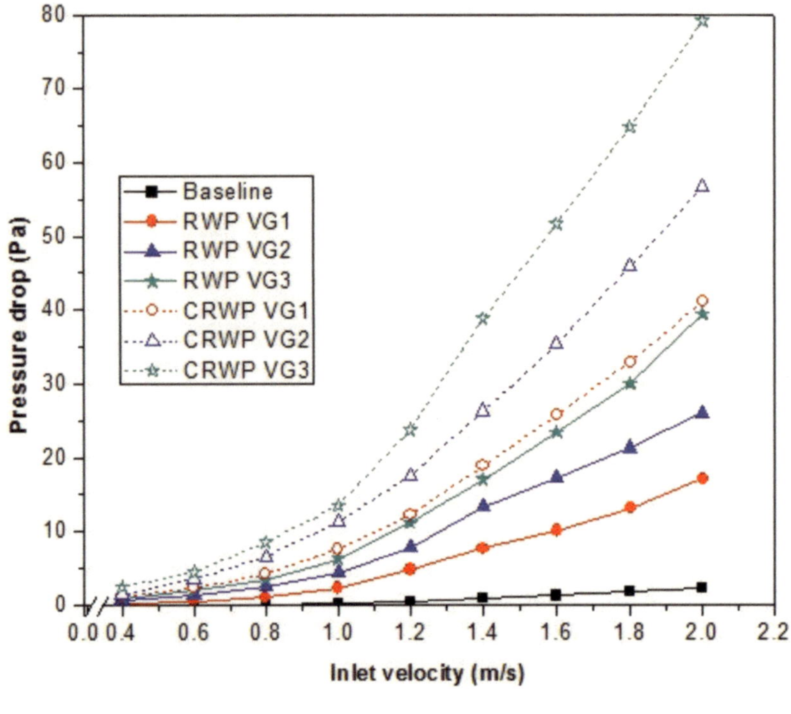

Figure 9. Pressure drop values with variations of inlet velocity in different a number of VG rows at 45° attack angle for RW and CRW VGs.

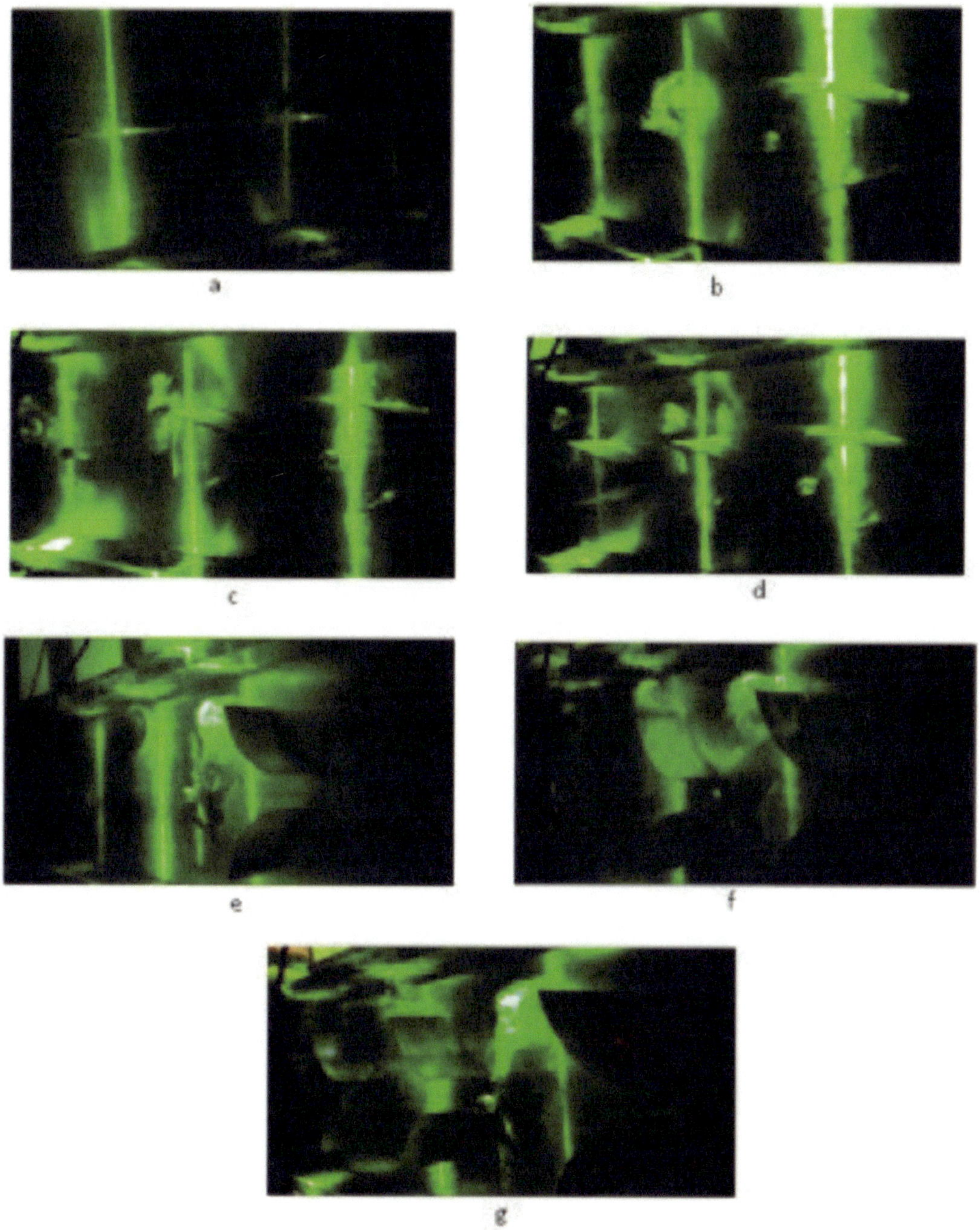

Figure 10. Flow structure of fluid flow passes through flat plate with and without VGs. (a) Baseline; (b) RWP VG1; (c) RWP VG2; (d) RWP VG3; (e) CRWP VG1; (f) CRWP VG2; and (g) CRWP VG3.

of fluid flow in the channel [34]. In general, the generation of vortex causes extra drag which leads to an increase in pressure drop [35].

3.4. Flow visualization

This flow visualization aims to know the flow structure as it passes through VGs. **Figure 10** shows the airflow structure through the plate for each case. In the baseline case, LV is not visible when the fluid passes through the plate because there are no VGs as can be seen in **Figure 10a**. LVs appear at two locations in the downstream of the VGs for the case of RWP VG1 as shown in **Figure 10b**. LV located on the third green laser cross section appears weaker than LV in front of it. This is due to viscous dissipation in the flow direction [36]. LVs are observed at three sites behind VGs with LV at the third location decreases in strength due to viscous dissipation as denoted in **Figure 10c**. Two vortices are observed in the second and third locations after the fluid passes through the VGs as shown in **Figure 10d**. When the fluid passes through the CRWP VG1, two vortices are formed in the wake region of the VGs and weaken their strength due to viscous dissipation as shown in **Figure 10e**. Vortices generated by CRW VG have a larger radius than those produced by RW VG. Vortices are also found when the fluid passes through CRWP VG2 and CRWP VG3 with a larger radius as shown in **Figure 10f** and **g**.

4. Conclusion

Experimental to know the effect of concave rectangular winglet vortex generator (CRW VG) has been completed. The heat transfer rate increased with increasing angle of attack. However, this increase in attack angle had a negative impact on the pressure loss penalty in which the pressure drop also increased with increasing angle of attack. Similar tendencies were also found when the number of VG pairs was increased, heat transfer increased with increasing pressure drop. The flow visualization was able to show the flow structure when the fluid passed through VGs. Longitudinal vortex (LV) was observed in the wake region of VGs.

Acknowledgements

This work was supported by the Fundamental Research Project of Indonesia (KEMENRISTEK DIKTI Number 699-21/UN7.P4.3/PP/2017). The authors are grateful to all laboratory members especially those of Thermofluid Laboratory of Mechanical Engineering of Diponegoro University, Indonesia.

Author details

Syaiful

Address all correspondence to: syaiful.undip2011@gmail.com

Mechanical Engineering Department, Diponegoro University, Semarang, Indonesia

References

[1] Fiebig M, Valencia A, Mitra NK. Wing-type vortex generators for fin-and-tube heat exchangers. Experimental Thermal and Fluid Science. 1993;**7**:287-295. DOI: 10.1016/0894-1777(93)90052-K

[2] Zhu JX, Fiebig M, Mitra NK. Numerical investigation of turbulent flows and heat transfer in a rib-roughened channel with longitudinal vortex generators. International Journal of Heat and Mass Transfer. 1995;**38**:495-501. DOI: 10.1016/0017-9310(94)00177-W

[3] Jacobi AM, Shah RK. Heat transfer surface enhancement through the use of longitudinal vortices: A review of recent progress. Experimental Thermal and Fluid Science. 1995;**11**:295-309. DOI: 10.1016/0894-1777(95)00066-U

[4] Biswas G, Torii K, Fujii D, Nishino K. Numerical and experimental determination of flow structure and heat transfer effects of longitudinal vortices in a channel flow. International Journal of Heat and Mass Transfer. 1996;**39**:3441-3451. DOI: 10.1016/0017-9310(95)00398-3

[5] Gentry MC, Jacobi AM. Heat transfer enhancement by delta-wing vortex generators on a flat plate: Vortex interactions with the boundary layer. Experimental Thermal and Fluid Science. 1997;**14**:231-242. DOI: 10.1016/S0894-1777(96)00067-2

[6] Guo ZY, Li DY, Wang BX. A novel concept for convective heat transfer enhancement. International Journal of Heat and Mass Transfer. 1998;**41**:2221-2225. DOI: 10.1016/S0017-9310(97)00272-X

[7] Lee SH, Ryou HS, Choi YK. Heat transfer in a three-dimensional turbulent boundary layer with longitudinal vortices. International Journal of Heat and Mass Transfer. 1999;**42**:1521-1534. DOI: 10.1016/S0017-9310(98)00290-7

[8] Chen Y, Fiebig M, Mitra NK. Heat transfer enhancement of finned oval tubes with staggered punched longitudinal vortex generators. International Journal of Heat and Mass Transfer. 2000;**43**:417-435. DOI: 10.1016/S0017-9310(99)00157-X

[9] Wang CC, Lo J, Lin YT, Mint WCS. Flow visualization of annular and delta winlet vortex generators in fin-and-tube heat exchanger application. International Journal of Heat and Mass Transfer. 2002;**45**:3803-3815. DOI: 10.1016/S0017-9310(02)00085-6

[10] Tiwari S, Maurya D, Biswas G, Eswaran V. Heat transfer enhancement in cross-flow heat exchangers using oval tubes and multiple delta winglets. International Journal of Heat and Mass Transfer. 2003;**46**:2841-2856. DOI: 10.1016/S0017-9310(03)00047-4

[11] Leu J-S, Ying-Hao W, Jang J-Y. Heat transfer and fluid flow analysis in plate-finand tube heat exchangers with a pair of block shape vortex generators. International Journal of Heat and Mass Transfer. 2004;**47**:4327-4338. DOI: 10.1016/j.ijheatmasstransfer.2004.04.031

[12] Sommers AD, Jacobi AM. Air-side heat transfer enhancement of a refrigerator evaporator using vortex generation. International Journal of Refrigeration. 2005;**28**:1006-1017. DOI: 10.1016/j.ijrefrig.2005.04.003

[13] Ferrouillat S, Tochon P, Garnier C, Peerhossaini H. Intensification of heat-transfer and mixing in multifunctional heat exchangers by artificially generated stream-wise vorticity. Applied Thermal Engineering. 2006;**26**:1820-1829. DOI: 10.1016/j.applthermaleng.2006.02.002

[14] Hiravennavar SR, Tulapurkara EG, Biswas G. A note on the flow and heat transfer enhancement in a channel with built-in winglet pair. International Journal of Heat and Fluid Flow. 2007;**28**:299-305. DOI: 10.1016/j.ijheatfluidflow.2006.03.030

[15] Wu JM, Tao WQ. Numerical study on laminar convection heat transfer in a rectangular channel with longitudinal vortex generator. Part A: Verification of field synergy principle. International Journal of Heat and Mass Transfer. 2008;**51**:1179-1191. DOI: 10.1016/j.ijheatmasstransfer.2007.03.032

[16] Wu JM, Tao WQ. Numerical study on laminar convection heat transfer in a channel with longitudinal vortex generator. Part B: Parametric study of major influence factors. International Journal of Heat and Mass Transfer. 2008;**51**:3683-3692. DOI: 10.1016/j.ijheatmasstransfer.2007.03.031

[17] Chu P, He YL, Lei YG, Tian LT, Li R. Three-dimensional numerical study on fin-and-oval-tube heat exchanger with longitudinal vortex generators. Applied Thermal Engineering. 2009;**29**:859-876. DOI: 10.1016/j.applthermaleng.2008.04.021

[18] Min C, Qi C, Kong X, Dong J. Experimental study of rectangular channel with modified rectangular longitudinal vortex generators. International Journal of Heat and Mass Transfer. 2010;**53**:3023-3029. DOI: 10.1016/j.ijheatmasstransfer.2010.03.026

[19] Wu JM, Tao WQ. Impact of delta winglet vortex generators on the performance of a novel fin-tube surfaces with two rows of tubes in different diameters. Energy Conversion and Management. 2011;**52**:2895-2901. DOI: 10.1016/j.enconman.2011.03.002

[20] Wu JM, Tao WQ. Effect of longitudinal vortex generator on heat transfer in rectangular channels. Applied Thermal Engineering. 2012;**37**:67-72. DOI: 10.1016/j.applthermaleng.2012.01.002

[21] He Y-L, Chu P, Tao W-Q, Zhang Y-W, Xie T. Analysis of heat transfer and pressure drop for fin-and-tube heat exchangers with rectangular winglet-type vortex generators. Applied Thermal Engineering. 2013;**61**:770-783. DOI: 10.1016/j.applthermaleng.2012.02.040

[22] Zhou G, Feng Z. Experimental investigations of heat transfer enhancement by plane and curved winglet type vortex generators with punched holes. International Journal of Thermal Sciences. 2014;**78**:26-35. DOI: 10.1016/j.ijthermalsci.2013.11.010

[23] Syaiful, Jalil R, Hambali I, Bae M-W. Numerical simulation of concave delta winglet vortex generator effect on heat transfer augmentation in airflow inside a channel. In: Proceedings of the Tenth International Symposium on Mechanics, Aerospace and Informatics Engineering (ISMAI X'15); 17-20 September 2015; Jinju, Korea. pp. 15-21.

[24] Syaiful, Reviansyah G, Bae M-W. Attack angle effect of concave delta winglet vortex generator on heat transfer augmentation in fin-and-tube heat exchanger for EGR cooler

application by numerical simulation. In: FISITA 2016 World Automotive Congress; 26-30 September 2016; Busan. Korea. F2016-ESYE-005.

[25] Oneissi M, Habchi C, Russeil S, Bougeard D, Lemenand T. Novel design of delta winglet pair vortex generator for heat transfer enhancement. International Journal of Thermal Sciences. 2016;**109**:1-9. DOI: 10.1016/j.ijthermalsci.2016.05.025

[26] Song KW, Xi ZP, Mei S, Wang LC, Wu X, Wang LB. Effect of geometric size of curved delta winglet vortex generators and tube pitch on heat transfer characteristics of fin-tube heat exchanger. Experimental Thermal and Fluid Science. 2017;**82**:8-18. DOI: 10.1016/j.expthermflusci.2016.11.002

[27] Syaiful, Ayutasari A, Soetanto MF, Siswantara AI, Bae M-w. Thermo-hydrodynamic performance analysis of fluid flow through concave delta winglet vortex generators by numerical simulation. International Journal of Technology. 2017;**7**:1276-1285. DOI: 10.14716/ijtech.v8i7.706

[28] Bergman TL, Lavine AS, Incropera FP, Dewitt DP. Fundamentals of Heat and Mass Transfer. 7th ed. USA: John Wiley & Sons; 2011. 467 p. ISBN 13: 978-0470-50197-9

[29] Malatesta V, Souza LF, Liu JTC, Kloker MJ. Heat transfer analysis in a flow over concave wall with primary and secondary instabilities. Procedia IUTAM. 2015;**14**:487-495. DOI: 10.1016/j.piutam.2015.03.077

[30] Riley S, Johnson MW, Gibbings JC. Boundary layer transition of strongly concave surfaces. In: Gas Turbine and Aeroengine Congress and Exposition; 4-8 June 1989; Toronto, Ontario, Canada: asmep/83350

[31] McCormagk PD, Welker H, Kelleher M. Taylor-Göertler vortices and their effect on heat transfer. In: ASME-AIChE Heat Transfer Conference; 3-6 August 1969; Minneapolis, USA: 69-HT-3

[32] Çengel YA, Cimbala JM. Fluid Mechanics: Fundamentals and Applications. 1st ed. USA: McGraw-Hill; 2006. 353 p. ISBN: 0-07-247236-7.

[33] Li H-Y, Chen C-L, Chao S-M, Liang G-F. Enhancing heat transfer in a plate-fin heat sink using delta winglet vortex generators. International Journal of Heat and Mass Transfer. 2013;**67**:666-677. DOI: 10.1016/j.ijheatmasstransfer.2013.08.042

[34] Xia HH, Tang GH, Shi Y, Tao WQ. Simulation of heat transfer enhancement by longitudinal vortex generators in dimple heat exchangers. Energy. 2014;**74**:27-36. DOI: 10.1016/j.energy.2014.02.075

[35] Saha P, Biswas G, Sarkar S. Comparison of winglet-type vortex generators periodically deployed in a plate-fin heat exchanger—A synergy based analysis. International Journal of Heat and Mass Transfer. 2014;**74**:292-305. DOI: 10.1016/j.ijheatmasstransfer.2014.03.015

[36] Arora A, Subbarao PMV, Agarwal RS. Numerical optimization of location of 'common flow up' delta winglets for inline aligned finned tube heat exchanger. Applied Thermal Engineering. 2015;**82**:329-340. DOI: 10.1016/j.applthermaleng.2015.02.071

Section 4

Applications

Heat Transfer of Helix Energy Pile: Part 1: Traditional Cylinder Helix Energy Pile

Guangqin Huang, Yajiao Liu, Xiaofeng Yang and Chunlong Zhuang

Additional information is available at the end of the chapter

http://dx.doi.org/10.5772/intechopen.76820

Abstract

Helix energy pile (HEP) is a new popular ground heat exchanger that has the advantages of large heat exchange rate and low initial cost. As for the traditional helix energy pile, the tube is wound on the cylindrical wall, which is called the cylinder helix energy pile (CyHEP). Further, both analytical solution model and numerical solution model for CyHEP are built to discuss the dynamic characteristics of thermal interferences and heat transfer performance. The results indicate that four heat exchange stages for the spiral pile geothermal heat exchanger along the fluid flow direction are revealed: inlet heat exchange stage, grout thermal short-circuiting stage, small temperature difference stage and outlet heat exchange stage. Each stage has corresponding heat transfer characteristics, and reducing the length of small temperature difference stage and increasing the other stages would enhance the heat exchange of spiral geothermal ground heat exchanger. As the pile diameter increases, the heat transfer per unit tube length decreases, and the heat exchange per unit pile depth increases. As the pile depth increases, the heat transfer per unit tube length and the heat exchange per unit pile depth are reduced. And as the pitch increases, the heat transfer per unit tube length increases, and the heat exchange per unit pile depth decreases.

Keywords: cylinder, helix energy pile, analytical, numerical, thermal interferences, stages

1. Introduction

Energy saving and environmental protection have been the symbols in modern development. In China, "Strategic Action Plan for Energy Development (2014—2020)" was published in November 2014 in order to achieve sustainable development. The plan points out that the non-fossil energy share of primary energy consumption would reach 15% by 2020 and the utilized quantity of geothermal energy would run up to 50 million tons of standard coal. Ground source heat pump (GSHP) is an efficient technology to exploit the shallow geothermal energy and has been applied extensively in building energy conservation.

GSHP utilizes the ground maintaining a nearly constant temperature as the heat source or sink, and the particular underground temperature is appropriate for efficient operation of the unit. The most important component for GSHP is the ground heat exchanger (GHE) and the borehole GHE with U-type tube. However, the high installation and drilling cost restrict the application of borehole GHE, especially for rocky areas. Therefore, energy piles [1] have been proposed and widely applied for its good economic and well-heated transfer performance. The primary structure forms of energy piles consisting of U type [2–4] (such as single U type, parallel double U type, triple U type and series U type) and helix type [5–7] and the helix energy piles (HEPs) are promising GHEs due to the characteristics of large heat exchange rate and no air choking in the pipes. However, due to the features of complex helix structures, shallow burial and a large diameter of the pile, the heat transfer characteristics of HEPs are completely different from the U-type borehole GHE and would not be described by classical Kelvin's line source. Hence, the heat transfer of HEP has formed an academic focus.

2. Heat transfer model and characteristics of the traditional cylinder helix energy pile (CyHEP)

Helix energy pile is a new kind of heat exchanger for ground source heat pump. It is usually buried in the building concrete pile foundation, combined with the building structure. Compared with the conventional U-type and W-type energy pile, helix energy pile has a larger heat transfer area under the same depth. In addition, helix energy pile can avoid the problem of air accumulation at the top of U-type and W-type ones. As for the traditional helix energy pile, the tube is wound on the cylindrical wall as shown in **Figure 1**, which is called the cylinder helix energy pile (CyHEP) in this section.

2.1. Heat transfer model

Through literature research and analysis, it was found that due to the complicated structure of the cylinder helix energy pile, and there is a lack of research on the heat transfer model of it. At present, the heat transfer model is mainly a thermal conductivity model. It is considered that the thermo-physical properties of the pile backfill are the same as those of the soil, which is different from the actual heat transfer process. Most of the numerical solution models are simulated with commercial CFD software. However commercial CFD software has modeling and meshing difficulties, poor flexibility and other shortcomings. Based on the above reasons, a new discrete method of cylinder helix energy pile is presented in this section, and a three-dimensional numerical heat transfer model is established based on a transient heat balance theory. It can provide a more accurate description of the heat transfer process of the cylinder helix energy pile, thus providing design guidance for the actual project.

http://dx.doi.org/10.5772/intechopen.76820

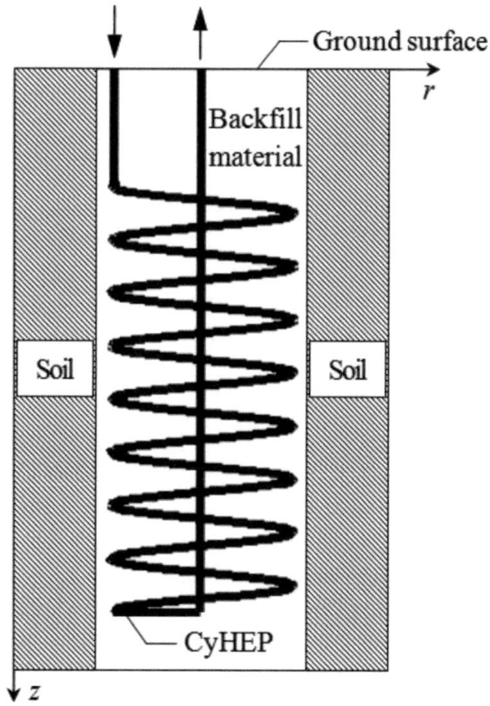

Figure 1. Cylinder helix energy pile (CyHEP).

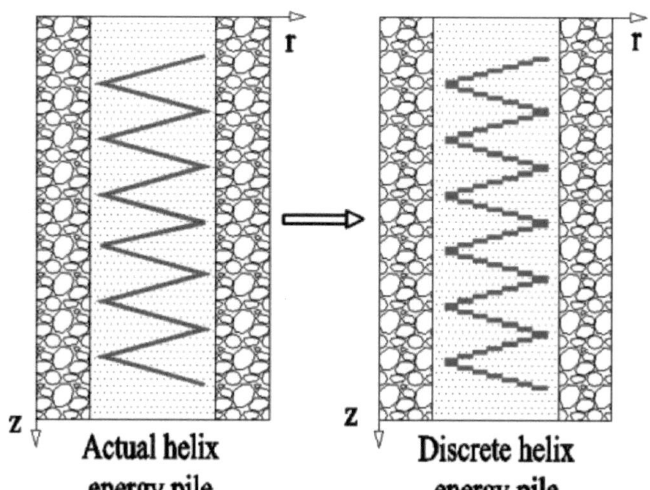

Figure 2. The sketch of discretized helix energy pile.

Figure 2 shows the sketch of a discretized helix energy pile. As shown in the figure, the continuous, circular cross-section helical fluid is discretely divided into a finite number of discrete, non-continuous stepped and square cross-section cells. The shape of a single fluid cell is shown in **Figure 3**, which is a micro-cell under the cylindrical coordinate system. In this way, the discrete fluid unit body can be better combined with the cylindrical coordinate system to facilitate the establishment of the heat balance equation of the fluid unit body, the backfill material and the geotechnical unit body.

Taking each discrete element as the research object, the amount of energy change per unit time of the element is equal to the net amount of heat transferred from the adjacent element, which is the essence of the transient heat balance principle of the element. As shown in **Figure 4**, within a certain period of time, the transient thermal equilibrium equation of the unit body is shown as the following:

$$q_{z+} + q_{z-} + q_{r+} + q_{r-} + q_{\theta+} + q_{\theta-} = \rho c_p V(i,j,k) \frac{T(i,j,k,\tau+1) - T(i,j,k,\tau)}{\tau_{sp}} \quad (1)$$

where q_{z+}、q_{z-} are heat flux transferred from the unit below and above in the z direction to the calculated unit, W/m^2; q_{r+}、q_{r-} are heat fluxes transferred from the unit in the radial positive normal direction and the negative normal direction to the calculated unit, W/m^2; $q_{\theta+}$、$q_{\theta-}$ are heat fluxes transferred from the unit in the positive normal of θ direction and the negative normal of θ direction to the calculated unit, W/m^2; ρ is the density of the calculated unit, kg/m^3; c_p is the specific heat of the calculated unit, J·m^{-3}·°C^{-1}; *V(i,j,k)* is the volume of the calculated

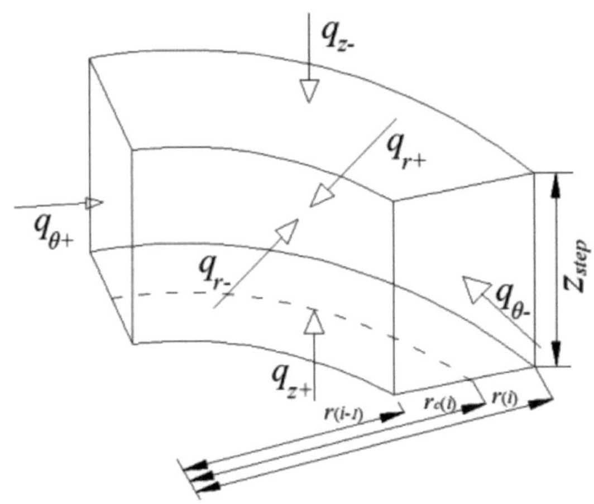

Figure 3. The sketch of the heat balance of unit body.

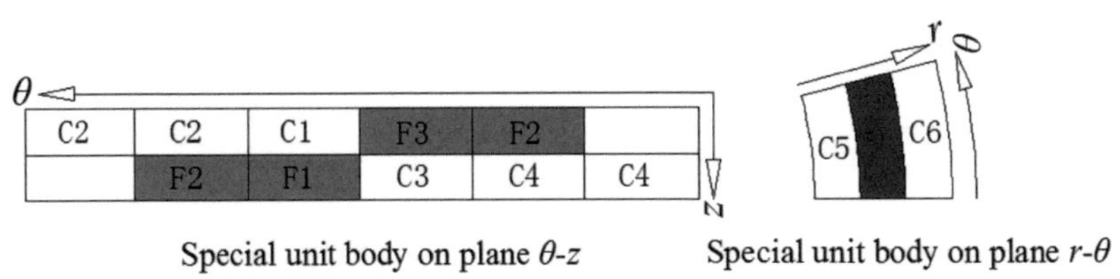

Figure 4. The sketch of the positional relation of special unit body.

unit, $V(i,j,k) = z_{sp}\cdot[0.5\theta_{sp}(j)\,(r^2(i) - r^2(i-1))]$, m^3; θ_{sp} is the increment in θ direction; τ_{sp} is the time step, s; T is the temperature, °C.

It is the key for the principle of transient heat balance to determine the heat transfer q between adjacent unit bodies in Eq. (1). And the heat transfer q between adjacent unit bodies is determined by the temperature difference and thermal resistance of adjacent unit bodies as shown in Eq. (2):

$$\begin{cases} q_{z+} = \dfrac{T(i,j,k+1) - T(i,j,k)}{R_{z+}} & q_{r+} = \dfrac{T(i+1,j,k) - T(i,j,k)}{R_{r+}} & q_{\theta+} = \dfrac{T(i,j+1,k) - T(i,j,k)}{R_{\theta+}} \\ q_{z-} = \dfrac{T(i,j,k-1) - T(i,j,k)}{R_{z-}} & q_{r-} = \dfrac{T(i-1,j,k) - T(i,j,k)}{R_{r-}} & q_{\theta-} = \dfrac{T(i,j-1,k) - T(i,j,k)}{R_{\theta-}} \end{cases} \tag{2}$$

where R_{z+}, R_{z-} is the heat transfer resistance between the unit below and above in the z direction and the calculated unit, °C·m^2·W^{-1}; R_{r+}, R_{r-} is the heat transfer resistance between the unit in the radial positive normal direction and the negative normal direction and the calculated unit, °C·m^2·W^{-1}; $R_{\theta+}$, $R_{\theta-}$ is the heat transfer resistance between the unit in the positive normal of θ direction and the negative normal of θ direction and the calculated unit, °C·m^2·W^{-1}.

In sum, for the three-dimensional heat transfer model proposed in this section, the key lies in the calculation of heat flow or thermal resistance, and the heat flow and thermal resistance for different types of unit bodies are determined, respectively.

(1) Thermal resistance calculation for unit body not in contact with the fluid: As for the unit body not in contact with the fluid, the heat transfer is a pure thermal conduction process; thus, the thermal resistance in the z direction, r direction and θ direction can be calculated as following:

$$R_{z+} = R_{z-} = \frac{2z_{sp}}{\lambda_s \theta_{sp}(j)(r^2(i) - r^2(i-1))} \tag{3}$$

$$R_{r+} = \frac{1}{\lambda_s \theta_{sp}(j) z_{sp}} \ln\left[\frac{r_c(i+1)}{r_c(i)}\right] \tag{4}$$

$$R_{r-} = \frac{1}{\lambda_s \theta_{sp}(j) z_{sp}} \ln\left[\frac{r_c(i)}{r_c(i-1)}\right] \tag{5}$$

$$R_{\theta+} = \frac{(\theta_{sp}(j+1) + \theta_{sp}(j)) r_c(i)}{2\lambda_s z_{sp}(r(i) - r(i-1))} \tag{6}$$

$$R_{\theta-} = \frac{(\theta_{sp}(j) + \theta_{sp}(j-1)) r_c(i)}{2\lambda_s z_{sp}(r(i) - r(i-1))} \tag{7}$$

where λ_s is the thermal conductivity of soil, W·m^{-1}·°C and $r_c(i)$ is the distance from center of calculated unit to center of the pile, m.

(2) Thermal resistance calculation for unit body in contact with fluid: For the unit body in contact with fluid, as shown in **Figure 4**, the heat transfer resistance of C1 and C6 units should include three items: fluid convection thermal resistance, thermal conductivity resistance of the

wall and thermal conductivity resistance of the backfills. Thus, for a typical unit body C2, the thermal resistance in the z direction can be calculated as the following and the other thermal resistances of unit body C2 can be calculated in the same way as thermal resistance calculation for unit body not in contact with the fluid. In the same way, the calculation methods of the thermal resistances of the unit bodies C1 and C3–C6 in contact with the fluid are similar to those of the unit body C2.

$$R_{z+} = \frac{2(0.5z_{sp} - \delta)}{\lambda_g \theta_{sp}(j)(r^2(i) - r^2(i-1))} + \frac{2\delta}{\lambda_p \theta_{sp}(j)(r^2(i) - r^2(i-1))} + \frac{2}{h \cdot \theta_{sp}(j)(r^2(i) - r^2(i-1))} \tag{8}$$

where λ_g is the thermal conductivity of the backfill, W·m^{-1}·°C; δ is the spiral wall thickness, m; h is the convection heat transfer coefficient inside the tube, W·m^{-2}·°C^{-1}.

(3) Heat flow calculation for fluid unit: Due to the discretized fluid cell body being in the form of stairs, in terms of geometry, the upper ladder cell body (cell F3 in **Figure 4**) and the lower ladder cell body (cell F1 in **Figure 4**) fail to ensure the continuity of flow and heat transfer. Therefore, in order to ensure continuity, it is assumed that the upper fluid unit F3 and the lower fluid unit F1 have a thermal effect, and the fluid unit F3 and the backfilling unit C1 and the fluid unit F1 and the backfilling unit C3 are insulated. Based on the above assumptions, the expression of heat flow in the θ direction of the fluid units F3 and F2 can be obtained as shown in Eq. (9). And heat transfer resistance in other directions is determined by the calculation of the thermal resistance of the unit in contact with the fluid. For the F1 fluid unit, in order to ensure the continuity of the fluid, $q_{\theta-}$ should be the heat transferred from unit F3 to unit F1, and it is calculated in the same way as unit F3 and unit F2.

$$\begin{cases} q_{\theta+} = -mc_{p,f}T(i,j,k,\tau) \\ q_{\theta-} = mc_{p,f}T(i,j-1,k,\tau) \end{cases} \tag{9}$$

where m is the fluid flow, m^3·s^{-1} and cp,f is the fluid specific heat, J·m^{-3}·°C^{-1}.

(4) Soil surface boundary condition: As the helix energy pile is shallow, pile depth is generally about 10 m; the soil surface environmental factors, such as air temperature, will affect the soil temperature of the pile foundation, thus affecting its heat transfer performance. For pile foundations installed in the basement, this section ignores the radiative heat transfer between the ground and other surfaces in the basement. And considering the convective heat transfer between the ground and the air, the boundary conditions of the soil surface are given by Eq. (10).

$$q_{z-} = (T_a(\tau) - T(i,j,0,\tau))h_a A(i,j,0) \tag{10}$$

where $A(i,j,0)$ is the top surface area of the unit $(i,j,0)$, $A(i,j,0) = \theta_{sp}(r(i)^2 - r(i-1)^2)$, m^2; h_a is the convection heat transfer coefficient, W·m^{-2}·°C^{-1} and T_a is the air temperature, °C.

Then the heat flow or thermal resistance formula of each type of unit body above is brought into the heat balance, Eq. (1), and the linear equations of the temperature field of helix energy pile at time τ can be obtained. The temperature field of fluid and heat transfer area is calculated with the C# programming language in the visual studio program developed platform, using the Gauss–Seidel iterative method. This is the three-dimensional numerical heat transfer model established in this section based on transient heat balance theory. It can provide more accurate description of the heat transfer process of the cylinder helix energy pile, thus providing design guidance for the actual project.

2.2. Heat transfer characteristic along the pipe

For helix energy piles, the heat transfer along the pipe and the geothermal temperature distribution characteristics are of great significance to the optimal design of the heat exchangers. The engineering significance of pitch, pile diameter, flow rate and other design parameters can be obtained from the perspective of efficient heat transfer. Based on the establishment of the three-dimensional numerical heat transfer model of the helix energy pile mentioned above, this section analyzes the fluid heat transfer along the pipe and geothermal temperature distribution characteristics and lays a theoretical foundation for its optimization design.

For helix energy piles, since the fluid flows downward along the tube length spirally in the spiral tube, the general direction of the fluid flow is vertically downward; there is no "reverse" flow between the fluids, the fluid temperature in the tube has the characteristic of gradually decreasing or increasing along the flow direction of the tube. And according to the geometric characteristics of the helix energy pile, the spiral pipe is divided into four stages of heat transfer along the process, as shown in **Figure 5**.

1. Entrance stage (L1): In the vicinity of the inlet part of the helix pipe, the top of the pipe is in direct contact with the covered area, and there is no heat exchange pipe in the covered area. In addition to transferring heat (cold) in the radial direction, fluid in the pipe also transfers heat (cold) to the covered area in the axial direction, so this stage has a larger heat exchange capacity. However, as the flow progresses, the distance between the helical fluid and the covered area is further away; the axial heat transfer gradually decreases, and the thermal short circuit between the adjacent spiral coils in the axial direction becomes stronger and stronger. The heat transfer efficiency gradually decreases, resulting in a gradual decline in heat transfer. Define the above stage as the entrance stage; the heat exchanger length is L1.

2. Thermal short circuit stage (L2): With the increase of the distance along the pipeline, the entrance stage is over; the phenomenon of thermal short circuit between adjacent spiral pipes in the axial direction is serious, and the amount of cold (heat) emitted from the spiral fluid in the axial direction to the soil is less and less. The amount of cold (heat) is mainly transmitted in the diameter direction. Due to the limited volume of the backfill area and the large temperature difference between the fluid in the initial stage of the spiral pipe and the surrounding soil/backfill, the intensity of the heat transfer is relatively large and the amount of cold (heat) emitted to the backfill area aggregates, resulting in the backfill area being short circuited seriously in the diameter direction, and this part is defined as the thermal short circuit stage; the length is L2. At this stage, it has the following salient

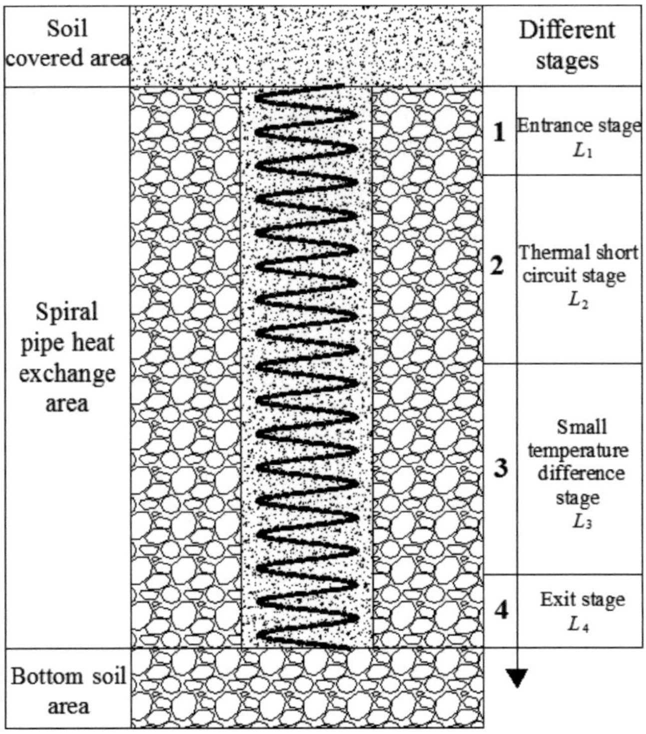

Figure 5. Heat transfer characteristic along the pipe.

features: the thermal short circuit effect in the diameter direction of the backfill area and the axial thermal short circuit effect are serious, but the thermal short circuit phenomenon in the diameter direction of the backfill area is gradually weakened and the heat exchange capacity gradually increases.

3. Small temperature difference stage (L3): When the thermal short circuit stage is over, the thermal short circuit phenomenon in the backfill area will be weak and the fluid temperature will decrease or increase further, resulting in lower heat transfer temperature difference between the fluid and the surrounding area; thus, the heat exchange capacity of the heat exchanger is gradually reduced. This stage is defined as the small temperature difference stage and the length is L3. This stage has the following salient features: temperature difference between fluid and rock and soil is small, capacity of geothermal storage (heat) is weak and heat transfer capacity gradually decreases.

4. Exit stage (L4): As the small temperature difference stage is over, the fluid is close to the outlet part of the helix pipe. This stage is defined as the exit stage and its length is L4. The exit stage is similar to the entrance stage. Since the spiral pipe near the exit is in direct contact with the soil at the bottom of the buried pipe, the axial thermal short circuit effect is weak. In addition to the heat exchange in the diameter direction, the fluid in the pipe also exchanges heat with the soil at the bottom of the ground tube. And as the fluid flows along

the pipe, the distance between the spiral fluid and the bottom of the soil gradually shortens, the axial thermal short circuit effect becomes weak; thus, the heat transfer capacity gradually increases.

Taking the case of summer heat release condition as an example, under the condition of constant inlet water temperature, the proposed three-dimensional numerical heat transfer model is used to simulate the fluid and soil temperature of the cylinder helix energy pile. And the four stages defined above are discussed. The main parameters of simulation are shown in **Table 1**. According to the thermo-physical properties of dense clay, the thermal conductivity of rock and soil is 1.2 W/(m·°C) and the specific heat of volume is 1680 kJ/(m^3·°C). The thermal conductivity of backfilling material is 1.5 W/(m·°C), the specific heat of volume is 2200 kJ/(m^3·°C) and the thermal conductivity of the spiral pipe is 0.45 W/(m·°C).

Figure 6 shows the temperature distribution of the fluid at τ = 48 h. It can be seen from the figure that the temperature of the fluid decreases gradually along the length of the pipe. The rate of change of fluid temperature at the entrance and exit stages is obviously larger, indicating that at the inlet and outlet heat transfer stage, the axial heat transfer is more prominent.

Parameters	Unit	Value
Inlet water temperature	°C	35
Water flow rate	m/s	0.3
Running time	h	48
Cover soil depth	m	2
Buried depth of the pile	m	8
Diameter of the pile	m	1.5
Pitch	m	0.15

Table 1. Main parameters of simulation.

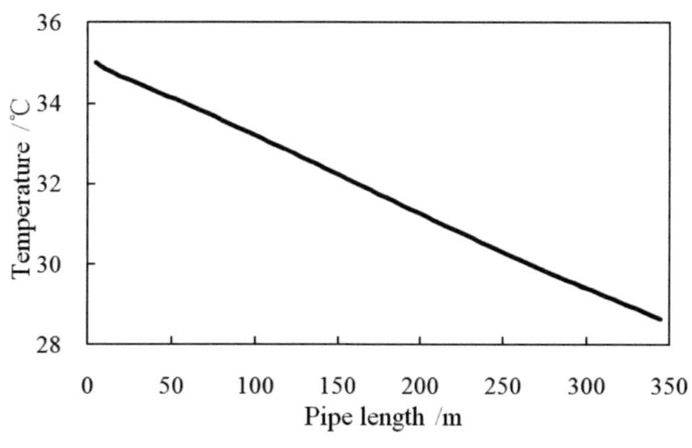

Figure 6. Fluid temperature distribution along the flow direction.

Figure 7 and **Figure 8** show the changes of fluid heat transfer along the pipe and the vertical geothermal temperature distribution; we can clearly distinguish the four heat transfer stages proposed above.

1. Entrance stage (0–30 m): According to the vertical geothermal temperature field given in **Figure 8**, it can be seen that the soil temperature in the affected area at this stage has a large temperature gradient in the axial and radial directions, resulting in a significantly larger

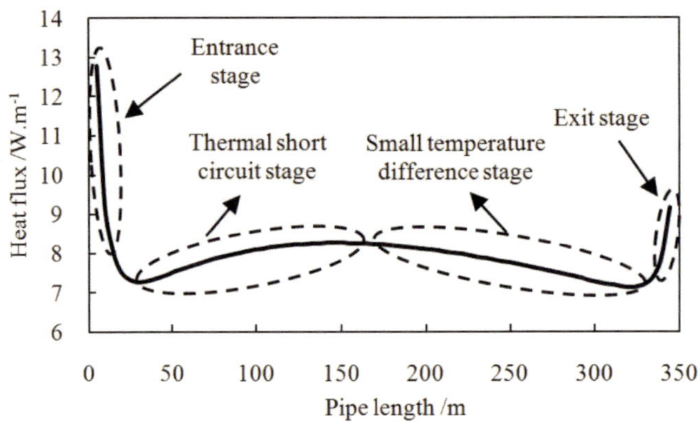

Figure 7. Fluid heat flux along the flow direction.

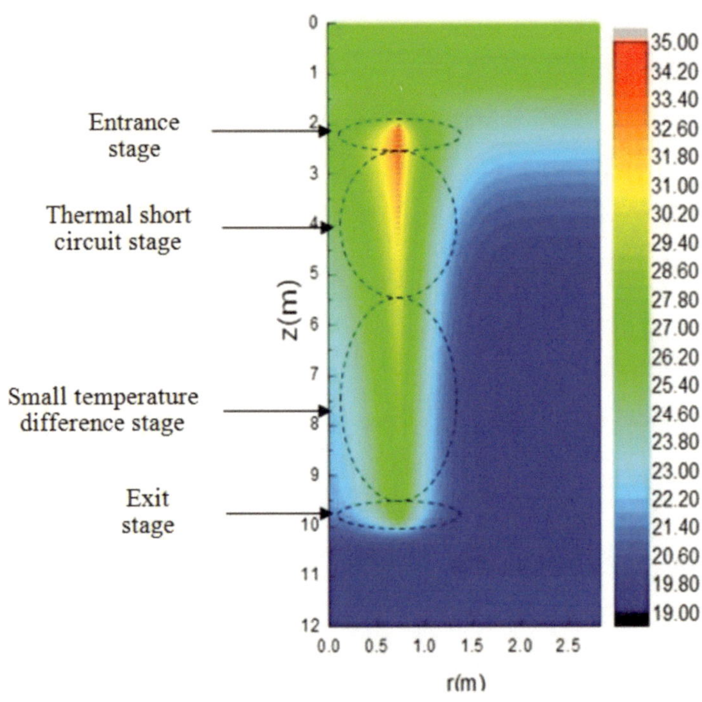

Figure 8. Temperature field on *r-z* plane.

heat transfer at this stage. However, as the distance between the fluid and the covered area increases, the axial thermal interference between the adjacent spiral coils increases and the amount of heat exchange decreases rapidly.

2. Thermal short circuit stage (30–150 m): Due to the high fluid temperature at this stage, a large amount of heat is transferred to the backfill area and the volume of the backfill area is limited, resulting in the accumulation of heat in the backfill area with high temperature and strong thermal short circuit effect in the diameter direction of the backfill area. However, as shown in **Figure 8**, with the gradual increase of the distance along the distance, the temperature in the backfill zone gradually decreases; the thermal influence in the backfill area decreases, the thermal effect radius decreases, the thermal short circuit effect gradually decreases and the heat transfer capacity gradually increases.

3. Small temperature difference stage (150–320 m): After the heat exchange at the entrance and thermal short circuit stages, the fluid temperature is lower and the heat transfer temperature difference between the fluid and the surrounding heat transfer area is gradually reduced, and the effect of cold (heat) storage in the soil is poor, resulting in the decrease of heat transfer capacity at this stage.

4. Exit stage (320–344 m): This stage is similar to the entrance stage. The fluid is in direct contact with the soil at the bottom of the spiral pipe. The soil temperature has a large temperature gradient both in radial direction and in axial direction, and the heat transfer effect is better. With the increase of the distance along the path, the closer the distance between the spiral fluid and the bottom soil, the more obvious the axial heat transfer effect is, and the heat transfer capacity gradually increases.

It can be seen from the analysis of the four stages that the entrance stage, thermal short circuit stage and exit stage belong to certain stages, and these three stages are beneficial for the whole heat exchange, especially for the entrance and exit stage; the heat transfer capacity is higher than the thermal short circuit and small temperature difference stages. For the small temperature difference stage, the heat transfer capacity decreases along the pipe length, so in the actual project design, it is necessary to reduce the proportion of the small temperature difference stage to the total heat exchange tube length and to maximize the other three stages, especially the entrance and exit stages.

2.3. Heat transfer capacity analysis

Based on the proposed three-dimensional numerical heat transfer model, this section analyzes and discusses the influence of different design parameters (pile diameter, pile depth and pitch) on cylinder helix energy pile, providing effective and direct theoretical guidance for the actual project.

2.3.1. Simulation conditions

The method of constant inlet water temperature is adopted to analyze and discuss the heat exchange capacity of the cylinder helix energy pile. The inlet temperature is 35°C and the flow rate is 0.3 m/s. According to the thermo-physical properties of dense clay, the thermal conductivity of rock and soil is 1.2 W/(m·°C) and the volumetric specific heat is 1680 kJ/(m^3·°C). The thermal conductivity of backfill material is 1.5 W/(m·°C), volumetric specific heat is 2200 kJ/(m^3·°C) and the thermal conductivity of the spiral pipe is 0.45 W/(m·°C).

2.3.2. *Effect of pile diameter on heat transfer performance*

Figure 9a shows the effect of pile diameter on heat flux per unit pipe length of the pile. The analysis shows that with the increase of pile diameter, the heat flux per unit pipe length decreases gradually, but the variation range is relatively small, which indicates that although the increase of pile diameter will cause the length of the spiral pipe to increase approximately proportionally, however, due to the increase of pile diameter, the phenomenon of thermal short circuit in the backfill area is weakened, which will weaken the influence on heat flux per unit pipe length. Taking the pile depth $H = 8$ m as an example, pitch b is, respectively, 0.1 and

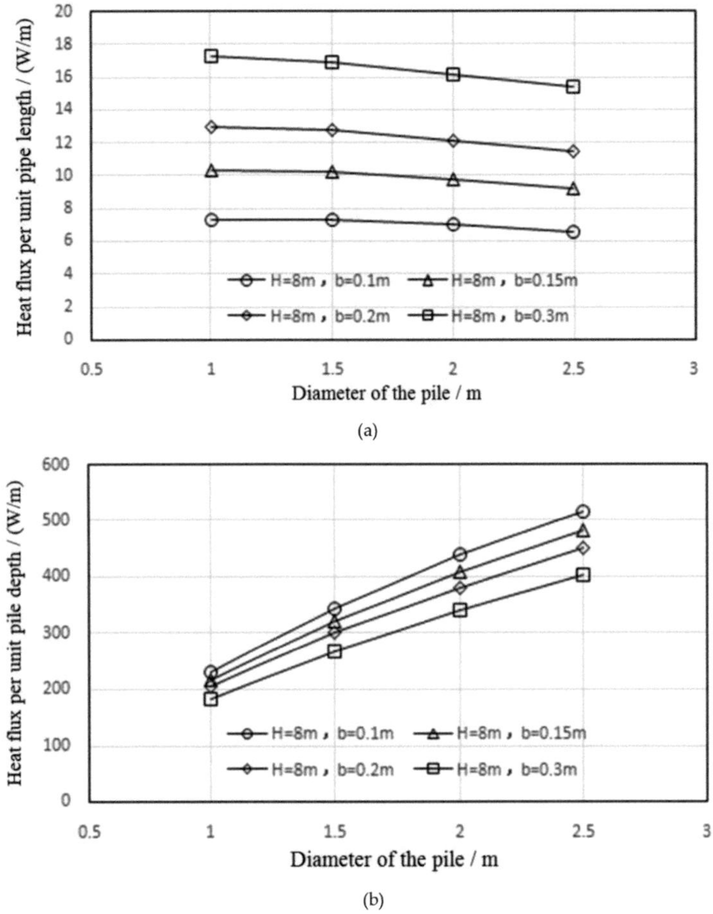

Figure 9. Effect of pile diameter on heat transfer performance. (a) Heat flux per unit pipe length. (b) Heat flux per unit pile depth.

0.3 m; when the pile diameter increases from 1 to 2.5 m, the average heat flux per unit pipe length decreases from 7.28 to 6.58 W/m, from 17.30 to 15.36 W/m, a relative decrease of 9.6 and 11.2%, respectively. The pile diameter increases by 0.1 m. The average heat flux per unit pipe length decreases by 0.047 and 0.129 W/m, respectively.

Figure 9b shows the effect of pile diameter on the heat flux per unit pile depth of the pile. It can be seen from the figure that with the increase of pile diameter, the heat flux per unit pile depth increases approximately linearly, which is due to the increase of buried pipe length, that is, the contact area between spiral buried pipe and surrounding area increases, resulting in the heat transfer amount of the pile greatly increasing; thus, the heat flux per unit pile depth increases approximately linearly in proportion. Taking the pile depth $H = 8$ m as an example, the pitch b is, respectively, 0.1 and 0.3 m; when the pile diameter increases from 1 to 2.5 m, the average heat flux per unit pile depth increases from 228.8 to 516.3 W/m, from 182.0 to 402.3 W/m, an increase of 1.26 times and 1.21 times, respectively. The pile diameter increases by 0.1 m. The average heat flux per unit pile depth increases by 19.2 and 14.7 W/m, respectively.

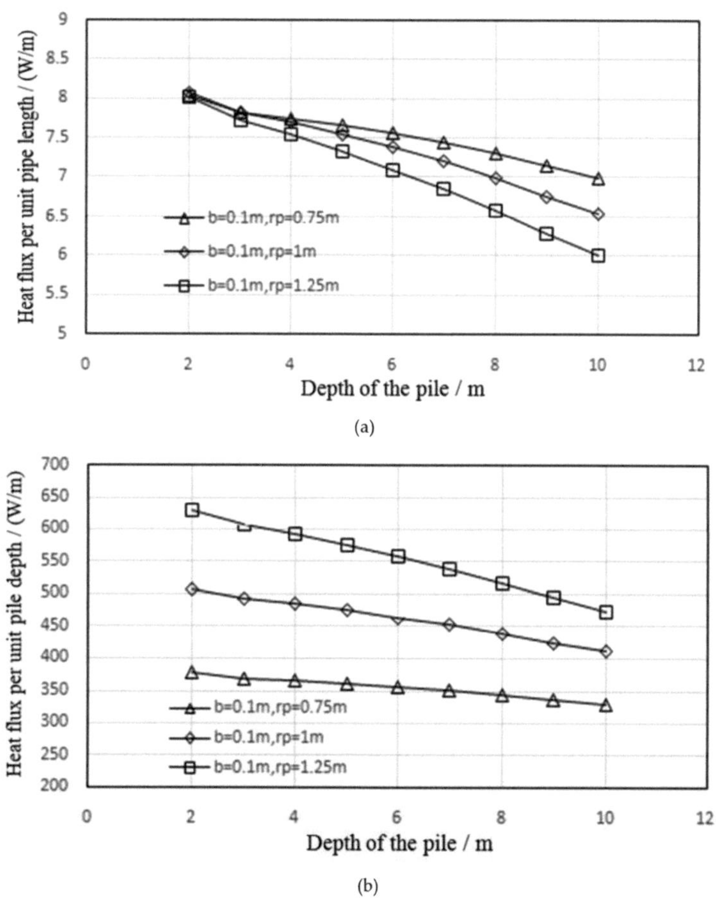

Figure 10. Effect of pile depth on heat transfer performance. (a) Heat flux per unit pipe length. (b) Heat flux per unit pile depth.

2.3.3. *Effect of pile depth on heat transfer performance*

Figure 10 shows the effect of different pile depths on the heat transfer performance under different pile diameters with pitch b = 0.1 m. As can be seen from the figure, with other conditions constant, the deeper the heat transfer pile, the longer the tube length, the lower the outlet temperature, the smaller the average heat transfer temperature difference and the smaller the heat flux per unit pipe length; at the same time, the increase of the pile depth will further increase the speed of the decrease of the heat flux per unit pipe length. Taking pile radius $r_p = 0.75$ and 1.2 m as an example, the heat flux per unit pipe length decreases from 8.03 to 6.99 W/m and 8.02 to 6.02 W/m when the pile depth H increases from 2 to 10 m, a decline of 12.9 and 25.0%, respectively.

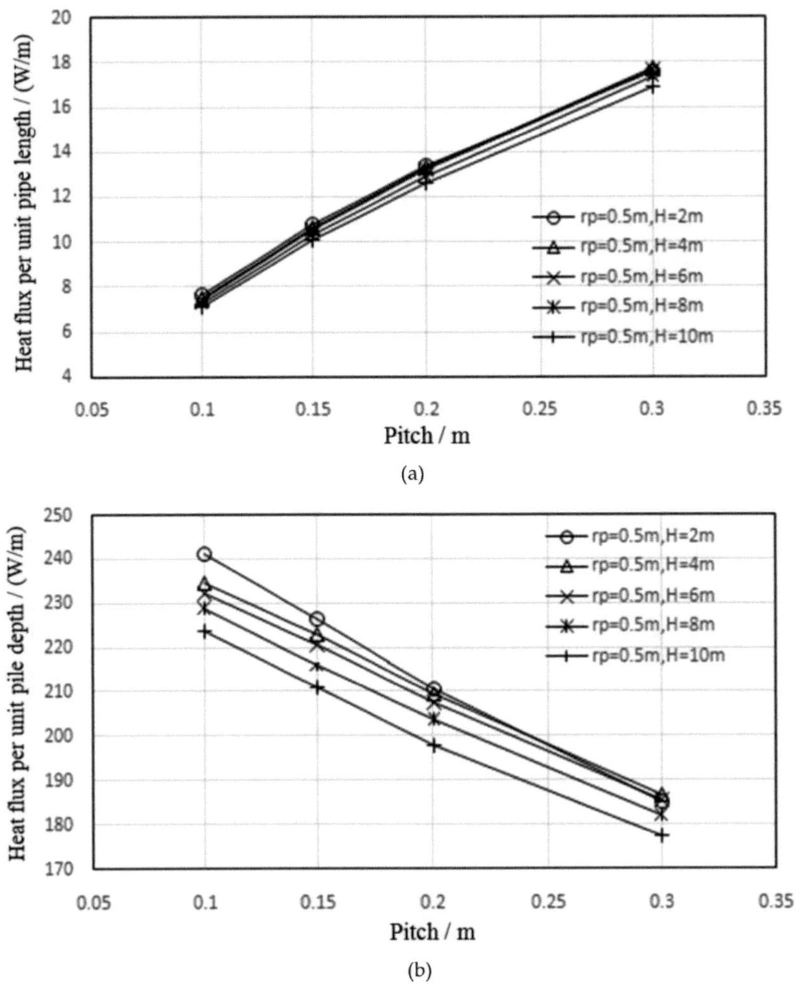

Figure 11. Effect of pitch on heat transfer performance. (a) Heat flux per unit pipe length. (b) Heat flux per unit pile depth.

Similarly, for the heat flux per unit pile depth, due to the increase of pile depth, resulting in lower average heat transfer temperature difference, heat flux per unit pile depth decreases. In addition, under different pile diameter conditions, the difference of heat flux per unit pile depth is larger, and the larger the pile diameter, heat flux per unit pile depth decreases. Taking $r_p = 0.75$ and 1.25 m as an example, when pile depth H increases from 2 to 10 m, the heat flux per unit pile depth decreases from 378.3 to 329.2 W/m and from 629.6 to 472.2 W/m, a decline of 12.9 and 25.0%, respectively.

2.3.4. *Effect of pitch on heat transfer performance*

Figure 11 shows the effect of pitch on heat transfer performance. Analysis finds that the pitch has greater impact on the heat flux per unit pipe length. Due to the increase of the pitch, the distance between the spiral tubes increases, and the axial short circuit phenomenon of the adjacent spiral tubes is weakened; thus, the heat flux per unit pipe length increases. Taking the pile radius $r_p = 0.5$ m as an example, when the pile depth $H = 8$ m, the pitch increases from 0.1 m to 0.3 m; the heat flux per unit pipe length increases from 7.28 to 17.31 W/m, an increase of 1.38 times. For each 0.1 m increase in pitch, the heat flux per unit pipe length increases by an average of 3.34 W/m.

Although the heat flux per unit pipe length increases approximately linearly with the increase of pitch, the heat flux per unit pile depth is approximately linearly reduced. The increase of pitch will reduce the axial thermal short circuit; thus, the effective heat transfer tube length is reduced for a certain size of the energy pile, that is, the effective heat transfer area of the fluid and the soil is reduced; thus, the total heat transfer amount of the pile is reduced. Taking the pile radius $r_p = 0.5$ m as an example, when the pile depth $H = 8$ m, the pitch increases from 0.1 to 0.3 m; the heat flux per unit pile depth decreases from 228.8 to 182.0 W/m, decreases by 20.5%. For every 0.1 m increase in pitch, the heat flux per unit pile depth has an average of 15.6 W/m decrease.

3. Conclusion(s)

Both the analytical solution model and numerical solution model for CyHEP are built to discuss their dynamic characteristics of thermal interferences and heat transfer performance. The following conclusions can be drawn.

1. Four heat exchange stages for the spiral pile geothermal heat exchanger along the fluid flow direction are revealed: inlet heat exchange stage, grout thermal short-circuiting stage, small temperature difference stage and outlet heat exchange stage. Each stage has corresponding heat transfer characteristics, and reducing the length of small temperature difference stage and increasing the other stages would enhance the heat exchange of spiral geothermal ground heat exchanger.

2. As the pile diameter increases, the heat transfer per unit tube length decreases, and the heat exchange per unit pile depth increases. For every 0.1 m increase in pile diameter, the

heat transfer per unit tube length decreases by an average from 0.047 to 0.129 W/m, and the unit heat exchange per unit pile depth increases by 14.7 to 19.2 W/m.

3. As the pile depth increases, the heat transfer per unit tube length and the heat exchange per unit pile depth are reduced. During the process of increasing H from 2 m to 10 m, the heat exchange per unit tube length and the heat exchange per unit pile depth are reduced from 1.04 to 2.00 W/m and 49.1–157.4 W/m, respectively.

4. As the pitch increases, the heat transfer per unit tube length increases, and the heat exchange per unit pile depth decreases. For a 0.1 m increase in pitch, the heat transfer per unit tube length increases by 3.34 W/m, and the unit heat exchange per pile depth decreases by 15.6 W/m.

Acknowledgements

I would like to express my gratitude to all those who helped me during the writing of this book. And I feel grateful to all the teachers in the Army Logistical University of PLA who once offered me valuable courses and advice during my study. Last, my thanks go to my beloved family for their loving considerations and great confidence in me all through these years. I also owe my sincere gratitude to my friends who gave me their help and time in listening to me and helping me work out my problems during the difficult course of the book.

This work received support from the Research Initiative for Basic Science and Frontier Technology of Chongqing, China [cstc2016jcyjA0496], Natural Science Foundation of China [51706243].

Conflict of interest

We declare that we have no financial and personal relationships with other people or organizations that can inappropriately influence our work; there is no professional or other personal interest of any nature or kind in any product, service and/or company that could be construed as influencing the position presented in, or the review of, our work.

Appendices and Nomenclature

x, y, z	Cartesian coordinate (m)
r, φ, z	cylindrical coordinate (m)
R, φ, Z	dimensionless cylindrical coordinate
h	height (m)
H	dimensionless height

http://dx.doi.org/10.5772/intechopen.76820

r	radial coordinate (m)
R	dimensionless radial coordinate
q_l	heating rate per length of pipe (W s^{-1})
τ	the time (s)
α_s	thermal diffusivity (m^{-2} s)
d_p	distance from the heat source point to the calculated point (m)
d_n	distance from the heat sink point to the calculated point (m)
D_p	dimensionless distance from the heat source point to the calculated point (m)
D_n	dimensionless distance from the heat sink point to the calculated point (m)
ρ	density (kg m^{-3})
c	specific heat (J kg^{-1} K^{-1})
λ	thermal conductivity (W m^{-1} K^{-1})
$\triangle T$	the temperature rise (K)
R	thermal resistance (m^{-1} K^{-1} W)
L	Length of helix pipe

Greek symbols

θ	cone angle (rad)
φ	spiral angle (rad)
Fo	Fourier number
η	influence coefficient
Θ	dimensionless temperature rise

Superscript

$'$	integration parameter
ave.	the average value
CyHEP	cylinder helix energy pile
CoHEP	truncated cone helix energy pile
t	top surface of pile
b	base surface of pile
mi	middle surface of pile

i	the index of arc
j	the index of coil
k, n	the index of time
p	Pipe
f	fluid
g	ground

Author details

Guangqin Huang*, Yajiao Liu, Xiaofeng Yang and Chunlong Zhuang

*Address all correspondence to: hgq880818@163.com

Department of Military Installation, Army Logistical University of PLA, Chongqing, China

References

[1] Hamada Y, Saitoh H, Nakamura M, Kubota H, Ochifuji K. Field performance of an energy pile system for space heating. Energy and Buildings. 2007;**37**:517-524. DOI: 10.1016/j.enbuild.2006.09.006

[2] Gao J, Xu Z, Liu J, Li KS, Yang J. Thermal performance and ground temperature of vertical pile-foundation heat exchangers: A case study. Applied Thermal Engineering. 2008;**28**: 2295-2304. DOI: 10.1016/j.applthermaleng.2008.01.013

[3] Cecinato F, Loveridge FA. Influences on the thermal efficiency of energy piles. Energy. 2015; **82**:1021-1033. DOI: 10.1016/j.energy.2015.02.001

[4] Park H, Lee S-R, Yoon S, Choi J-C. Evaluation of thermal response and performance of PHC energy pile: Field experiments and numerical simulation. Applied Energy. 2013;**103**:12-24. DOI: 10.1016/j.apenergy.2012.10.012

[5] Zarrella A, De Carli M. Heat transfer analysis of short helical borehole heat exchangers. Applied Energy. 2013;**102**:1477-1491. DOI: 10.1016/j.apenergy.2012.09.012

[6] Zarrella A, De Carli M, Galgaro A. Thermal performance of two types of energy foundation pile: Helical pipe and triple U-tube. Applied Thermal Engineering. 2013;**2**:301-310. DOI: 10.1016/j.applthermaleng.2013.08.011

[7] Zarrella A, Capozza A, De Carli M. Analysis of short helical and double U-tube borehole heat exchangers: A simulation-based comparison. Applied Energy. 2013;**112**:358-370. DOI: 10.1016/j.apenergy.2013.06.032

Chapter 13

Heat Transfer of Helix Energy Pile: Part 2—Novel Truncated Cone Helix Energy Pile

Guangqin Huang, Yajiao Liu, Xiaofeng Yang and Chunlong Zhuang

Additional information is available at the end of the chapter

http://dx.doi.org/10.5772/intechopen.76821

Abstract

Owing to the fact that severe thermal interferences exist in the radial and generatrix directions of the traditional cylinder helix energy pile due to the limited thermal heat capacity of the pile and small ratio between coil pitch and radius of pile, therefore, a novel truncated cone helix energy pile (CoHEP) is presented to weaken the thermal interferences and improve the heat transfer efficiency. Further, both the analytical solution model and numerical solution model for CoHEP are built to discuss the dynamic characteristics of thermal interferences and heat transfer performance. The results indicate that the thermal interference of CoHEP is dynamic. The thermal interference in the upper part of the CoHEP is much smaller than the traditional CyHEP. And in general the heat flux per unit pipe length of the novel CoHEP is larger than that of the traditional CyHEP. Heat flux per unit pipe length of the CoHEP increases linearly with inlet water temperature. For the same inlet water temperature, the thermal short circuit is serious at the bottom of the CoHEP, and it's weak in the upper part of CoHEP. Also it's obvious that as the inlet water temperature increases, the thermal short circuit becomes more serious.

Keywords: truncated cone, helix energy pile, analytical, numerical, thermal interferences, stages

1. Introduction

Although the CyHEP has the characteristics of large heat transfer area and large heat transfer, the results of Park et al. [1, 2] show that the traditional CyHEP has a close distance between the adjacent tubes axially, which lead to a serious thermal interference phenomenon. Even when the pitch is small enough, the relative heat transfer efficiency is lower than the U-type ground heat exchanger. In addition, Yang et al. [3] conducted a laboratory investigation of the traditional CyHEP and found that the reduction in pitch would increase the total heat transfer but reduce the unit length of heat transfer. Thus it can be speculated that the distance between the adjacent tubes axially determines the thermal interference of the exchanger which plays an important role in the heat transfer efficiency. It can be found that at present researchers mainly study the mathematical models of the traditional CyHEP; there is little work about the thermal interference and heat transfer characteristics. Thus, in this section, for the sake of reducing thermal interference and improving heat transfer efficiency, a novel truncated cone helix energy pile (CoHEP) was proposed as shown in **Figure 1**.

As shown in **Figure 2**, the structure of novel CoHEP is characterized by the fact that the spiral heat transfer tube is wound on the wall of the truncated cone, and its spiral radius decreases linearly with the increase of the depth. **Figure 2** shows the physical mode of CoHEP. The cone angle is θ, the pitch in the depth direction is b, the distance between the adjacent tube is d; thus, the relationship between b and d is: $b = d{\cdot}\cos\theta$. What's more, the bottom radius is r_b, the top radius is r_t, $r_t > r_b$. The distance from the top of the structure to the ground surface is h_t and the distance from the bottom of the structure to the ground surface is h_b. And the spiral heat transfer tube is distributed according to the following equation:

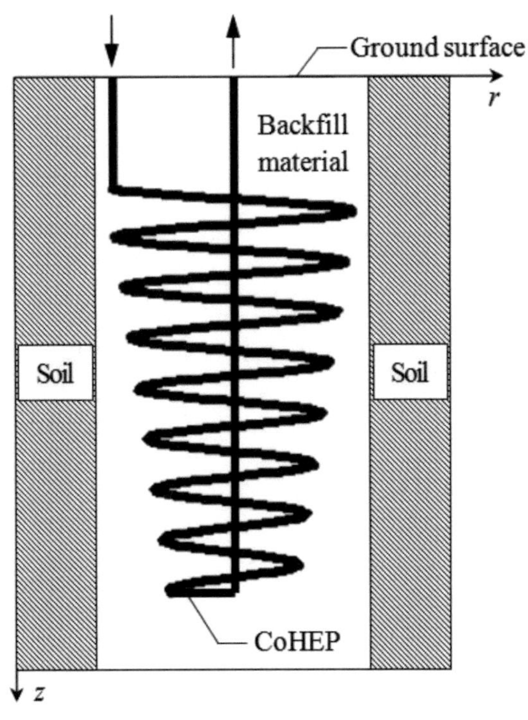

Figure 1. Truncated cone helix energy pile (CoHEP).

$$\begin{cases} r = [r_b + (h_b - z) \cdot \tan\theta] \\ z = b \cdot \varphi/2\pi \end{cases} \tag{1}$$

where (r, ϕ, z) is the coordinate of a certain point of the truncated cone spiral coil structure in the column coordinate system.

Compared with the traditional CyHEP, the novel CoHEP proposed in this chapter has the following advantages:

1. Due to the fact that the spiral heat transfer tube of the novel CoHEP is wound on the wall of the truncated cone, there is a certain angle between the generatrix of the truncated cone and the vertical direction. Compared with the vertical generatrix of the cylinder, this can effectively reduce the risk of drilling wall collapse during construction.

2. When the fluid enters the heat transfer tube, the temperature difference between the top fluid and the soil is large and the radial thermal interference is strong which will weaken the heat transfer efficiency. Therefore, the novel CoHEP is designed with a larger top radius compared with the traditional CyHEP. This will effectively alleviate the radial thermal interference and improve thermal efficiency.

3. In the case where the pitch in the depth direction (b) is the same, the distance between the adjacent tube of CoHEP (d) can be calculated as: $d = b/\cos\theta$, which is obviously larger than the distance between the adjacent tube of CyHEP (b). It can effectively reduce the axial thermal interference and improve heat transfer efficiency.

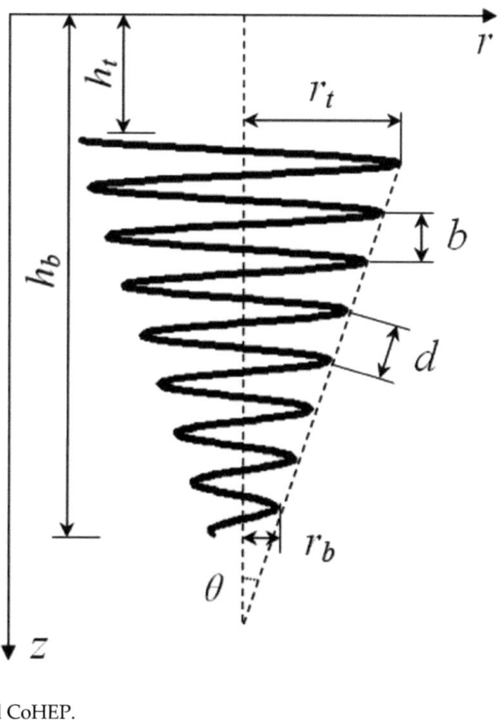

Figure 2. Structure of the novel CoHEP.

2. Heat transfer model and characteristics of the novel truncated cone helix energy pile (CoHEP)

2.1. Analytic solution model of CoHEP

2.1.1. Analytic solution model

To investigate the thermal performance of CoHEP, an analytical solution model of CoHEP based on Green's function is presented and the thermal interference of CoHEP is investigated in the manuscript.

For modeling the heat transfer of CoHEP, the following assumptions are made:

1. The medium is assumed to be a homogeneous infinite medium and the difference between soil and pile is ignored. The medium thermal properties do not change with the variation of temperature.
2. The medium has a uniform initial temperature, T_0.
3. The ground surface at $z = 0$ maintains a constant temperature, T_0.
4. Helix tube is regarded as a concentric spiral coil line with heating rate per length of pipe as ql. Mass, heat capacity, and thickness of the spiral coils' heat source are neglected.

The method of images for conduction questions is employed to maintain a constant temperature at the ground surface $z = 0$. Imagine that a mirror-image truncated cone spiral line of heat sink exists with heat rate $-q_l$, as shown in **Figure 3**.

On the condition of the above assumptions, the Green's function in the cylindrical coordinates could be written as Eq. (2) in the infinite medium. The Green's function represents the temperature rise by effect of an instantaneous point heat source of unit strength generated at time τ' and at point (r', ϕ', z').

$$G\left(r',,\varphi',,z',r,\varphi,z,,\tau',\tau\right) = \frac{1}{8\left[\sqrt{\pi\alpha_s(\tau-\tau')}\right]^3}\exp\left[-\frac{r^2+r'^2-2rr'\cos(\varphi-\varphi')+(z-z')^2}{4\alpha_s(\tau-\tau')}\right] \quad (2)$$

where G is the Green's function; α_s is the thermal diffusivity; τ' and τ is the time when heat is first emitted and the calculated time, respectively; (r', ϕ', z') are the heat sources or sink-point coordinates; (r, ϕ, z) are the coordinates at the calculated point in the medium except the heat source. In fact, the term of $r^2 + r'^2 - 2rr'\cos(\phi-\phi') + (zi-z')^2$ is the square of distance from the calculated point to the point in the line of truncated cone spiral heat source or sink.

Based on the information given in **Figure 2**, the relationship among r', ϕ', z' would be described as Eq. (3) for truncated cone spiral heat source line with heat flux q_l:

$$\begin{cases} r' = \left[r_{mi} + \left(h_{mi} - b\varphi'/2\pi\right)\cdot \mathrm{tg}\,\theta\right] \\ z' = b\varphi'/2\pi \end{cases} \quad (3)$$

http://dx.doi.org/10.5772/intechopen.76821

Similarly, the relationship among coordinates of the mirror-image truncated cone spiral line of heat sinks is written as Eq. (4).

$$\begin{cases} r' = \left[r_{\text{mi}} + \left(h_{\text{mi}} - b\varphi'/2\pi\right) \cdot \text{tg}\,\theta\right] \\ z' = -b\varphi'/2\pi \end{cases} \tag{4}$$

Hence, the distance dp from the heat source point (r', ϕ', z') to the calculated point (r, ϕ, z) could be described as Eq. (5) based on Eq. (3) and Eq. (4).

$$d_{\text{p}} = \sqrt{\begin{array}{l} r^2 + \left[r_{\text{mi}} + \left(h_{\text{mi}} - b\varphi'/2\pi\right) \cdot \text{tg}\,\theta\right]^2 - 2r\left[r_{\text{mi}} + \left(h_{\text{mi}} - b\varphi'/2\pi\right) \cdot \text{tg}\,\theta\right] \\ \times\cos\left(\varphi - \varphi'\right) + \left(z - b\varphi'/2\pi\right)^2 \end{array}} \tag{5}$$

Accordingly, the distance dn from the heat sink point to the calculated point could be written as Eq. (6).

$$d_{\text{n}} = \sqrt{\begin{array}{l} r^2 + \left[r_{\text{mi}} + \left(h_{\text{mi}} - b\varphi'/2\pi\right) \cdot \text{tg}\,\theta\right]^2 - 2r\left[r_{\text{mi}} + \left(h_{\text{mi}} - b\varphi'/2\pi\right) \cdot \text{tg}\,\theta\right] \\ \times\cos\left(\varphi - \varphi'\right) + \left(z + b\varphi'/2\pi\right)^2 \end{array}} \tag{6}$$

The temperature rise in the heat transfer medium is induced by the effects of spiral heat source and heat sink. Therefore, when the CoHEP emits heat at the intensity of q_l from time $\tau' = 0$, the

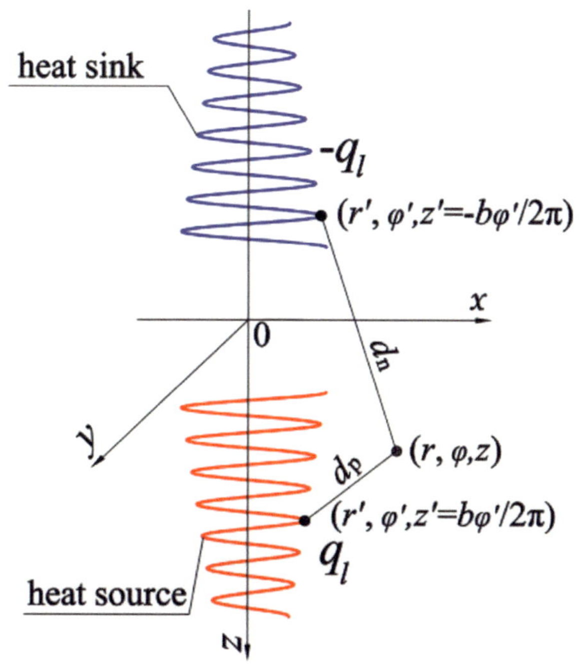

Figure 3. Schematic diagram of mirror heat source method.

temperature rise Δt can be obtained by integrating Green's function Eq. (2) with respect to the time τ' and the spiral line of heat source and sink.

$$\begin{aligned}\Delta T(r,\varphi,z,\tau) &= \frac{q_l}{\rho c}\int_0^{\tau}\int_L G\left(r',\varphi' z' = b\varphi'/2\pi r\varphi z, \tau' \tau\right) - G\left(r',\varphi' z' = -b\varphi'/2\pi r\varphi z, \tau' \tau\right)\mathrm{d}\tau \mathrm{d}l \\ &= \frac{q_l}{\rho c}\int_0^{\tau}\frac{1}{8\left[\sqrt{\pi\alpha_s(\tau-\tau')}\right]^3}\mathrm{d}\tau' \int_L \left\{\exp\left[-\frac{d_p{}^2}{4\alpha_s(\tau-\tau')}\right] - \exp\left[-\frac{d_n{}^2}{4\alpha_s(\tau-\tau')}\right]\right\}\mathrm{d}l\end{aligned} \tag{7}$$

where $\triangle T$ is the temperature rise; ρ is the density; c is the specific heat; L stands for the spiral line; and d_p and d_n are the above-defined ones.

Denoting $u = 1/\left[2\sqrt{\alpha_s(\tau-\tau')}\right]$ to make Eq. (7) concise, τ' is a function of u:

$$\tau' = \tau - 1/\left(4\alpha_s u^2\right) \tag{8}$$

Differential form of Eq. (8) would be achieved:

$$\mathrm{d}\tau' = 1/\left(2\alpha_s u^3\right)\mathrm{d}u \tag{9}$$

Thus, $u = 1/\left(2\sqrt{\alpha_s\tau}\right)$ when $\tau' = 0$ and $u = \infty$ when $\tau' = \tau$. Then, Eq. (7) can be transformed into another expression as follows.

$$\Delta T(r,\varphi,z,\tau) = \frac{q_l}{2\pi^{3/2}\lambda}\int_{1/(2\sqrt{\alpha\tau})}^{\infty}\mathrm{d}u\int_L\left\{\exp\left[-d_p{}^2u^2\right] - \exp\left[-d_n{}^2u^2\right]\right\}\mathrm{d}l \tag{10}$$

where λ is the thermal conductivity.

Eq. (10) is an expression of integrating with respect to the spiral coil line and is improper for computation. Thus, the transformation from the integration variable l to spiral angle ϕ' is necessary. The expression of the truncated cone spiral coil line in the Cartesian coordinates can be written as Eq. (11).

$$\begin{cases} x = \left[r_{mi} + \left(h_{mi} - b\varphi'/2\pi\right)\cdot \mathrm{tg}\,\theta\right]\cos\varphi' \\ y = \left[r_{mi} + \left(h_{mi} - b\varphi'/2\pi\right)\cdot \mathrm{tg}\,\theta\right]\sin\varphi' \\ z = \pm b/2\pi \end{cases} \tag{11}$$

Thus, the differential $\mathrm{d}l$ can be calculated by Eq. (12).

$$\mathrm{d}l = \sqrt{\mathrm{d}x^2 + \mathrm{d}y^2 + \mathrm{d}z^2} = \sqrt{\begin{matrix}\left[r_{mi} + \left(h_{mi} - b\varphi'/2\pi\right)\cdot \mathrm{tg}\,\theta\right]^2 \\ +b^2/4\pi^2\left(1 + (\mathrm{tg}\,\theta)^2\right)\end{matrix}}\,\mathrm{d}\varphi' \tag{12}$$

Denoting $f(\varphi') = \sqrt{\left[r_{mi} + \left(h_{mi} - b\varphi'/2\pi\right)\cdot \mathrm{tg}\,\theta\right]^2 + b^2/4\pi^2(1 + \mathrm{tg}\,\theta^2)}$ and then $\mathrm{d}l = f(\varphi')\mathrm{d}\varphi'$, Eq. (10) can be converted to Eq. (13).

$$\Delta T(r,\varphi,z,\tau)=\frac{q_l}{2\pi^{3/2}\lambda}\int_{2\pi h_t/b}^{2\pi h_b/b} f(\varphi')\mathrm{d}\varphi'\int_{1/(2\sqrt{\alpha\tau})}^{\infty}\left\{\exp\left[-{d_p}^2u^2\right]-\exp\left[-{d_n}^2u^2\right]\right\}\mathrm{d}u \tag{13}$$

Denoting $d_p u = x_p$ and $d_n u = x_n$, Eq. (14) can be obtained.

$$\Delta T(r,\varphi,z,\tau)=\frac{q_l}{2\pi^{3/2}\lambda}\int_{2\pi h_t/b}^{2\pi h_b/b} f(\varphi')\mathrm{d}\varphi'\left\{\begin{array}{l}\frac{1}{d_p}\cdot\int_{d_p/(2\sqrt{\alpha_s\tau})}^{\infty}\exp\left[-{x_p}^2\right]\mathrm{d}x_p\\ -\frac{1}{d_n}\cdot\int_{d_n/(2\sqrt{\alpha_s\tau})}^{\infty}\exp\left[-{x_n}^2\right]\mathrm{d}x_n\end{array}\right\} \tag{14}$$

According to the characteristics of the error function, the more concise expression can be achieved.

$$\Delta T(r,\varphi,z,\tau)=\frac{q_l}{4\pi\lambda}\int_{2\pi h_t/b}^{2\pi h_b/b} f(\varphi')\left[\frac{1}{d_p}\cdot\operatorname{erfc}\left(\frac{d_p}{2\sqrt{\alpha_s\tau}}\right)-\frac{1}{d_n}\cdot\operatorname{erfc}\left(\frac{d_n}{2\sqrt{\alpha_s\tau}}\right)\right]\mathrm{d}\varphi' \tag{15}$$

In which, $\operatorname{erfc}(x)=1-\frac{2}{\sqrt{\pi}}\int_0^x\exp(-u^2)\mathrm{d}u$ is the complementary error function. The final analytical solution of CoHEP can be obtained by employing $f(\phi')$ into Eq. (15).

$$\Delta T(r,\varphi,z,\tau)=\frac{q_l}{4\pi\lambda}\int_{2\pi h_t/b}^{2\pi h_b/b}\sqrt{\left[r_{mi}+\left(h_{mi}-\frac{b\varphi'}{2\pi}\right)\cdot\operatorname{tg}\theta\right]^2+\frac{b^2}{4\pi^2}\left(1+(\operatorname{tg}\theta)^2\right)}\left[\begin{array}{l}\frac{1}{d_p}\cdot\operatorname{erfc}\left(\frac{d_p}{2\sqrt{\alpha_s\tau}}\right)\\ -\frac{1}{d_n}\cdot\operatorname{erfc}\left(\frac{d_n}{2\sqrt{\alpha_s\tau}}\right)\end{array}\right]\mathrm{d}\varphi' \tag{16}$$

To reduce the number of parameters and simplify the calculation, the corresponding non-dimensional parameters should be denoted: $\Theta = \lambda\Delta T/q_l$, $H_{mi} = h_{mi}/r_{mi}$, $B = b/ r_{mi}$, $R = r/ r_{mi}$, $R_b = r_b/ r_{mi}$, $Z = z/ r_{mi}$, $Fo = \alpha_s\tau/ {r_{mi}}^2$. Thus, the dimensionless expression of CoHEP can be obtained:

$$\Theta=\frac{1}{4\pi}\int_{2\pi H_t/B}^{2\pi H_b/B}\sqrt{\left[1+\left(H_{mi}-\frac{B\varphi'}{2\pi}\right)\cdot\operatorname{tg}\theta\right]^2+\frac{B^2}{4\pi^2}(1+\operatorname{tg}\theta^2)}\cdot\left[\begin{array}{l}\frac{1}{D_p}\cdot\operatorname{erfc}\left(\frac{D_p}{2\sqrt{Fo}}\right)\\ -\frac{1}{D_n}\cdot\operatorname{erfc}\left(\frac{D_n}{2\sqrt{Fo}}\right)\end{array}\right]\mathrm{d}\varphi' \tag{17}$$

In which

$$D_p=\sqrt{R^2+[1+(H_{mi}-B\varphi'/2\pi)\cdot\operatorname{tg}\theta]^2-2R[1+(H_{mi}-B\varphi'/2\pi)\cdot\operatorname{tg}\theta]\cos(\varphi-\varphi')+(Z-B\varphi'/2\pi)^2}$$

$$D_n=\sqrt{R^2+[1+(H_{mi}-B\varphi'/2\pi)\cdot\operatorname{tg}\theta]^2-2R[1+(H_{mi}-B\varphi'/2\pi)\cdot\operatorname{tg}\theta]\cos(\varphi-\varphi')+(Z+B\varphi'/2\pi)^2}$$

In fact, the dimensionless expression Eq. (17) will become Eq. (18) when the cone angle θ equals to zero. Further, we found that the form of Eq. (18) is also the analytical solution of ScM [4, 5], which is indicated that the ScM is a particular case of the CoHEP model presented in the section.

$$\Theta = \frac{1}{4\pi} \int_{2\pi H_t/B}^{2\pi H_b/B} \sqrt{1 + \frac{B^2}{4\pi^2}} \cdot \left[\frac{1}{D_p} \cdot \operatorname{erfc}\left(\frac{D_p}{2\sqrt{Fo}} \right) - \frac{1}{D_n} \cdot \operatorname{erfc}\left(\frac{D_n}{2\sqrt{Fo}} \right) \right] d\varphi' \tag{18}$$

In which

$$D_p = \sqrt{R^2 + 1 - 2R\cos(\varphi - \varphi') + (Z - B\varphi'/2\pi)^2}$$

$$D_n = \sqrt{R^2 + 1 - 2R\cos(\varphi - \varphi') + (Z + B\varphi'/2\pi)^2}$$

The term of $\sqrt{1 + B^2/4\pi^2}$ in Eq. (18) is induced by the different definitions of heating rate q_l in the ScM and the proposed model in the manuscript. The ScM treats the ql as heating rate per depth of pile while the ql is regarded as heating rate per length of pipe in the model of CoHEP.

2.1.2. Calculation of pipe wall temperature rise

The temperature rise on the pipe wall is an important parameter and could reflect the heat transfer characteristic of CoHEP, which also directly affects the energy efficiency of GSHP systems. To calculate the average temperature on the pipe wall accurately, the helical pipe is divided into numbers of spiral arc sections, and the number of spiral sections is n, the length of each spiral section is L_j, as shown in **Figure 4**. According to Eq. (11), $L_j = f(\phi_j)\cdot\triangle\phi_j$. Therefore, the entire pipe wall's dimensionless temperature rise Θ_{ave} can be obtained by the way of weighted average among the spiral arc sections as Eq. (19).

$$\Theta_{\text{ave}} = \frac{\sum\limits_{j=1}^{n} L_j \Theta_{\text{ave},j}}{\sum\limits_{j}^{n} L_j} = \frac{\sum\limits_{j=1}^{n} f(\varphi_j)\Delta\varphi_j \Theta_{\text{ave},j}}{\sum\limits_{j}^{n} f(\varphi_j)\Delta\varphi_j} \tag{19}$$

where ϕ_j and $\triangle\phi_j$ is the spiral angle and increment of spiral angle, respectively and $\Theta_{ave,j}$ is the average dimensionless temperature rise of the jth spiral section of helical pipe.

The cross-section of pipe with a radius of r_s is divided into four circular arcs to meet the demands of accuracy and the value of angle for each circular arc is $\pi/2$. So the average temperature raise of the jth spiral section of helical pipe can be determined:

$$\Theta_{\text{ave},j} = \sum_{i=1}^{4} \Theta_i/4 \tag{20}$$

where i stands for the index of the divided circular arc.

Ultimately, the influence coefficient η is employed to discuss the heat transfer efficiency of CoHEP compared with CyHEP, quantitatively.

$$\eta = \left(\frac{1}{\Theta_{\text{ave, CoHEP}}} - \frac{1}{\Theta_{\text{ave, CyHEP}}}\right) / \frac{1}{\Theta_{\text{ave, CyHEP}}} \tag{21}$$

It's known that Θ_{ave} could stand for the thermal resistance of the ground. Therefore, the reciprocal of Θ_{ave} can be the representative for the heat transfer coefficient. Hence, the employed influence coefficient η can reflect the thermal performance of CoHEP, especially for analysis of the influence of cone angle on heat transfer.

2.2. Numerical solution model of CoHEP

As the geometry of CoHEP is complex, the three-dimensional model of prototype is too complex and difficult to calculate. In order to improve the efficiency of simulation calculation, a laboratory-scale model was built based on the similarity principle. This will help achieve a long-term simulation of CoHEP efficiently, with lower computer configuration requirements and also a shorter calculation cycle. Innovatively, in addition to calculating the temperature of outlet water and the measuring point in the soil, the temperature of the fluid along the spiral tube is also calculated in order to study the heat transfer characteristics along the novel CoHEP.

2.2.1. CoHEP model

Similarity principle was used to shrink the prototype model in this chapter. According to the similarity principle, the similarity index of two similar models must be equal to 1 for the sake of thermal performance of the prototype and miniature models. When the miniature model of

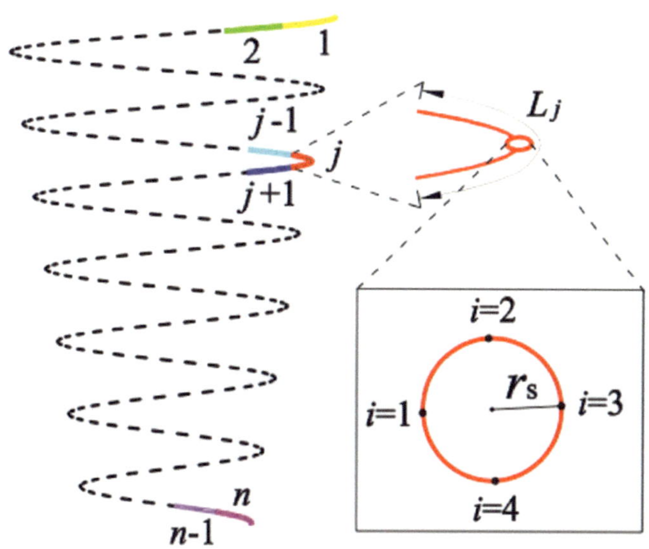

Figure 4. Schematic diagram of calculation of entire average temperature rise on the wall of pipe.

the CoHEP was designed, the heat transfer model was similarly converted using the length-scale coefficient on the prototype basis, and the soil physical parameters and the fluid in the miniature model were the same as the prototype one. Thus, according to the theory of heat transfer similarity, the Nu, Fo, and Re numbers in the prototype and miniature models were equal. Under the conditions set above, the relationship between the calculation time of the miniature and prototype system is that:

$$\tau_{min} = C_l^2 \cdot \tau_{pro} \tag{22}$$

Therefore, by scaling the prototype model the calculation results can be obtained in a relatively short period of time comparing the long-time calculation cycle of the prototype model. And it will certainly reduce the requirements for computer configuration. Through the analysis above, we can get the corresponding parameters of the prototype and miniature models, and they are shown in **Table 1**.

An example of the novel CoHEP whose θ is 10° is given in **Figure 5**. The detailed dimensions are shown in the figure. In this chapter the model was built by Solidworks, and imported to

Parameters	Spiral pitch in the depth direction/m	Bottom diameter/m	Inside diameter of coil pipe/m	Total length/m	Water flow/L/h	Time/h	Re
Prototype	0.17	0.5	0.025	100	175	300	3611
Miniature	0.034	0.1	0.005	20	35	12	3611

Table 1. The main parameters for the miniature and prototype models.

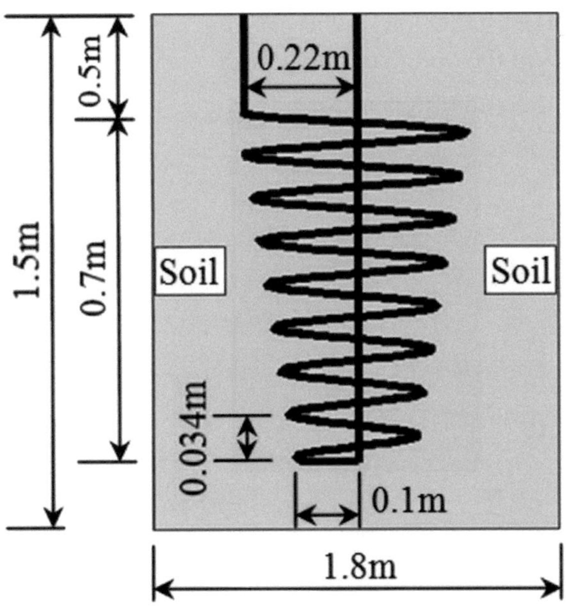

Figure 5. CoHEP of 10° cone angle.

Workbench for grid partitioning. The three-dimensional unstructured grid and CoHEP positioning are shown in **Figure 6**. As the temperature change around the spiral tube is more intense and the calculation accuracy requirement is higher, the grid around the tube is partially encrypted. At the same time, the soil temperature changes more and more slowly as the radius increases, so the spacing of the grid can be increased in the radial direction. What's more, owing to the fact that the inlet and outlet sections of the CoHEP are wrapped in insulation material, there is no heat exchange in those sections. Thus for the convenience of grid partitioning, the outlet section was moved from the center of the CoHEP to the bottom of the model when the model was built. In the simulation process, three sets of different grids (3,233,475, 3,907,586, 4,590,437) for the same geometric model were built, respectively. By comparison, the increase in the number of grids results in almost no effect on the calculation results. Thus the grid number of 3,233,475 was selected as the computational grid of the model. The main parameters of the numerical model are listed in **Table 2** that are obtained from the verification test.

2.2.2. Governing equations

As for the ground heat exchanger, fluid in the pipe relies on the pump for forced circulation, so it's the forced convection heat transfer between the fluid and pipe wall. In this chapter, soil is treated as homogeneous and isotropic solid and the effect of groundwater seepage is ignored; thus, there is only thermal conductivity in the soil.

In the simulation, the k-ε two-equation model is chosen for the flow of the fluid in the CoHEP. The general form of the governing equations is shown below:

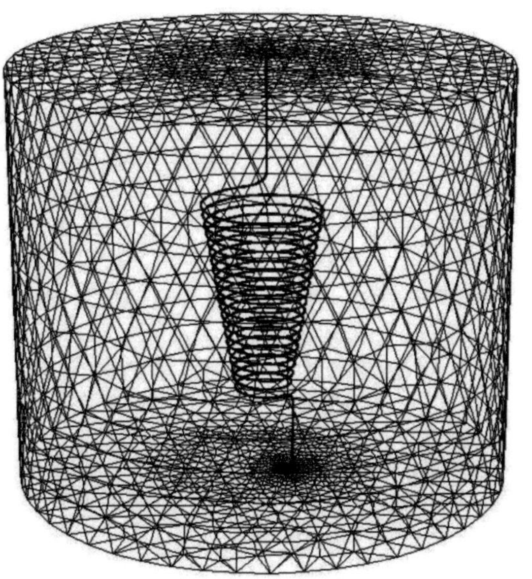

Figure 6. 3-D unstructured grid of CoHEP.

Parameter	Component	Symbol	Value	Unit
Density	Water	ρ_f	993.45	kg/m^3
	Pipe	ρ_p	900	kg/m^3
	Soil	ρ_s	2500	kg/m^3
Heat conductivity	Water	λ_f	0.72	W/(m·K)
	Pipe	λ_p	0.35	W/(m·K)
	Soil	λ_s	0.4	W/(m·K)
Gravimetric heat capacity	Water	c_f	4182	J/(kg·K)
	Pipe	c_p	1920	J/(kg·K)
	Soil	c_s	920	J/(kg·K)
Inlet water temperature	Water	T_{in}	38	°C
Water flow rate	Water	V_f	35	L/h

Table 2. The main parameters of the numerical model.

$$\frac{\partial\left(\rho_f\phi\right)}{\partial\tau}+\frac{\partial\left(\rho_f u\phi\right)}{\partial x}+\frac{\partial\left(\rho_f v\phi\right)}{\partial y}+\frac{\partial\left(\rho_f w\phi\right)}{\partial z}=\frac{\partial}{\partial x}\left(\Gamma\frac{\partial\phi}{\partial x}\right)+\frac{\partial}{\partial y}\left(\Gamma\frac{\partial\phi}{\partial y}\right)+\frac{\partial}{\partial z}\left(\Gamma\frac{\partial\phi}{\partial z}\right)+S \quad (23)$$

$$\frac{\partial\left(\rho_f\phi\right)}{\partial\tau}+\operatorname{div}\left(\rho_f\,\vec{V}\,\phi\right)=\operatorname{div}(\Gamma\cdot\operatorname{grad}(\phi))+S \quad (24)$$

Table 3 lists the governing equations for the k-ε two-equation model in the Cartesian coordinate system, and the turbulent kinetic energy generation term G_k from the mean velocity gradient is:

$$G_k=\frac{\eta_t}{\rho_f}\left\{2\cdot\left[\left(\frac{\partial u}{\partial x}\right)^2+\left(\frac{\partial v}{\partial y}\right)^2+\left(\frac{\partial w}{\partial z}\right)^2\right]+\left(\frac{\partial u}{\partial y}+\frac{\partial v}{\partial x}\right)^2+\left(\frac{\partial u}{\partial z}+\frac{\partial w}{\partial x}\right)^2+\left(\frac{\partial v}{\partial z}+\frac{\partial w}{\partial y}\right)^2\right\} \quad (25)$$

where u, v, w, respectively, represents the velocity in the x, y, z direction in the Cartesian coordinate system. k is the turbulent kinetic energy. ε is the turbulent energy dissipation rate. η is the molecular viscosity coefficient. η_t is the turbulent viscosity coefficient. η_{eff} is the effective viscosity coefficient. σ_k, σ_ε, σ_T, respectively, represents the turbulent Prandtl number of the turbulent kinetic energy k, the turbulent kinetic energy dissipation rate ε, and the temperature T_f. ρ_f is the density of the fluid. P_f is the pressure of the fluid.

As soil is treated as homogeneous and isotropic solid and the effect of groundwater seepage is ignored, thus, there is only thermal conductivity in the soil. The thermal governing equation in the soil can be expressed by the following formula:

$$\frac{\partial(\rho_s T_s)}{\partial\tau}=\frac{\partial}{\partial x}\left(\frac{\lambda_s}{c_s}\cdot\frac{\partial T_s}{\partial x}\right)+\frac{\partial}{\partial y}\left(\frac{\lambda_s}{c_s}\cdot\frac{\partial T_s}{\partial y}\right)+\frac{\partial}{\partial z}\left(\frac{\lambda_s}{c_s}\cdot\frac{\partial T_s}{\partial z}\right) \quad (26)$$

Equation	ϕ	Γ	S
Continuity equation	1	0	0
X-momentum equation	u	$\eta_{eff} = \eta + \eta_t$	$-\frac{\partial p_f}{\partial x} + \frac{\partial}{\partial x}\left(\eta_{eff}\frac{\partial u}{\partial x}\right) + \frac{\partial}{\partial y}\left(\eta_{eff}\frac{\partial v}{\partial x}\right) + \frac{\partial}{\partial z}\left(\eta_{eff}\frac{\partial w}{\partial x}\right)$
Y-momentum equation	v	$\eta_{eff} = \eta + \eta_t$	$-\frac{\partial p_f}{\partial y} + \frac{\partial}{\partial x}\left(\eta_{eff}\frac{\partial u}{\partial y}\right) + \frac{\partial}{\partial y}\left(\eta_{eff}\frac{\partial v}{\partial y}\right) + \frac{\partial}{\partial z}\left(\eta_{eff}\frac{\partial w}{\partial y}\right)$
Z-momentum equation	w	$\eta_{eff} = \eta + \eta_t$	$-\frac{\partial p_f}{\partial z} + \frac{\partial}{\partial x}\left(\eta_{eff}\frac{\partial u}{\partial z}\right) + \frac{\partial}{\partial y}\left(\eta_{eff}\frac{\partial v}{\partial z}\right) + \frac{\partial}{\partial z}\left(\eta_{eff}\frac{\partial w}{\partial z}\right)$
Turbulent kinetic energy	k	$\eta + \frac{\eta_t}{\sigma_k}$	$\rho_f G_k - \rho_f \varepsilon$
Turbulence energy dissipation rate	ε	$\eta + \frac{\eta_t}{\sigma_\varepsilon}$	$\frac{\varepsilon}{k}\left(c_1 \rho_f G_k - c_2 \rho_f \varepsilon\right)$
Energy equation	T	$\frac{\eta}{\Pr} + \frac{\eta_t}{\sigma_T}$	0

Table 3. The governing equations for the k-ε two-equation model in the Cartesian coordinate system.

where c_s is the specific heat capacity of the soil. λ_s is the thermal conductivity of the soil. T_s is the temperature.

2.2.3. Initial and boundary conditions

As the depth of the CyHEP is shallow, it's usually buried 10–20 m below from the surface of the ground. Thus the heat transfer of the buried pipe is greatly affected by the environmental conditions of the soil surface. Particularly, as the top diameter of the novel CoHEP is bigger than the traditional CyHEP, there is more area of pipe close to the ground surface. Thus the effects of dynamic surface conditions on the heat transfer of the novel CoHEP cannot be ignored. However, for the sake of computational convenience most models of the traditional CyHEP including the analytical solution models and the numerical solution models always make a certain simplification by setting the ground surface temperature as a constant one which will lead to the deviation from actual situation.

Figure 7 shows the dynamic near-surface air temperature recorded in the verification test. The following functional Eq. (27) is obtained by fitting the recorded data. And the corresponding fitting curve is shown in **Figure 5**.

$$\begin{cases} T_{air} = 24.3 - 0.026\tau + 3.3\times 10^{-4}\tau^2 - 1.6\times 10^{-6}\tau^3, & 0\le\tau\le 50 \\ T_{air} = 23.6 + 4.7\times 10^{-4}\tau, & 50 < \tau\le 720 \end{cases} \quad (27)$$

where T_{air} is the near-surface air temperature. τ is the system running minutes.

The initial soil temperature distribution is affected by air temperature and solar radiation, which varies with depth and time, and has a great influence on the heat transfer performance of underground tube heat exchangers. It's one of the most basic parameters in theoretical calculations. Especially for the CoHEP, due to its shallow depth, the initial soil temperature distribution is not uniform. Considering the initial soil temperature gradient in the depth direction, the model will be more suitable for the actual condition. **Figure 8** shows the initial soil temperature distribution in the depth direction recorded in the verification test. The following functional Eq. (28) is obtained by fitting the recorded data. And the corresponding fitting curve is shown in **Figure 8**.

$$T_{initial} = 24.3 - 1.44h + 1.05h^2 - 0.27h^3 \tag{28}$$

where $T_{initial}$ is the initial soil temperature and h is the depth in the soil.

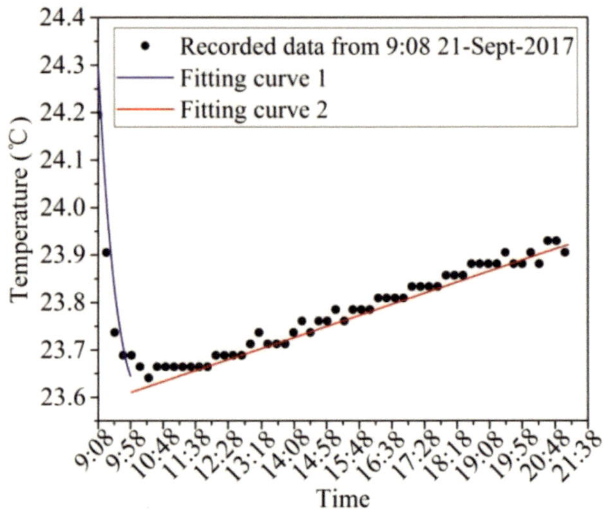

Figure 7. Recorded dynamic near-surface air temperature.

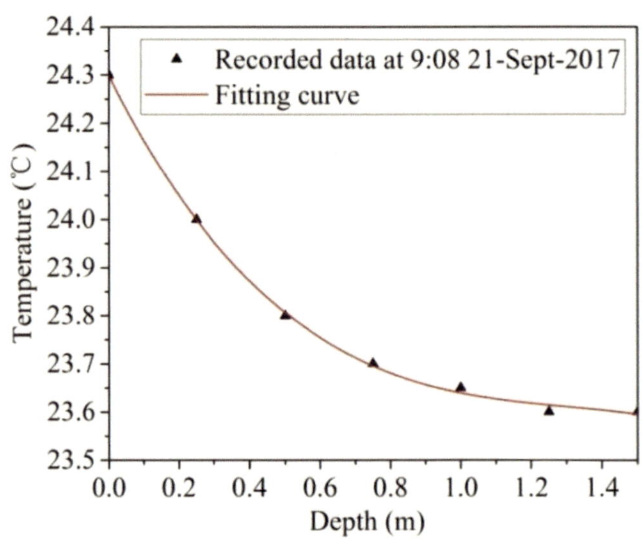

Figure 8. Recorded initial soil temperature distribution in the depth direction.

Both the dynamic surface condition and the initial soil temperature are modified with user defined functions (UDF) written in C language according to the functions above and then incorporated to the calculation model. As for the model's other boundary conditions, the top surface, bottom surface and the surrounding surface of the cylindrical soil model, are defined as the wall boundary. The inlet of the tube is set to velocity-flow-inlet and the outlet is set to pressure-outlet. Buried pipe and soil areas are solid; the area inside the tube is fluid.

2.2.4. CFD simulation setup

The commercial CFD software ANSYS_FLUENT is used to simulate the heat transfer of the CoHEP. Because the length of the tube is 20 m, the inside diameter of the tube is 0.005 m; thus, the model belongs to the slender region heat transfer model and the solver selected in the calculation is a three-dimensional double-precision solver. What's more, k-ε two-equation model is selected for calculating the turbulent flow in the tube. The physical parameters of fluid in the tube, buried pipe and soil areas are defined according to the verification test experiment. The top surface of the soil model is defined as the heat transfer boundary according to the actual situation. The bottom surface and surrounding surface of the cylindrical soil model are defined as adiabatic boundary conditions. The momentum equation, turbulent kinetic energy equation, turbulence energy dissipation rate and the energy equation all use the second-order upwind discretization format for calculation.

In order to help analyze the thermal performance of CoHEP, three-series monitoring points (a total of 18) shown in **Figure 9(a)** are set to monitor the soil temperature. Innovatively, a series of monitoring points are set along the tube of CoHEP to investigate the water temperature drop along the tube, and the separation distance of the points is shown in **Figure 9(b)**.

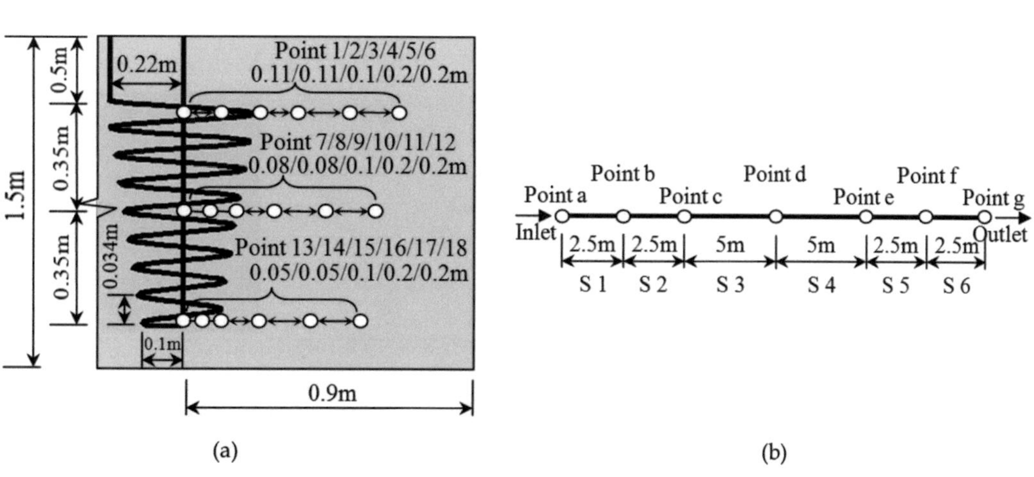

Figure 9. Arrangements of series monitoring points. (a) Monitoring points in the soil, (b) monitoring points in the pipe of CoHEP.

2.3. Heat transfer characteristics of CoHEP

2.3.1. The thermal interference of CoHEP by analytic solution model

Significant radial thermal interferences (RTI) and generatrix thermal interferences (GTI) exist in the cylinder helix energy pile. **Figure 10(a)** shows the contour of Θ in the *Z-R* plan of CyHEP with $\phi = 180°$, $B = 1$ and Fo = 10. It is indicated that GTI is serious duo to the small pitch and also RTI is significant because of the limited thermal heat capacity of pile, especially in the middle of pile. However, the remarkable 3-D effect weakens both RTI and GTI at the upper and lower part of pile, which leads to the reduction of Θ at the top and base of CyHEP compared to the middle of CyHEP.

The contours of Θ in the *Z-R* plan of CoHEP with $\theta = 5°$ and $\theta = 10°$ are shown as **Figure 10(b)** and **Figure 10(c)**, respectively, when the length of heat exchange pipe is consistent. The figures show that bigger the cone angle, lower the dimensionless temperature at the bottom of HEP. The upper part of CyHEP has the same situation. It is indicated that the structure of CoHEP can weaken the GTI in the bottom of pile due to the local large ratio between coil pitch and radius of pile. Besides, the degree of RTI also reduces in the upper parts of pile because of the local large radius.

2.3.2. The dynamic thermal interference of CoHEP by analytic solution model

The CoHEP can effectively reduce the thermal interference, and further investigations discover that the thermal interference is dynamic by analyzing the *Z-*Θ curve on the generatrix of

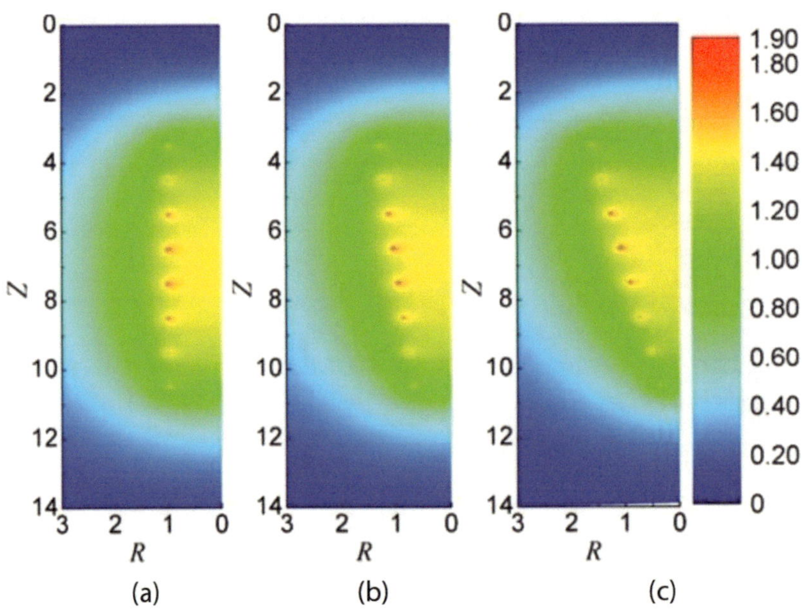

Figure 10. Contours of dimensionless temperature rise in the *Z-R* plan at $\phi = 180°$($B = 1$, Fo = 10). (a) $\theta = 0°$, (b) $\theta = 5°$, (c) $\theta = 10°$.

CyHEP and CoHEP with variable Fo from 0.01 to 100. The calculation precondition is that length of pipe and coil pitch are consistent. The results are shown in **Figure 11**.

With the increasing Fo, the ground surface boundary thermal interference (BTI) appears, in addition to the RTI and GTI. BTI is induced by the first kind of boundary condition and affected by distance from the top surface of CoHEP to the ground surface ht. Besides, according to the expression of Green's function, the heat affected zone is continuously growing with the increasing Fo, which results in the aggravation of three thermal interferences. However, each kind of thermal interference has a spatial inconsistence with variation of Fo.

In order to explain the dynamic characteristics revealed by the Z-Θ curve from Fo = 0.01 to Fo = 100 plotted in **Figure 11**, six different degrees are qualitatively defined to describe the thermal interference including "none," "little," "mild," "moderate," "serious," and "severe." The defined degree gradually increases from "none" to "severe." According to both the structure features of HEP and the plotted Z-Θ curve, the specific degrees of RTI, GTI, and BTI have been deduced, which has been noted at the top and bottom of figure for each case of Fo.

Figure 11 shows that Θ in the middle of the adjacent coils is zero when Fo is very small (such as Fo = 0.01), which indicates that the degrees of thermal interference are "none" and the curves of Z-Θ for CyHEP and CyHEP overlap. With the increase of Fo, RTI at the base of CoHEP appears firstly due to the small base radius when cone angle θ equals to 10° and then, the GTI in the CyHEP and the top of CoHEP begins to emerge because of the small ratio between coil pitch and local radius of pile. Next, the RTI in the CyHEP and the top of CoHEP tends to appear in succession due to the large radius of pile. Finally, BTI starts to show in the top of CyHEP and CoHEP due to the distance of ht. It is worth mentioning that the Θ in the top of HEP would be severely influenced by BTI when Fo is big (such as Fo = 10 and 100).

In general, the GTI in CoHEP tends to weaken with the increasing of depth, because the ratio between coils pitch and local radius of the pile gradually increases while the local radius of pile linearly decreases. Therefore, the GTI in the base of CoHEP is weaker than other place of CoHEP and CyHEP, and the dimensionless temperature rise Θ in the base of CoHEP is lower when Fo > 1. However, the Θ in the base is higher than that of CyHEP when Fo < 1 due to the local smaller radius of CoHEP, which means that the RTI of CyHEP is weaker than that of CoHEP in the base of the pile for a small Fo (such as Fo = 0.1 and Fo = 0.5).

Besides, the Θ in the top of CoHEP is lower than that of CyHEP with range of Fo from 0.1 to 5 due to a large local radius of CoHEP, which indicates that the RTI of CoHEP is weaker than that of CyHEP in the top of the pile. However, the curves in the top of CyHEP and CoHEP overlaps while Fo > 5 because the BTI tends to be severe.

2.3.3. *Heat transfer characteristics of CoHEP by the numerical solution model*

Ground source heat pump heat exchanger relies on the temperature difference between the circulating medium in the pipe and the soil around the pipe to absorb or release energy, thus heating or cooling the circulating medium in the pipe. The temperature of the circulating medium in the pipe has a significant effect on the heat transfer of the ground heat exchanger,

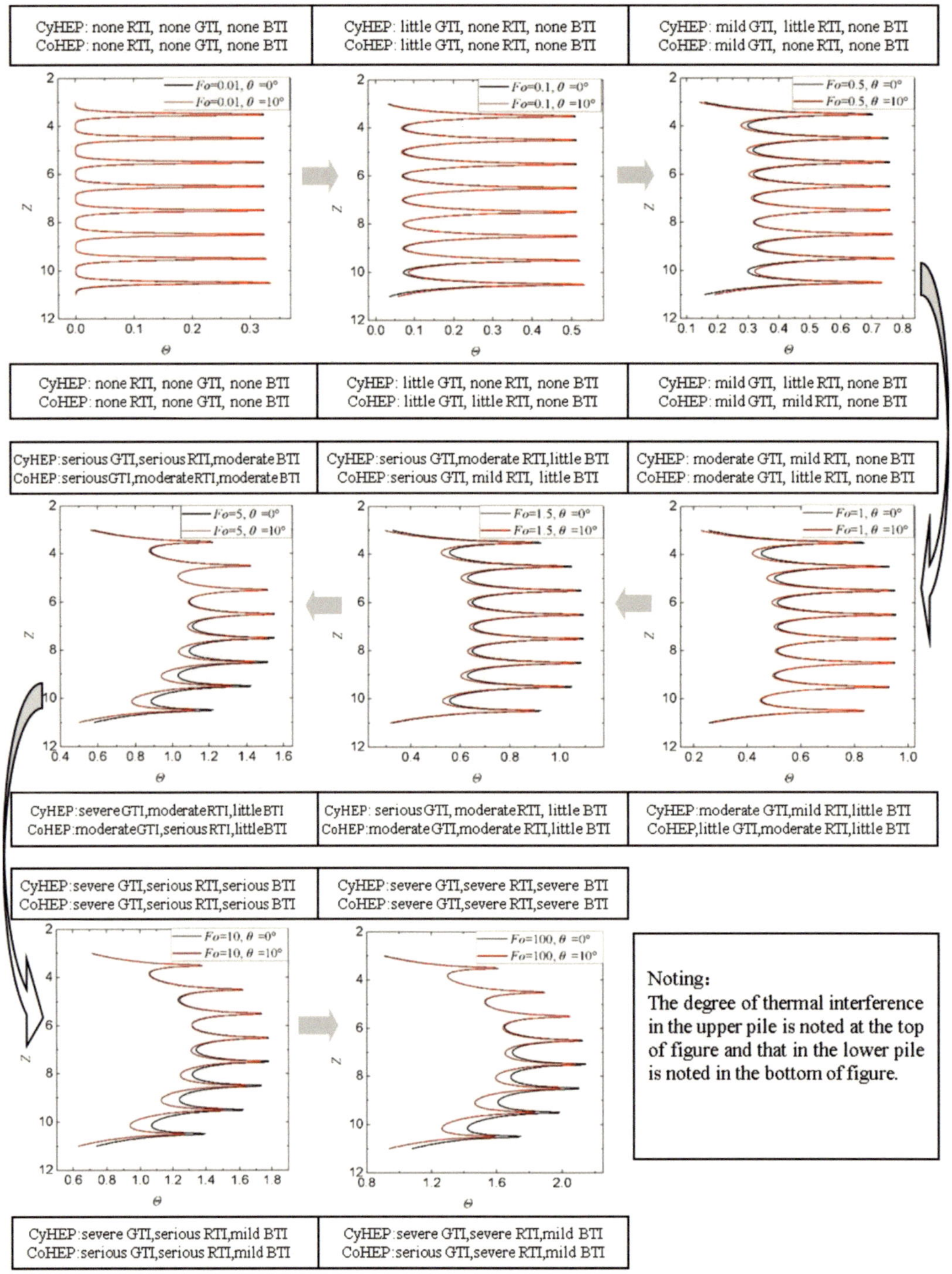

Figure 11. Dynamic characteristic of thermal interference of HEP ($B = 1$, $\phi = 180°$).

while the temperature of the circulating medium in the pipe is controlled by the inlet temperature of the ground heat exchanger.

In order to investigate the influence of inlet water temperature on the heat transfer performance of the novel CoHEP, three different inlet water temperatures, 33, 38, 43°C are chosen to study in this chapter. These three conditions are simulated with the established numerical model verified above. All the other parameters in addition to the inlet water temperature are the same as shown in **Table 2**.

Figure 12 shows the simulated outlet water temperature changing with operation time in three different inlet water temperature conditions. We can easily find that the outlet water temperature increases with time and eventually gradually reaches a steady state. Correspondingly, **Figure 13** shows the simulated heat flux per unit pipe length changing with operation time in these three conditions. And it shows that the heat flux per unit pipe length decreases with time and eventually reaches a steady state. The main reason for this trend is: In the early stage of operation, the temperature difference between the fluid and soil is large, and the heat transfer is sufficient. Therefore the system has large inlet and outlet water temperature difference and large heat transfer flux. However, after that, the heat-affected zone increases in radius and the thermal resistance of the soil becomes the main contradiction of the heat transfer process, resulting in the decrease of the heat transfer rate. After a period of heat exchange, with the heat accumulation in the soil, the temperature difference between the fluid and soil is small; thus, the heat flux per unit pipe length sharply reduces. And then later, when the soil temperature field tends to be stable, the heat transfer performance also is gradually stabilized. As shown in **Figure 14**, the heat flux per unit pipe length increases linearly as the inlet water temperature increases. When the system running time is 4, 8, and 12 h, respectively, the heat

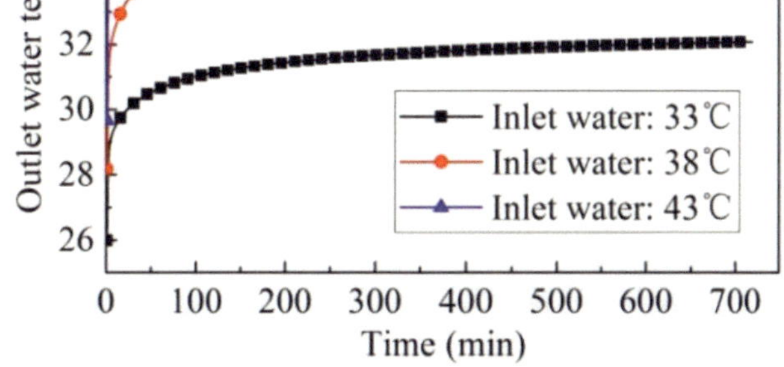

Figure 12. Outlet water temperature with operation time under different inlet water temperatures.

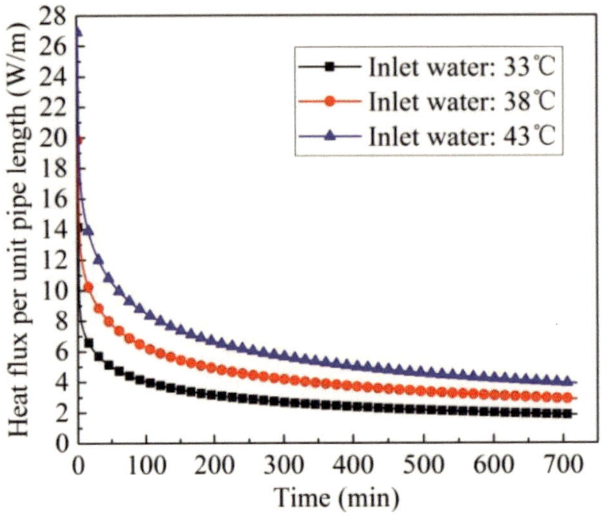

Figure 13. Heat flux per unit pipe length with operation time under different inlet water temperatures.

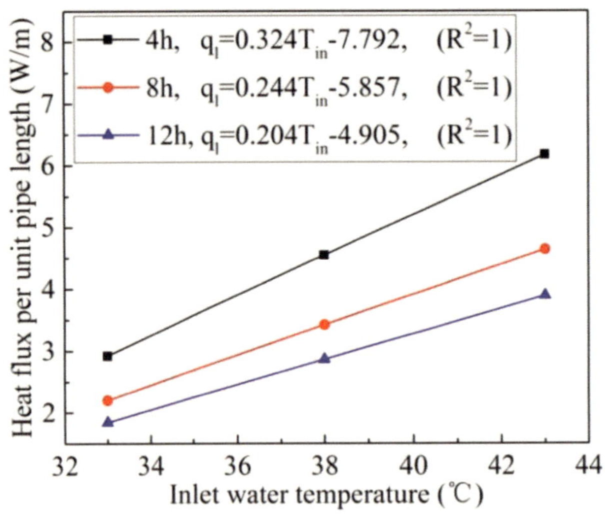

Figure 14. Heat flux per unit pipe length with various inlet water temperatures at different operation time.

fluxes per unit pipe length are 2.9, 2.2, and 1.83 W/m in the condition of the 33°C inlet water temperature. The heat fluxes per unit pipe length are 4.52, 3.42, and 2.83 W/m in the condition of 38°C inlet water temperature. Moreover, the heat fluxes per unit pipe length are 6.14, 4.64, and 3.87 W/m in the condition of 43°C inlet water temperature. The larger the inlet water temperature, the greater the heat flux per unit pipe length.

Figure 15 shows that with the increase of running time, the heat flux per unit pipe length decreases nonlinearly. And the fitting equations for three different inlet water temperature conditions are shown in the figure.

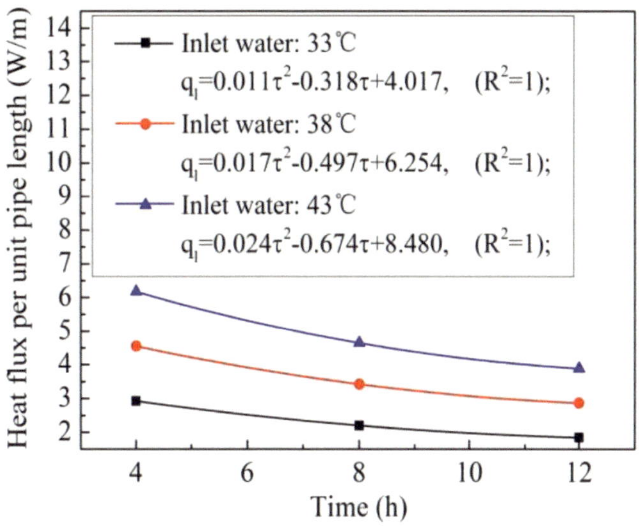

Figure 15. Heat flux per unit pipe length decreasing with operation time under different inlet water temperatures.

Figure 16 shows the fluid temperature distribution in the flow direction along the pipe length with the inlet water temperature of 33 38 and 43°C. It can be seen from the figure that the fluid temperature gradually decreases along the pipe length, and the fluid temperature reduction is more obvious in the entrance and exit stages of the CoHEP. **Figure 17** shows the heat flux per unit pipe length in the flow direction along the pipe length with the inlet water temperature of 33, 38 and 43°C. According to the geometry of CoHEP, as shown in **Figure 17**, the whole pipe of CoHEP is divided into four stages along the flow direction of the pipe length:

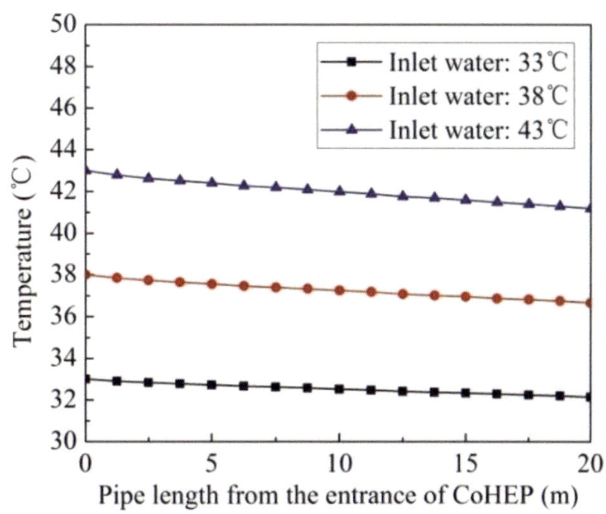

Figure 16. Water temperature distribution in the flow direction along the pipe length.

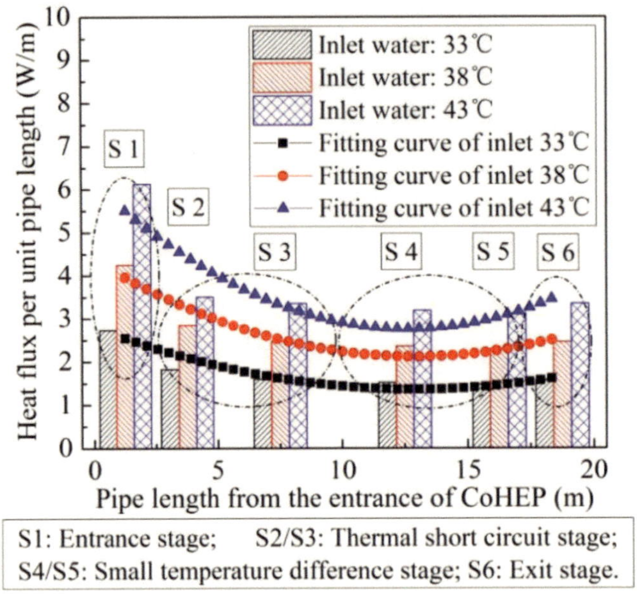

Figure 17. Heat flux per unit pipe length in the flow direction along the pipe length.

1. The entrance stage (S1): This stage is at the top of CoHEP which is in direct contact with the soil-covering area, and there is no heat transfer pipe in this area. Therefore, the heat transfer process in this stage has the following features: The fluid in this stage of the pipe not only radiates heat in the direction of diameter but also in the axial direction to the soil-covering area above the CoHEP, resulting in greater heat exchange capacity at this stage.

2. The thermal short circuit stage (S2 + S3): As the flow distance along the pipe length increases, the entrance stage ends. Due to the accumulation of heat in the soil, the thermal short circuit between the spiral pipes in the axial direction is serious, and the heat is mainly transferred in the diameter direction. Thus, the heat exchange capacity at this stage weakens.

3. The small temperature difference stage (S4 + S5): As the thermal short circuit stage ends, the fluid in the pipe decreases further, resulting in a decrease of the temperature difference between the fluid and the soil around. At the same time, due to the small diameter of the lower part of CoHEP, there is thermal short circuit both in the diameter direction and the in the axial direction. Therefore the heat exchange capacity reduces further at this stage.

4. The exit stage (S6): As the small temperature difference stage ends, the fluid is close to the exit of the pipe and this stage is defined as the exit stage. The exit stage is similar to the entrance stage. Since the pipe at the exit stage is in direct contact with the soil at the bottom of the buried CoHEP, the axial thermal short circuit effect is weak. The fluid in this stage of the pipe not only radiates heat in the direction of diameter but also in the axial direction to the soil at the bottom of the buried CoHEP. Thus, the heat exchange capacity increases at this stage.

3. Conclusions

A novel truncated cone helix energy pile (CoHEP) is presented to weaken the thermal interferences and improve the heat transfer efficiency. Further, both analytical solution model and numerical solution model for CoHEP are built to discuss their dynamic characteristics of thermal interferences and heat transfer performance. The following conclusions can be drawn.

1. Four heat exchange stages for the spiral pile geothermal heat exchanger along the fluid flow direction are revealed: inlet heat exchange stage, grout thermal short-circuiting stage, small temperature difference stage and outlet heat exchange stage. Each stage has corresponding heat transfer characteristics, and reducing the length of small temperature difference stage and increasing the other stages would enhance the heat exchange of the spiral geothermal ground heat exchanger.

2. The thermal interference of CoHEP is dynamic. When Fo < 1, the RTI of CyHEP is weaker than that of CoHEP in the bottom and the dimensionless temperature rise on the pile wall is lower than that of CoHEP. However, The GTI in the base of CoHEP is weaker than the other place of CoHEP and CyHEP, and the dimensionless temperature rise in the base of CoHEP is lower when Fo > 1.

Besides, the dimensionless temperature rise on the upper pile wall of CoHEP is lower than that of CyHEP when Fo < 5 and the RTI of CoHEP is weaker than that of CyHEP in the top of the pile. However, when Fo > 5, the temperature difference between CyHEP and CoHEP in the upper parts is nothing because the BTI tends to severe.

3. The thermal interference in the upper part of the CoHEP is much smaller than the traditional CyHEP. Moreover, under the same pitch in the depth direction (b) condition, the distance between the adjacent tube of CoHEP (d) is obviously larger than that of the traditional CyHEP, which can effectively reduce the axial thermal interference. Thus in general the heat flux per unit pipe length of the novel CoHEP is larger than that of the traditional CyHEP.

4. Heat flux per unit pipe length of the CoHEP increases linearly with the inlet water temperature. Thus the thermal performance of the CoHEP can be enhanced by increasing the inlet water temperature. But the inlet water temperature's increase will also lead to the increase of the outlet water temperature. It will cause a high condensation temperature and reduce the efficiency of the system. There is an optimal inlet water temperature, which needs to be analyzed together with the overall system's heat exchange efficiency.

For the same inlet water temperature, the thermal short circuit is serious at the bottom of the CoHEP, and it's weak in the upper part of the CoHEP. Also it's obvious that as the inlet water temperature increases, the thermal short circuit becomes more serious.

Acknowledgements

I would like to express my gratitude to all those who helped me during the writing of this book. And I feel grateful to all the teachers in the Army Logistical University of PLA who once offered me valuable courses and advice during my study. Last, my thanks go to my beloved family for their loving considerations and great confidence in me all through these years. I also owe my sincere gratitude to my friends who gave me their help and time in listening to me and helping me work out my problems during the difficult course of the book.

This work received support from the Research Initiative for Basic Science and Frontier Technology of Chongqing, China [cstc2016jcyjA0496], Natural Science Foundation of China [51706243].

Conflict of interest

We declare that we have no financial and personal relationships with other people or organizations that can inappropriately influence our work; there is no professional or other personal interest of any nature or kind in any product, service, and/or company that could be construed as influencing the position presented in, or the review of, our work.

Nomenclature

x, y, z	Cartesian coordinate (m)
r, ϕ, z	cylindrical coordinate (m)
R, ϕ, Z	dimensionless cylindrical coordinate
h	height (m)
H	dimensionless height
r	radial coordinate (m)
R	dimensionless radial coordinate
q_l	heating rate per length of pipe (W s^{-1})
τ	the time (s)
α_s	thermal diffusivity (m^{-2} s)
d_p	distance from the heat source point to the calculated point (m)
d_n	distance from the heat sink point to the calculated point (m)
D_p	dimensionless distance from the heat source point to the calculated point (m)
D_n	dimensionless distance from the heat sink point to the calculated point (m)
ρ	density (kg m^{-3})
c	specific heat (J kg^{-1} K^{-1})

λ	thermal conductivity (W m^{-1} K^{-1})
$\triangle T$	the temperature rise (K)
R	thermal resistance (m^{-1} K^{-1} W)
L	length of helix pipe
Greek symbols.	
θ	cone angle (rad)
ϕ	spiral angle (rad)
Fo	Fourier number
η	influence coefficient
Θ	dimensionless temperature rise
Superscript.	
$'$	integration parameter
ave.	the average value
CyHEP	cylinder helix energy pile
CoHEP	truncated cone helix energy pile
t	top surface of pile
b	base surface of pile
mi	middle surface of pile
i	the index of arc
j	the index of coil
k, n	the index of time
p	pipe
f	fluid
g	ground

Author details

Guangqin Huang*, Yajiao Liu, Xiaofeng Yang and Chunlong Zhuang

*Address all correspondence to: hgq880818@163.com

Department of Military Installation, Army Logistical University of PLA, Chongqing, China

References

[1] Park S, Lee D, Choi H-J, Jung K, Choi H. Relative constructability and thermal performance of cast-in-place concrete energy pile: Coil-type GHEX (ground heat exchanger). Energy. 2015;**81**:56-66. DOI: 10.1016/j.energy.2014.08.012

[2] Park S, Sung C, Jung K, Sohn B, Chauchois A, Choi H. Constructability and heat exchange efficiency of large diameter cast-in-place energy piles with various configurations of heat exchange pipe. Applied Thermal Engineering. 2015;**90**:1061-1071. DOI: 10.1016/j.applthermaleng.2015.05.044

[3] Yang W, Lu P, Chen Y. Laboratory investigations of the thermal performance of an energy pile with spiral coil ground heat exchanger. Energy and Buildings. 2016;**128**:491-502. DOI: 10.1016/j.enbuild.2016.07.012

[4] Park SK, Lee S-R, Park H, Yoon S, Chung J. Characteristics of an analytical solution for a spiral coil type ground heat exchanger. Computers and Geotechnics. 2013;**49**:18-24. DOI: 10.1016/j.compgeo.2012.11.006

[5] Li M, Lai Alvin CK. Heat-source solutions to heat conduction in anisotropic media with application to pile and borehole ground heat exchangers. Applied Energy. 2012;**96**:451-458. DOI: 10.1016/j.apenergy.2012.02.084

Chapter 14

Techniques for the Fabrication of Super-Hydrophobic Surfaces and Their Heat Transfer Applications

Hafiz Muhammad Ali, Muhammad Arslan Qasim,
Sullahuddin Malik and Ghulam Murtaza

Additional information is available at the end of the chapter

http://dx.doi.org/10.5772/intechopen.72820

Abstract

Super-hydrophobic surfaces are surfaces that have extreme water-repellent properties and show contact angle greater than 150° and sliding angle less than 5°. These surfaces play a significant role in different processes like icing delay, anti-frosting, boiling, condensation, drag reduction, self-cleaning, etc. The present study comprises of different techniques for the fabrication of super-hydrophobic surfaces. These techniques include chemical etching, solution immersion, laser electrodeposition, template deposition, spray coating, various others. Important characteristics of super-hydrophobic surfaces like durability, storability, corrosion resistance, etc. are achieved differently by different methods. Also, some methods are simple, rapid, cost-effective and versatile. Moreover, various heat transfer applications of super-hydrophobic surfaces like boiling, condensation, icing delay, drag reduction, etc. have also been discussed in this chapter.

Keywords: super-hydrophobic surfaces, contact angle, sliding angle, surface roughness, micro-nanostructures, surface energy

1. Introduction

It has already been found that leaves of some plants like Lotus flower, Nasturtium, Prickly pear, etc. and wings of some insects like Pamassius glacialis, etc. have micro-nanostructures on their surfaces which have ability to repel water excellently [1]. These structures lead to minimum surface energy (i.e. wettability) due to which the contact area fraction between liquid and solid surface also becomes minimum. Moreover, surface tension of liquid causes its droplets to get shaped into spherical shape which has minimal surface area. So, the combined effect of high surface tension of liquid and low surface energy can lead to the formation of spherical

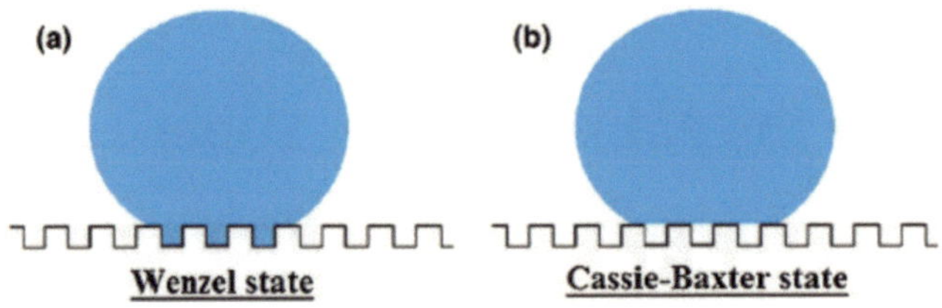

Figure 1. Water droplet in (a) Wenzel state (b) Cassie-Baxter state [2].

drops on the surface. The contact angle of these drops can be more than 150° on such surfaces. Such surfaces are named as super-hydrophobic surfaces. Water droplets can actually bounce on these surfaces. By mimicking the nature, these surfaces can be synthesized.

A liquid droplet can be in two different states when resting on a solid surface which is hydrophobic. In Wenzel state, the liquid droplets penetrate into the asperities of surface resulting in increased contact area fraction of liquid with solid surface. This leads to the reduced mobility of droplets on surface. On the other hand, in Cassie-Baxter state, droplet stays on the hierarchical structures and air gets entrapped in the cavities of surface due to which it has minimum contact area fraction with solid surface. This helps in the formation of spherical droplets which can roll on the surface. Water droplets stay on the super-hydrophobic surface in Cassie-Baxter state [2].

Figure 1 shows the water droplet in Wenzel and Cassie-Baxter state.

First section of this chapter summarizes different techniques for the fabrication of super-hydrophobic surfaces. These surfaces must possess some important characteristics like durability, corrosion resistance, etc. which are required in various applications. The challenge now-a-days is the fabrication of super-hydrophobic surfaces which have long-term stability. Second section gives an idea of different heat transfer applications of super-hydrophobic surfaces. Major applications like condensation, boiling, anti-icing and drag reduction with experimental work of different authors have also been discussed.

2. Fabrication techniques

Fabrication of super-hydrophobic surfaces usually requires the roughening of surface to get micro-nanostructures followed by surface modification which leads to low surface energy. Some methods like chemical etching, solution-immersion process, spray coating use coating material for surface modification after surface roughening, while some methods like laser electrodeposition, template deposition do not need to modify the surface. Simplicity, least time consumption, cost-effectiveness and versatility are the important parameters during fabrication process. Moreover, characteristics like durability, corrosion resistance and storability of super-hydrophobic surfaces formed are achieved at different levels from each method.

2.1. Chemical etching

Esmaeilirad et al. [3] developed physically and thermally stable SHS by a simple and cost-saving method. Chemical etching method was used for developing micro-nanostructures on the surface of Al alloy which was used as a substrate. After surface roughing of substrate with abrasive paper and ultrasonically cleaning with deionized water, it was treated with NaOH first and then with solution of HCl and CH_3COOH resulting in micro-nanostructures on surface. Then surface energy of etched substrate was lowered by immersing it in the solution of silanes. The maximum water contact angle (WCA) obtained was 165° and contact angle hysteresis (CAH) was 3°. SHS formed remained stable after different tests and was storable for more than a month. Qi et al. [4] developed a rapid fabrication method of SHS. Metal-assisted chemical etching was done on zinc plate which was used as substrate. Pure zinc plate was first ultrasonically cleaned and dried in oven. For surface roughing, it was put in an aqueous solution of nitric acid and metal nitrate. Metal nitrates used were $AgNO_3$, $Cr(NO_3)_3$ and Cu $(NO_3)_2$. After etching, sample was immersed in a solution of ethanol and flouroalkyl silane (FAS) for surface modification. WCA of pure zinc without etching was found to be 102 ± 3° after modification by FAS only and maximum of 135 ± 2° with just etching. The maximum WCA was obtained to be 140 ± 2° for $Cr(NO_3)_3$ after 5 s, 157 ± 2° for $Cu(NO_3)_2$ after 1 s and 162 ± 2° for $AgNO_3$ after 1 s. In case of Cu^{2+}-assisted chemical etching, etching time, concentration of $cu(NO_3)_2$ and concentration of HNO_3 were optimized.

Figure 2 shows the schematic illustration of Zn SHS fabrication by metal-assisted chemical etching.

Chu and Wu [5] fabricated SHS on Al and Cu substrates simultaneously using chemical etching method. Surfaces were first cleaned and then solution of HCl and $Cu(NO_3)_2$ was used for Al plate etching, while for copper plate etching, solution of HNO_3 and $AgNO_3$ was used. Surface modification was done using aqueous solution of FAS. The results showed that for Al and Cu substrates, the micro-nanostructures were similar to that of lotus leaves and moss, respectively. The WCA was found to be 164 ± 1° for Al and 157 ± 1° for Cu. The measured rolling angles were 2 ± 1° and 6 ± 1° for Al and Cu, respectively. Moreover, condensation experiment was also performed over both SHS resulting in lower droplet density, higher droplet jumping probability, slower droplet growth rate and lower surface coverage for Al as compared to Cu. Yin et al. [6]

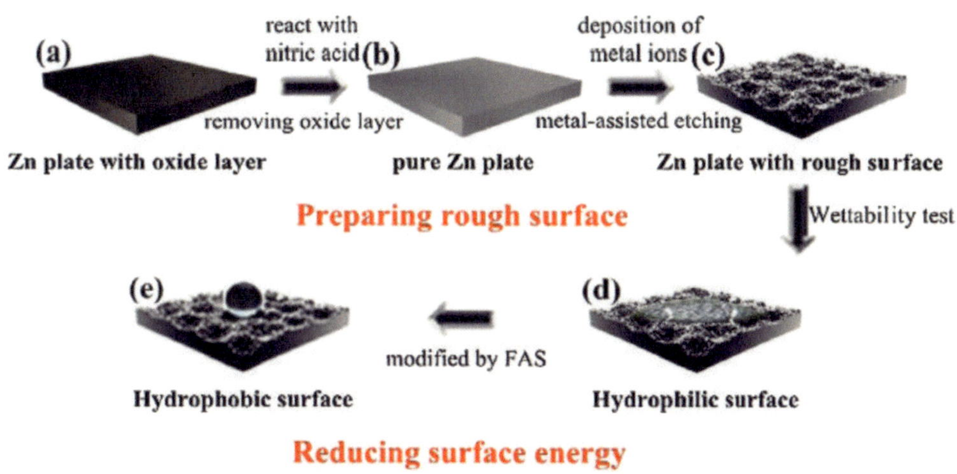

Figure 2. Schematic illustration of fabrication of Zn SHS by metal-assisted chemical etching [4].

developed SHS using chemical etching method. Al plate was used as the substrate material. For etching process, a solution of HF and HCl in deionized water was prepared. Later, Al substrate was immersed in that solution for roughing the surface. For surface modification, three different coatings such as perfluoroalkyltriethoxysilane (PFO), PA and room temperature vulcanized (RTV) coating were separately used to determine the super-hydrophobic nature of surfaces. WCA measurements showed the maximum CA of 162, 161.7 and 158.3° for PFO, PA and RTV coatings, respectively. Environmental factors varied, that is, temperature in range of −10 to 30°C and RH values of 30, 60 and 90%, during condensation experiment to determine any changes in super-hydrophobic nature of surfaces. CA, SA and contact area fraction were calculated and it was found that with increasing RH and lowering temperature, CA and SA values decreased and increased, respectively, while contact area fraction increased showing increase in wettability. Super-hydrophobicity of RTV coating was greatly affected during condensation at low temperature which was recovered simply by drying.

To counter the problem of ice accumulation on Al substrate which has excessive use in transmission lines, Liao et al. [7] made SHS on Al which showed excellent anti-icing property. A simple continuous chemical etching method was used for this purpose. The roughness on Al foil was increased by using emery paper and then washed and cleaned properly. For etching process, it was first immersed in $CuCl_2$ solution and then in HCl solution. Both resulted in increased surface roughness of Al foil. Finally, hexadecyl-trimethoxy silane was used for surface modification which reduced surface energy resulting in super-hydrophobic Al surface. WCA measurements showed that bare Al surface was transformed into SHS with contact angle increasing from 101.1° to 161.9°. It was found after thermal stability tests that WCA reduced a little with increase in temperature but surface was still super-hydrophobic. It was determined that the fabricated SHS could mitigate freezing process. Moreover, stability of SHS was also confirmed against water-drop impact and sand-impact abrasion. Nguyen et al. [8] fabricated super-hydrophobic and oleophobic surfaces using electroless etching method. Silicon surface was used as substrate which was first dipped in $AgNO_3/NaBF_4$ solution. This resulted in etching of silicon surface creating nanostructures. This etched surface was then coated by Ag nanoparticles which modified the surface topology. Third and last step was the lowering of surface energy, first by C_4F_8 plasma deposition and second by SiO_x overlayers chemically modified with PFTS. Contact angle of water and hexadecane droplets were measured. When C_4F_8 plasma deposition was used for lowering surface energy, then nano-Si structure showed maximum CA of 160 and 87° for water and hexadecane, respectively. Results were 160 and 110°, respectively, for water and hexadecane when SiO_x overlayers chemically modified with PFTS were used for lowering surface energy. It was found that deposition of Ag particles led to improved wetting properties.

Yin et al. [9] developed a novel way for fabrication of SHS. Al alloy was used as substrate which was properly polished and then washed. It was then anodized under constant current in a particular solution. The next step was chemical etching resulting in leaf-like structure on surface. Last step was the modification of sample with solution of KH-832 in ethanol which decreased the surface energy of sample. Untreated sample was found to be hydrophilic with WCA of 79.4°, sample after pre-treatment and anodization showed super-hydrophilic nature with 0° WCA while sample after pre-treatment, anodization as well as chemical modification was found to be super-hydrophobic with WCA of 167°. SHS formed was also found stable with contact angle decreasing only 4–7° after 3 months. It also possessed great corrosion protection. Lin et al. [10] used Ti6Al4V as substrate material on which SHS having 3D porous structures, was fabricated. The process of fabrication started with sand blasting of samples resulting in microscale pattern structures. After that, etching was done in two steps. In first step, above samples were immersed in a solution of H_2SO_4 to get microscale pits. Second step of etching was done using alkaline solution and it resulted in porous structures on surface. The samples were then annealed at a temperature of almost 500°C followed by lowering surface energy by modification with FAS-17 solution. NaOH concentration varied during fabrication process and it was found that the best concentration was in range of 2.5–3.5 M. WCA was determined to reach maximum of 160° and SA was 3°. SHS formed by this process also showed very good corrosion protection behavior based on electrochemical corrosion results.

Choi et al. [11] developed a facile and cost-effective approach to fabricate SHS from easily available and environment-friendly materials. Fabrication process was started by pouring polydimethyloxysilane (PDMS) solution on substrate followed by spin coating to make it flat. Then salt particles were introduced on hot embedded PDMS surface. In the last step, dissolution of salts was done by dipping sample into water bath. This resulted in etching of only salt particles and not PDMS that produced micro/nano-hierarchical structures. Surface was modeled as in Cassie-Baxter state with WCA of 151° as compared to 102° for flat PDMS surface without etching. Dust removal experiments showed that SHS formed were capable of self-cleaning. Moreover, SHS had great corrosion resistance properties with droplets of acid/alkali solutions showing contact angles in range of 147–152°.

Table 1 summarizes the work of different researchers who have used chemical etching method for SHS fabrication.

2.2. Solution immersion

Xu et al. [12] developed a simple one-step method of SHS fabrication. Solution-immersion process was used in which copper foams were used as substrate. Pure copper foams after properly washed, ultrasonically cleaned and dried were immersed in ethanolic stearic acid solution for 4 h, 2 days and 4 days. With increase in immersion time, clusters formed became denser, covered the surface more, the skeleton of 3D porous structure got thicker and rougher while pore size decreased. WCA kept increasing with immersion time to 156° for 4 days and SA decreased to minimum of 4°. This fabrication process resulted in robust and mechanically stable SHS. Zhao et al. [13] developed a simple and pure chemical approach of fabricating SHS. SiO_2-coated SiC nanowires were used as substrate material which already contained nanostructures on surface. Fabrication process started with the preparation of ethanol solution of FAS. The substrate was then immersed in the solution for a day which resulted in the surface with super-hydrophobic nature. The treatment process did not produce any change in the surface morphology, micro-texture and crystal phase of substrate. WCA of 5 μL droplet was measured to be 153°. To check for durability, SHS fabricated was irradiated by UV lamps of 100 W which resulted in the great durability of surface. Zheng et al. [14] fabricated SHS on glass substrate by dip coating sol-gel process. At first, glass plate was properly washed and cleaned. Then, two solutions named as 'A' and 'B' were first prepared. Solution 'A' contained glycidoxypropyl trimethoxysilane (GPTS) modified with silica sol while solution 'B' consisted of PTFE emulsion.

Ref. no	Substrate	Materials used	Coating materials for lowering surface energy	Max. water contact angle/ min. sliding angle	Stability	Remarks
[3]	Aluminum alloy	(1) Cast aluminum alloy (2) Hexane (3) (TCODS) (4) (TCDS) (5) (TCOS) (6) Sodium hydroxide (7) Hydrochloric acid (8) Acetic acid (9) Deionized water	Solution of (TCODS + TCDS + TCOS) & hexane	165°/3°	Remained stable after (1) 100 h immersing in water (2) 30 min ultrasonication (3) (3) Heated to 375°C for 20 min	Cost-effective method to create physically stable, thermally stable and storable SHS
[4]	Zinc	(1) Zinc plate (2) FAS (3) Silver nitrate (4) Copper nitrate (5) Chromium nitrate (6) Anhydrous ethanol (7) Nitric acid	Solution of ethanol and FAS	(1) When $AgNO_3$: 162 ± 2° (2) When Cu $(NO_3)_2$: 157 ± 2° (3) (3) When Cr $(NO_3)_3$: 140 ± 2°	WCA remained almost constant for 7 months in air	A fast method to fabricate SHS, good for metals
[7]	(1) Aluminum (2) Copper	(1) Aluminum foils (2) Copper foils (3) Fine sand paper (4) Absolute ethyl alcohol (5) $Cu(NO_3)_2{\cdot}3H_2O$ (6) Pure HCl (7) $AgNO_3$ (8) Pure HNO_3 (9) 1H,1H, 2H, 2H-Perfluorodecyltriethoxysilane (10) Deionized water	1H,1H, 2H, 2H-Perfluorodecyltriethoxysilane	(1) For aluminum: 164° (2) For copper: 157°	–	A simple chemical etching method for fabrication of SHS on Aluminum and copper substrate
[13]	Aluminum	(1) Aluminum plates (2) Methanol (3) Acetone (4) Toluene (5) Xylene (6) HF (7) HCl	(1) PFO coating (2) PA coating (3) RTV coating	(1) When PFO coated: 162.0 ± 2.2°/0.4 ± 0.1° (2) When PA coated: 161.7 ± 1.9°/2.5 ± 1.0° (3) When RTV		Effect of environmental factors on super-hydrophobic behavior of surfaces was also studied

Ref. no	Substrate	Materials used	Coating materials for lowering surface energy	Max. water contact angle/ min. sliding angle	Stability	Remarks
		(8) Perfluoroalkyltriethoxysilanes (9) Palmitic acid (10) Silicon rubber (11) RTV		coated: 158.3 ± 1.5°/3.6 ± 1.4°		
[20]	Aluminum foil	(1) Aluminum foil (2) $CuCl_2{\cdot}2H_2O$ (3) Emery paper no. 400, no. 600, and no. 1000 (4) Hydrochloric acid (5) Hexadecyl-trimethoxysilane	2 wt.% hexadecyl-trimethoxy silane	161.9 ± 0.5°/6.8°	The as-prepared SHS can effectively reduce the ice formation and is stable under indoor and ambient environments and shows good stability against water drop impact and sand-impact abrasion	The SHS fabricated by this method can mitigate the freezing process and is stable against water drop impact
[26]	Silicon	(1) Cleaning reagents (2) Acetone (3) Ethanol (4) Iso-propanol (5) $NaBF_4$ (6) $AgNO_3$ (7) PFTS (8) Water (9) Hexadecane	(1) C_4F_8 (2) SiO_x overlayers chemically modified with PFTS	(1) NanoSi + C_4F_8: (i) 160° with H_2O, (ii) 87° with hexadecane (2) Nano-Si+ PFTS: (i) 160° with H_2O, (ii) 110° with hexadecane	–	SHS formed are highly oleophobic by this simple, electroless etching method
[29]	Aluminum alloy	(1) 2024-Al (2) KH-832 (3) Ethanol (4) H_2SO_4, HNO_3, H_3PO_4 (5) $Al_2(SO_4)_3{\cdot}18H_2O$ (6) CrO_3, NaOH, Na_2CO_3 (7) $Na_2SiO_3{\cdot}9H_2O$ (8) $Na_3PO_4{\cdot}12H_2O$	Solution of KH-832 in ethanol	167.7°/5°	Surface was excellent corrosion resistant and showed stable super-hydrophobicity with WCA decreased by only 4° to 7° when kept in air for 3 months	Very stable and corrosion resistant SHS was formed
[31]	Ti6Al4V	(1) Ti6Al4V (2) FAS-17 solution (3) Acetone (4) Alcohol	FAS-17 solution	160°/3°	Corrosion resistance test showed that i_{corr} decreased by 3 times while E_{corr} increased by 0.26V	SHS formed could be used in corrosion environment

Ref. no	Substrate	Materials used	Coating materials for lowering surface energy	Max. water contact angle/ min. sliding angle	Stability	Remarks
		(5) Deionized water (6) H_2SO_4 (7) NaOH (8) $NaHCO_3$			making the SHS excellent corrosion resistant	
[32]		(1) PDMS + curing agent (10:1) (2) NaCl (3) KOH, HCl (4) Deionized water (5) KCl (6) Carbon nano-powder	(No coating material used)	$168 \pm 7°/6°$	Excellent corrosion resistance showed when SHS was put in different pH solutions with WCA always found in range of 147°–152°	Inexpensive, easy and environment-friendly method of fabricating SHS with good self-cleaning and corrosion-resistant characteristics

Table 1. Chemical etching.

Both solutions were mixed into each other. After then, glass was immersed in the mixture by dip-coating process. Later, drying and heating of modified glass resulted in SHS. Both micro and nanostructures were produced on surface and WCA and SA were found to be 156 and 6°, respectively. Weight contents of SiO_2 and PTFE were also varied and then optimized to be 20:9. To check the thermal stability of coating, the temperature of reacting process varied as 100, 200, 250 and 300°C. Best super-hydrophobic features were found at 250°C after which WCA and SA kept decreasing and increasing, respectively. Fabrication procedure used was very simple and of low cost.

Fleming and Zou [15] used dip-coating method to fabricate stable super-hydrophobic and super-hydrophilic surfaces. Titanium plate was used as substrate material. Fabrication was done in three steps. In first step, grade 5 titanium substrate was sandblasted to have micro roughness on surface. Second step consisted of dipping sample into colloidal silica solution by dip-coater to generate nanoscale roughness on surface. In last step, sample was put in DRIE chamber in which C_4F_8 gas was introduced resulting in low surface energy film over the surface. WCA was found to be about 152°. Stability tests of wetting/de-wetting cycles showed that both super-hydrophilic and super-hydrophobic surfaces were stable for about 54 days. Super-hydrophilic surface fabricated by this process could be stored for about 25 months in ambient atmosphere. Liu et al. [16] used a simple one-step sol-gel process to fabricate super-hydrophobic coatings which were transparent and possessed excellent self-cleaning property due to very high WCA. The fabrication process was done on glass substrate which was initially cleaned properly. A coating solution was prepared which contained trimethoxysilane in ethanol, ammonia and deionized water. Glass substrate was immersed in that solution. Different deposition times were taken as 10, 100, 300 and 600 s and surface morphology as well WCA were observed at each deposition time. The coatings at those deposition times were named as T-1, T-2, T-3 and T-4 coatings, respectively. Surface roughness increased with increase in deposition time but decreased at 600 s deposition time. Similarly, WCA also increased with deposition time with maximum of 169° for T-3 coating and then decreased due to decrease in surface roughness. The coating showed good behavior under water-jet impact. Mechanical durability of coating was found to be poor due to softness of FAS layer. Coating also showed plastron stability.

Mahadik et al. [17] fabricated SHS by a simple and inexpensive sol-gel method. Super-hydrophobic features were achieved without using any surface modification material. Quartz substrate was used in this study which was first properly cleaned. The alcohol was prepared consisting of MTMS, methanol, oxalic acid and ammonium hydroxide. Finally, quartz substrate was dipped in the alcohol. Later on, it was annealed and dried resulting in SHS. Contact angles were measured for droplets of water, oils, acids and organic liquids. CA measurements showed very high WCA of 168° and sliding angle of 3° for double distilled water. During thermal stability tests, temperature of surface was increased to 560°C but it remained super-hydrophobic. Ultra-high water-repellent SHS was also found to be very durable. With petroleum oils, it behaved as super-oleophilic surface. Huang and Lin [18] fabricated transparent SHS by simple sol-gel process in which temperature was kept low to about 80°C. Transparent plain glass was the substrate material which was initially cleaned properly. Silicic acid was prepared from TEOS and HCl aqueous solution. It was then added to the silica/ethanol solution. Glass substrate was then dipped in the final solution for 1 min. This coated substrate was then dipped in PFOTS ethanol solution for surface modification lowering surface energy. Silica powders used were of two sizes such as 15 and 40 nm. Surface coating varied with silica particle loading and silicic acid concentration. Best super-hydrophobic features were obtained with sample code 15B containing 0.09 M silicic acid and 6 g 15 nm silica particles. CA measurements were taken for the droplets of water as well as CH_2I_2. For sample code 15B, contact angles for water and CH_2I_2 were found to be $160.4 \pm 4.8°$ and $152.1 \pm 2.8°$, respectively. Moreover, coatings fabricated in this study were also found to be stable after ultrasonication test and water-drop impact.

Ramezani et al. [19] fabricated transparent and thermally stable SHS by sol-gel process. Initially glass substrate was washed and ultrasonically cleaned. Alcohol was obtained from the mixture of ethanol, ETES and NH_4OH in right proportions. Substrate was vertically dipped in this transparent silica sol. Heating and drying resulted in the deposition of silica film over glass substrate. In last step, iso-OTMS in hexane was used as surface modification agent and its concentration varied from 0 to 15% to determine its effect on super-hydrophobicity of surface. When its concentration was 0%, WCA was found to be 108°. With increase in its concentration, WCA also increased reaching to its maximum value of 160° at 15% concentration after which no more WCA increased. The size of silica nanoparticles increased from 26 to 42 nm after modification. To determine thermal stability of SHS, temperature was increased from room temperature to 600°C but super-hydrophobicity was retained to 440°C. Moreover, SH films were highly transparent and had uniform size. Chen et al. [20] developed a simple and green methodology to fabricate SHS which have significant role in industries now-a-days. Al foil was used as a substrate followed by zinc sheet in second phase. Al foil was properly washed, cleaned and dried before applying fabrication methodology. A solution of stearic acid was made with water followed by addition of HCl in that solution. The etching process was started by dipping Al foil into the solution for 90 min, and 75 min for zinc sheet. After drying the sample, ladder-like microstructures and wave-like structures were formed on Al-etched and zinc-etched samples, respectively. WCA for both substrates was more than 150° showing super-hydrophobicity. Anti-corrosion property of SHS was also improved as compared to bare samples. The results of steam-freezing experiment showed that SHS possessed icing delay. Effect of pH on WCA of SHS was also determined and it was found that during whole range of pH 1–14, it remained higher than 150° showing stability in all conditions.

Figure 3 shows the process of solution immersion.

Wu et al. [21] fabricated SHS by one-step method which was green as well as versatile, that is, it could be applied on different types of substrates. In this method, emulsion was created in different steps beginning by dispersion of silica nanoparticles in ethanol followed by modification with PTES. Epoxy resin and curing agent were later mixed in that solution by stirring. Three different types of substrates such as wood, paper and glass were used on which SHS was fabricated. Emulsion on substrate was also coated by three methods such as brushing, spraying and dipping. The results showed super-hydrophobicity with CA around 152° and SA less than 10° on all substrates using all three methods. This acted as a proof that fabrication method is versatile and not affected by the method of coating. Epoxy resin appeared to be as a necessary ingredient for fabrication of SHS as the one formed without epoxy resin was not mechanically stable. Durability of SHS was tested by sand abrasion, knife scratching and adhesive tape tests resulting in almost constant values of CA.

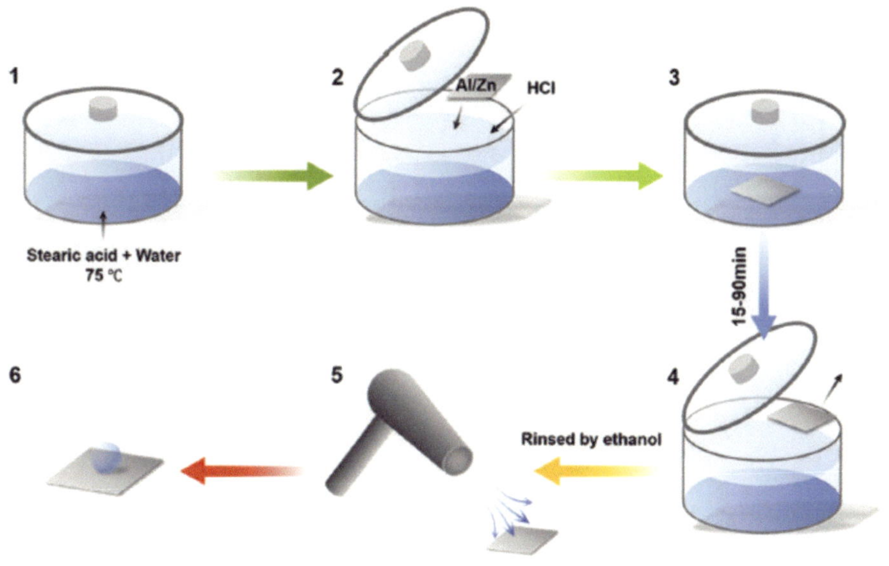

Figure 3. Solution-immersion process [20].

Table 2 gives the summary of different researchers' work who have used solution immersion method for SHS fabrication.

2.3. Laser electrodeposition

Tang et al. [22] used laser electrodeposited composite method to fabricate SHS. Copper channel was used as substrate. First surface was roughed by laser, then electrodeposition was performed to obtain microstructures on surface and then dried. No coating material was used for lowering surface energy. Static WCA was found to be decreasing by increasing the channel width and the maximum WCA was 156°. Rolling angle was also determined and it was less than 5°. Moreover, the effect of microstructures on pressure drop was also studied. Pressure drop increased with increase in channel width and was also smaller in super-hydrophobic channel as compared to smooth channel. Friction factor was reduced in super-hydrophobic micro-channels by maximum of 48.8%. Li et al. [23] used a new one-step and simple idea for the fabrication of super-hydrophobic and hydrophobic surfaces. Zinc sheets were used as a substrate material. Laser ablation process was used to increase surface roughness of zinc sheet dipped in aqueous solution of H_2O_2. Two types of laser ablation were used in the study such as nanosecond (ns) and femtosecond (fs) laser ablation. ZnO and $Zn(OH)_2$ were generated on the zinc surface as a result of laser ablation. Both types of laser ablation created different microstructures on surface. Clustered flower-like microstructures were formed on ns-laser ablated sample while non-directional flaky nanostructures were formed on fs-laser ablated sample. Roughness was also found to be higher for ns-laser ablated sample due to which it was super-hydrophobic with WCA and WSA of 158.5 and 4.3°, respectively. On the other hand, fs-laser ablated sample was highly hydrophobic with WCA of 145.7° and WSA of 12.5°.

Ref. no	Substrate	Materials used	Coating materials for lowering surface energy	Max. water contact angle/min. sliding angle	Stability	Remarks
[9]	Copper foams	(1) Pure copper foams (2) Acetone (3) Ethanol (4) HCl (5) Deionized water (6) Nitrogen gas (7) Ethanolic stearic acid solution	(No coating material)	156°/4°	Robust and mechanically stable	A simple one-step method, but can take longer time for fabrication of SHS
[12]	SiO_2-coated SiC nanowires	(1) Ethanol solution of FAS (2) Glacial acetic acid (3) SiO_2-coated SiC nanowires (4) Ethanol	Fluoroalkylsilane	153°	Possessed ultra-violet durability	A pure chemical approach to fabricate SHS with no change in morphology, microstructure and crystal phase
[16]	Glass substrate	(1) Glass substrate (2) Silica sol (3) Mixture of water and ethanol (4) Acetic acid (5) GPTS (6) PTFE emulsion (7) Deionized water	Modified SiO_2 sol and PTFE emulsion	156°/6°	Coating gets destroyed as temperature is increased above 250°C	A simple and cost-effective procedure for nano-composite coating, but coating gets destroyed as temperature is increased above 250°C
[18]	Titanium surface	(1) Grade 5 titanium (2) 165-micron (3) Alumina media (4) Acetone (5) Isopropyl alcohol (6) (DI) water (7) Nitrogen (8) Colloidal silica solution (9) C_4F_8 gas	C_4F_8 gas	152°	High degree of stability, as surfaces retain their wetting properties for at least 54 days under multiple wetting/de-wetting cycles	An easy method of fabrication, gives long-term wetting stability and storability
[19]	Glass	(1) Glass substrates (2) Trimethoxysilane (3) Ethanol	Trimethoxysilane	169°/5°	The super-hydrophobic wetting state was preserved under the impact of a high-speed water jet,	A simple one-step process to fabricate optically transparent SHS with very high contact angle

http://dx.doi.org/10.5772/intechopen.72820

Ref. no	Substrate	Materials used	Coating materials for lowering surface energy	Max. water contact angle/min. sliding angle	Stability	Remarks
		(4) Ammonia (5) Deionized water			however, the coating showed poor mechanical durability	of 169° but mechanical durability is poor
[21]	Quartz substrates	(1) Quartz substrates (2) Methyl-trimethoxysilane (MTMS) (3) Methanol (99%) (4) Hexane (85%) (5) Oxalic acid (6) Liquor ammonia (7) pH indicator papers 1–14	(No coating material used)	$168 \pm 2°/3 \pm 1°$	When MTMS based coating was exposed with outdoor temperature, wind, air and humidity variations over six months environment, the coating showed reduced water contact angle by 27°, and increased sliding angle $8 \pm 1°$	A simple and inexpensive method to fabricate SHS with very high WCA on quartz substrate without surface modification
[22]	Transparent plain glass	(1) Transparent plain glass (2) Tetraethoxysilane (TEOS) $C_8H_{20}OSi$ (3) Hydrochloric acid (HCl) (4) Ethanol (C_2H_5OH) (5) Silica powder, 40 nm and 15 nm, SiO_2 (6) 1H,1H,2H,2H-perfluorooctyltrichlorosilane, PFOTS, $CF_3(CF_2)_5(CH_2)_2SiCl_3$	1H, 1H, 2H, 2H-perfluorooctyltrichlorosilane, PFOTS, $CF_3(CF_2)_5(CH_2)_2SiCl_3$	(1) For water, $160.4 \pm 4.8°/7.1 \pm 2.0°$ (2) For CH_2I_2, $152.1 \pm 2.8°/84.5 \pm 9.3°$	Coating was chemically and mechanically stable as tested by ultrasonic damage and water dropping impact	SHS fabrication process was conducted by a wet chemical technique at relatively low temperature conditions
[23]	Glass substrates	(1) Glass substrates (2) Ethanol and hexane (3) Ethyl-triethoxysilane (4) Isooctyl-trimethoxysilane (5) Ammonia (6) Deionized water	Isooctyl-trimethoxysilane (iso-OTMS)	160°	The SH silica films are thermally stable up to 440°C and above this temperature, the silica films lose super-hydrophobicity	Transparent and thermally stable films are fabricated by this method
[33]	(1) Aluminum foil (2) Zinc sheet	(1) Al foil (2) Zn sheet (3) Ethanol (4) Acetone (5) Hydrochloric acid (6) Stearic acid (7) Deionized water	(No coating material used)	>150° for both Al foil & Zn sheet	SHS showed excellent stability when put in acidic and basic conditions with WCA always more than 150°. Also great corrosion protection was found	A facile and green method to prepare very stable and corrosion resistant SHS

Ref. no	Substrate	Materials used	Coating materials for lowering surface energy	Max. water contact angle/min. sliding angle	Stability	Remarks
[35]	(1) Wood (2) Paper (3) Glass	(1) Wood sample (2) Filter paper (3) Glass (4) Silica nanoparticles, 12 nm (5) PTES (6) Epoxy resin & curing agent (E-44) (7) Absolute ethanol	PTES	~152°/10° for all SHS	SHS formed were mechanically very durable when tested by finger wiping, tape peeling and sandpaper abrasion and maintained their super-hydrophobicity in solutions of different pH	A green, versatile, one-step method independent of substrate to fabricate very durable SHS

Table 2. Solution immersion.

Chun et al. [24] developed fabrication method of SHS on copper substrate and the focus of this study was kept to the fast fabrication of SHS. First, copper substrate was exposed to the nanosecond laser beam originating from laser beam machining of 3.3 watt power and 20 Hz frequency. Earlier before this study, it had already been found that laser beam produced layer of CuO on sample which was hydrophilic and it took a long time about 27 days to be converted into SHS which had Cu_2O on the surface. So, instead of keeping the laser textured surface in ambient conditions, it was made super-hydrophobic using annealing at low temperature, that is, 100°C. This accelerated the conversion of CuO into Cu_2O resulting in the formation of SHS after 13 h. As copper is also stable with ethanol, so additional reduction in time to get SHS was achieved when ethanol was used by which SHS was formed in less than 5 h. WCA was also found to be maximum of 165° showing super-hydrophobicity.

Figure 4 shows the schematic illustration for fabrication of SHS by laser beam machining.

Table 3 summarizes the work of different researchers who have used electrodeposition technique for SHS fabrication.

2.4. Template deposition

Wang et al. [25] used template deposition method for the fabrication of SHS using copper sheet as substrate. The fabrication method consisted of three steps. First step was to prepare template from polystyrene microspheres powders. Second step was to put copper plate in polystyrene colloidal microspheres with $CuSO_4$ solution as an electrolyte. The electrodeposition potential was about −0.5 V and SEM images were taken after deposition time of 19 s. During this process, copper atoms filled the voids in the template. In final step, surface was modified using fluorosilane solution resulting in a SHS. WCA of 5 μL droplet was measured to be 156.3°. WCA increased with increase in deposition time to 19 s and then decreased after that. Bhagat and Gupta [26] developed a purely physical process of fabricating super-hydrophobic polycarbonate surfaces (PC). Silicon wafers were used as substrate which along-with PC were initially washed and cleaned properly. In first step of fabrication, micro-textures were produced on substrate by high power laser and in second step, those microstructures were thermally replicated on PC surface by sandwiching them between two hot plates having different temperatures. Upon cooling, both surfaces were then separated from each other. Effect of replication temperature was such that when temperature varied from 155 to 175°C,

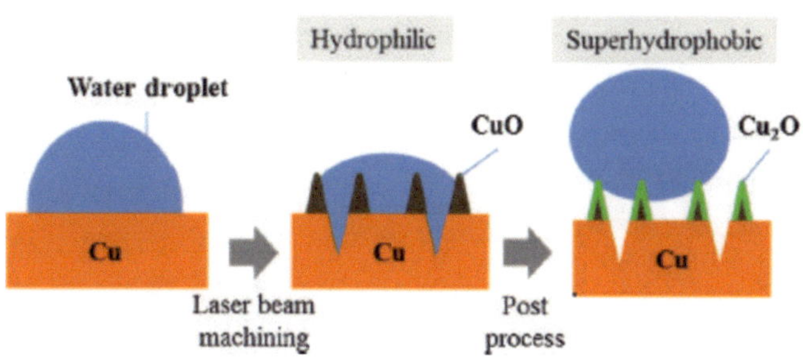

Figure 4. Schematic illustration of SHS with laser beam machining and post-process [24].

Ref. no	Substrate	Materials used	Coating materials for lowering surface energy	Max. water contact angle/ min. sliding angle	Stability	Remarks
[5]	Copper	(1) Copper channel (2) Solution of $CuSO_4$ and HCl as electrolyte (3) Flat copper as anode (4) Ethyl glycol	(No coating material used)	156°/<5°	–	A method to fabricate SH copper channels without the use of coating material
[30]	Zinc sheet	(1) Zinc sheet (2) Hydrogen peroxide aqueous solution	(No coating material used)	(1) ns-laser ablated sample: 158.5°/4.3° (2) fs-laser ablated sample: 145.7°/12.5°	–	One-step and simple method with small number of materials needed
[37]	Copper	(1) Pure copper plate (2) Ethanol	(No coating material used)	165°/9°	–	Method to fabricate SHS with reduced time to gain super-hydrophobicity after laser texturing

Table 3. Laser electrodeposition.

micro-pillars formed on PC surface varying in the average height from 1.34 μm to 6.68 μm. During this, WCA also varied from 82° to maximum 155°. Super-hydrophobic PDMS surfaces were also fabricated by this process with maximum CA of 162°. This physical process of fabrication resulted in robust SHS with good mechanical properties.

Figure 5 shows schematic diagram for the fabrication of super-hydrophobic PC by electrodeposition.

Liu et al. [27] prepared SHS by simple, rapid and facile method. SiC paper was first sonicated, rinsed with ethanol and then baked. PDMS pre-polymer was pored over SiC paper. It was peeled from PDMS cast after cooling. As rough PDMS was templated from SiC paper, flat PDMS was also templated in the same way on Si wafer. The WCA for both cases were measured and it was found that rough PDMS behaved as super-hydrophobic while flat PDMS showed hydrophobic behavior with WCA of 154 and 113°, respectively. SHS formed was also found to possess excellent adhesive properties described by hybrid wetting state.

Table 4 gives the summary of different researchers' work who have used template deposition method for fabrication of SHS.

2.5. Spray coating

Momen and Farzaneh [28] fabricated exceptionally stable SHS using a simple and cost-effective approach of spray coating. Glass substrate used was first ultrasonically cleaned followed by preparation of three separate samples. First sample 'a' consisted of RTV silicon

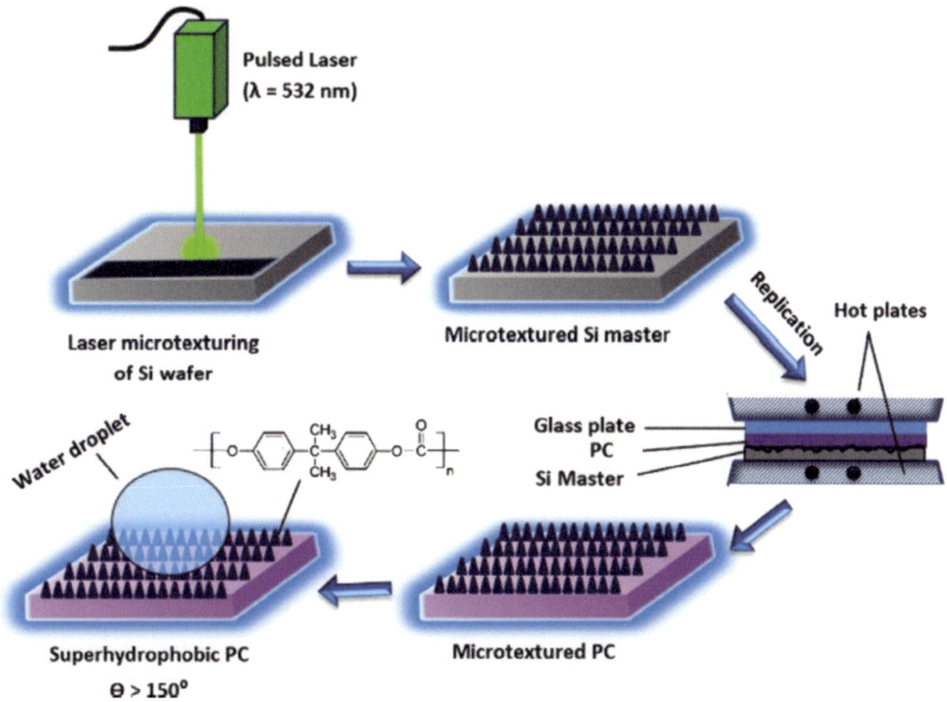

Figure 5. Schematic diagram for fabrication of super-hydrophobic PC by laser micro-texturing of Si wafer and its subsequent replication [26].

rubber (SR) in hexane, sample 'b' of ZnO nanoparticles in hexane and SR while sample 'c' contained both ZnO and SiO_2 nanoparticles in hexane and SR. Then all three samples were separately spray-coated on glass slides by using spray gun. SEM investigations showed that sample 'a' coating resulted in smooth surface. Sample 'b' coating caused nanscale roughness on surface while sample 'c' coating resulted in both micro and nanostructures. WCAs for all samples a, b and c were measured to be 117.3, 132.5 and 162.7°, respectively. Different stability tests carried out on SHS resulted in slight decrease in WCA after 10 days by immersing in different pH solutions. UV and humidity had also little effect on WCA. Against heating treatment at 150°C for almost a month, WCA was decreased just a little showing the stability of SHS. Ipekci et al. [29] developed one-step spray-coating method of fabricating SHS which had improved mechanical robustness and durability. A matrix was required for better dispersing of silica nanoparticles functionalized with fluorinated silanes. Hydroxyl-terminated polystyrene was used as a matrix which reacted with substrate to form polymer brushes through covalent bonding. Glass slides were used as a substrate. WCA measurements showed that CA was greater than 170° and SA was approaching 0°. WCA was highest for weight-ratio of 1 between PS-OH and FNP. Effect of curing temperature was also determined and WCA was found to be maximum at 190°C. Abrasion tests also showed that WCA decreased after grain velocity of 10 km/h. SHS formed were also transparent to 85%. Enhancement of mechanical robustness was different for different substrates and polymers.

Ref. no	Substrate	Materials used	Coating materials for lowering surface energy	Max. water contact angle/ min. sliding angle	Stability	Remarks
[8]	Copper	(1) Copper sheet (2) Acetone (3) Styrene (4) Sodium dodecyl sulfate (5) Ethanol (6) Deionized water	Fluoroalkylsilane	156.3°	–	The procedure consists of three steps, that is, template preparation, electrodeposition and surface modification
[11]	Polycarbonate surface	(1) Silicon wafers (2) Acetone (3) Methanol (4) Deionized water (5) Polycarbonate sheet (6) PDMS sheet	(No coating material used)	(1) For micro-textured PC: 155°/<10° (2) For micro-textured PDMS: 162°	The micro-textured PC surfaces are robust enough and exhibit very low damage to the master	A simple physical method to make SH PC surface without any chemical change
[25]	PDMS	(1) PDMS (2) SiC paper (3) Deionized water (4) Formamide, 99.5% (5) dimethyl sulfoxide (DMSO), 99.8% (6) Ethanol	(No coating material used)	153°/23°	Water drop test, ultrasonic treatment and water immersion tests show the high durability of SHS, also thermally stable below 500°C	A simple, inexpensive and modification-free process of fabricating highly durable, transparent, thermally stable SHS

Table 4. Template deposition.

Wang et al. [30] fabricated SHS by cost-effective and environment-friendly method. The method used was salt-spray method which increased surface roughness, followed by modification of surface reducing its surface energy. Two substrate materials such as steel and Mg alloy were separately used. Both substrates were first properly polished, cleaned and then dried. These were placed for about 2 h in the neutral spray chamber which contained NaCl solution creating humid and fog condition. This generated needle-like structures on steel and flower-like structures on Mg alloy surface. Finally, super-hydrophobicity was achieved when above samples were put in solution of FAS-17. WCA measurements showed that CA of 156.05 and 152.65° while SA of 2 and 5° were achieved for steel and magnesium alloy samples, respectively. Mechanical durability of SHS formed was tested by water fall test in which kinetic energy of falling water varied. The results showed that WCA decreased after 20 minutes but water-repellent properties still remained in the solution. Although durability of SHS formed by this method was not excellent, it is still good.

Figure 6 shows the schematic diagram for fabrication of SHS by salt-spray method.

Table 5 summarizes the work of different researchers who have used spray coating technique for SHS fabrication.

http://dx.doi.org/10.5772/intechopen.72820

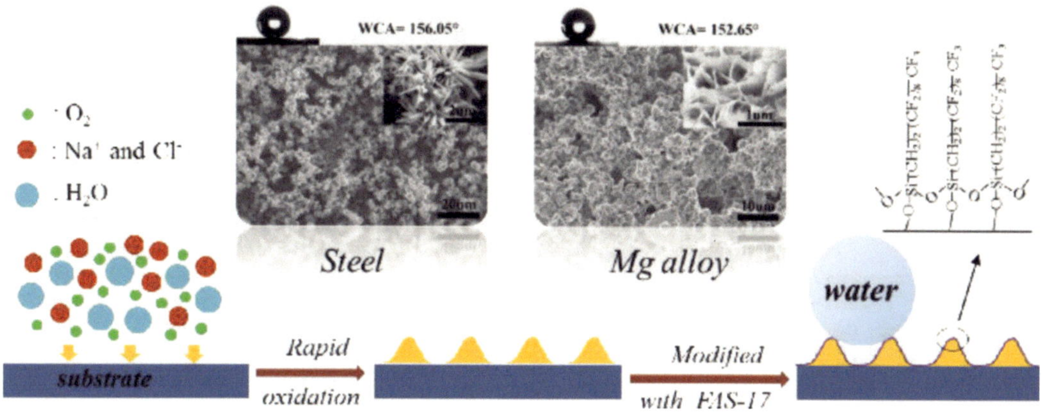

Figure 6. Schematic diagram for fabrication of SHS by salt-spray method [30].

2.6. Other techniques

Feng et al. [31] fabricated SHS using a simple method in which copper plate was used as a substrate material. Fabrication process was completed in three steps. First, copper powder was oxidized with $AgNO_3$ solution after it was ultrasonically cleaned with nitric acid and distilled water. Second, surface hydrophilicity was changed into hydrophobicity by lauryl mercaptan (DM). In last step, mold compression was performed on copper powder under 80 MPa to obtain super-hydrophobic characteristics. Tests showed maximum WCA of 155.2° and SA was also less than 5°. Super-hydrophobicity of surface was also dominant for other fluids. Copper-based SHS was also found to be very long-term durable, storable, regenerable and of excellent cleaning property.

Sun et al. [32] developed composite electro-brush flow plating technology for the fabrication of SHS over copper substrate. Experimental device was made at first which consisted of tank, linear motor, pump, substrate cathode, steel anode, plating solution, etc. Copper plate was polished mechanically, ultrasonically cleaned and then fixed in experimental device. Ni plating solution consisted of thoroughly mixed nano-Al_2O_3 particles was prepared. Electro-brush was moved reciprocately by motor to rub the substrate surface. Finally, n-Al_2O_3/Ni coating was obtained on copper surface with micro-nanostructures. The resulting sample was immersed in FAS solution for surface modification. The results showed WCA of 162° and SA of less than 10°. The WCA increased with increasing plating voltage to 12 V and then decreased after it. Similarly, WCA increased with plating time to 2 s and then remained almost constant. Final SHS obtained possessed excellent mechanical properties and stability.

Huang and Leu [33] used spin coating method to fabricate SHS. Substrate used was copper heat sink. Fabrication process was completed in three steps. In first step, chemical polishing was done on substrate to remove oxides and impurities. In second step, CuO nanostructures were produced on substrate surface by immersion in H_2O_2 solution. In last step, electronic liquid EGC-1720 was prepared by fluorosilane polymer and hydrofluoroether solvents. Sample was placed on acrylic cylinder resulting in SHS after spin coating. WCA was measured to be 153.43 ± 5.13°. Comparison of pure copper heat sink and super-hydrophobic copper heat

Ref. no	Substrate	Materials used	Coating materials for lowering surface energy	Max. water contact angle/ min. sliding angle	Stability	Remarks
[17]	Glass slide	(1) Glass substrate (2) RTV silicon rubber (3) Hexane (4) ZnO nanoparticles (5) SiO_2 nanoparticles (6) Acetone (7) Water (8) Nitrogen gas	SR/ZnO/ SiO_2 composite	162°/7°	Coating possesses (1) Medium-good stability against a wide range of pH solutions (2) No significant effect of UV and humidity (3) Retained its super-hydrophobic and roll-off properties against the heating treatment	A simple spray-coating method in one-step process, ultra-water-repellent properties in several environmental factors
[27]	Glass slide	(1) Glass slides (2) Polystyrene (PS) (3) hydroxyl-terminated polystyrene (PS-OH) (4) Toluene (99%) (5) Cyclohexane (HPLC grade, 99.9%) (6) Silica NPs (12 nm diameter) (7) FDTS	Flourinated silanes	170°/0°	(1) For impact velocities lower than 10 km/h during sand abrasion tests, surfaces retained their super-hydrophobicity (2) Surfaces were transparent with 85% light passing capability	Mechanically robust and transparent SHS was formed by one-step spray-coating method
[36]	(1) Steel (2) Magnesium alloy	(1) Mg alloy (2) Low carbon steel (3) Ethanol (4) Acetone (5) Sand paper (6) NaCl (7) Deionized water (8) 1 wt% FAS-17	1 wt% FAS-17	(1) For Mg: 152.65°/5° (2) For steel: 156.05°/2°	Waterfall jet test at 100 kPa showed that WCA and sliding angle started decreasing and increasing, respectively, after 20 min, so moderate durability was found	Cost-effective and environment-friendly method to fabricate SHS with moderate durability

Table 5. Spray coating.

sink was also carried out by performing CHT experiments. Surface coating was degraded during CHT experiments due to continuous heating by condensate water droplets.

Zhang et al. [34] fabricated anti-corrosive and durable SHS by dual-layer method. Glass slide was used as substrate material which was properly cleaned. The fluorinated silica was prepared and mixed with PFTS under stirring to get final fluorinated silica. Then PDMS solution was mixed with fluorinated silica and layer 1 was obtained by its spray coating on glass substrate. After then, layer 2 was obtained by spray coating of fluorinated silica on top. Micro-protrusion structured surface was obtained whose WCA was found to be maximum at 157°. Positive results from the surface abrasion test and ultrasonication test showed mechanical stability of super-hydrophobic coating. SHS also showed anti-corrosive behavior when copper substrate was used. Moreover, immersion in different pH solutions also did not change super-hydrophobicity of surface fabricated in this study.

Stanton et al. [35] prepared SHS by simple and cost-effective method. Glass substrate was used which was dip-coated into the already prepared PDMS solutions. These PDMS films were exposed to the candle flame. Candle soot made the films black. Removal of candle soot then resulted into the formation of nanostructures on the films and transparency was also obtained. Then, calcination was performed in muffle furnace at 450°C to obtain SHS. Transmittance of SHS was affected by calcination temperature and dip-coating times. Transmittance was found to be maximum at 450°C. Three tests including water-drop test, ultrasonic treatment and water immersion were conducted to check the durability of obtained SHS. Tests showed that SHS were very durable and could easily recover super-hydrophobicity after heat treatments. WCA and SA were measured to be as 163 and 1°, respectively.

Fabrication of SHS requires roughness of surface followed by lowering surface energy. Song et al. [36] used silicon substrate for fabrication of SHS. Surface was made textured by the Al-induced crystallization of amorphous silicon. Lowering of surface energy was done by the deposition of OTS coating. WCA increased by both roughing surface and then by lowering its energy. Only lowering the surface energy of silicon without modifying surface structure resulted in less increase in WCA. OTS-modified textured surface showed maximum CA angle of 155° while OTS-modified smooth surface showed only 112°. Adhesion and friction performance of SHS also decreased by texturing the surface. So, it was obvious that although both texturing as well as surface modification caused increase in WCA along-with decrease in adhesion and friction but texturing played a greater role in this. Sun et al. [37] fabricated SHS on Al substrate by a facile and scalable method. Al foils were first pre-treated in which oxide layer over their surface was completely removed. Then sample was cleaned, rinsed and finally dried. Chemical etching was done by dipping pre-treated samples in HCl solution for 5 min. The chemically etched sample was then placed over stainless steel mesh in a crucible pot containing PDMS solution. The calcination of PDMS at 300°C resulted in decomposition and oxidation reactions on PDMS producing organic SiO_2 nanoparticles. These PDMS vapors grew on chemically etched Al surface leading to surface roughness as well as lowering of surface energy. This combined methodology of chemical etching and chemical vapor deposition produced SHS on Al with WCA reaching 158.7°. The polarization curves showed increased corrosion resistance of super-hydrophobic Al surface. SHS formed were also capable of self-cleaning.

Table 6 gives the summary of different researchers' work who have used other techniques for fabrication of SHS.

3. Heat transfer and other applications

Super-hydrophobic surfaces play a significant role in various heat transfer applications in which surface morphology is very important. In boiling, super-hydrophobic spots along-with hydrophilic spots can result in very high critical heat flux and heat transfer coefficient (HTC) values. Similarly, SHS in condensation heat transfer (CHT) experiments leads to drop-wise condensation in which more than 10 times heat transfer coefficients can be achieved as compared to that in film-wise condensation. Long term-stability and storability are the key parameters to focus on while using SHS for condensation. Another important application of SHS is

Ref. no	Technique used	Substrate	Materials used	Coating materials for lowering surface energy	Max. water contact angle/ min. sliding angle	Stability	Remarks
[6]	Compression molding	Copper	(1) Copper powder (2) $AgNO_3$ (3) HNO_3 (4) Lauryl mercaptan (DM) (5) Ethanol	Lauryl mercaptan (DM)	155.2°/<5°	Remained stable when (1) Placed in the air for ten months (2) Soaked in water for half a month (3) For over a wide pH range (i.e. pH = 6–14)	A simple method to fabricate SHS with long-term durability, stability, storability regenerability and self-cleaning property
[10]	Flow plating technology	Copper plate	(1) Pure copper plate (2) Fluoroalkylsilane (3) Nano-particle n-Al_2O_3 (4) Nickel plating solution reagent	Fluoroalkylsilane	162°/ <10°	Outstanding mechanical properties and stability	A method to fabricate very stable SHS and greater contact angle but also greater sliding angle
[14]	Spin coating	Pure copper heat sink	(1) Pure copper heat sink (2) HNO_3 solution (3) H_2O_2 aqueous solution (4) EGC – 1720 (fluorosilane polymer + hydrofluoroether)	EGC – 1720 (fluorosilane polymer + hydrofluoroether)	153.43 ± 5.13°	EGC-1720 can withstand a harsh vapor environment and is durable for 100 h	A concise and low-cost surface modification method but not much durable
[15]	Dual-layer method	Copper surface	(1) Silicon dioxide nano-powder (2) PFTS (3) Polydimethyl-siloxane		158.7°	SHS were mechanically durable with WCA always greater than 150° after abrasion, ultrasonication and immersion in different pH solution tests. SHS was also found very corrosion resistant	Highly stable and corrosion-resistant SHS was produced by this method
[24]	Calcination process	Glass substrate	(1) Glass slides (2) Wax (3) Cotton threads (4) α, ω-dihydroxypolydimethylsiloxane (5) TEOS	(No coating material used)	163°/1°	–	A rapid, reproducible technique to fabricate very adhesive SHS

http://dx.doi.org/10.5772/intechopen.72820

Ref. no	Technique used	Substrate	Materials used	Coating materials for lowering surface energy	Max. water contact angle/ min. sliding angle	Stability	Remarks
			(6) Dibutyltindilaurate (DBTDL) (7) Ethanol and n-hexane (8) *n*-dodecane				
[28]	Chemical vapor deposition	Silicon	(1) Silicon wafers (2) Acetone (3) Iso-propanol (4) Deionized water (5) Aluminum etchant—type D (6) Octadecyltrichlorosilane (OTS)	Octadecyltrichlorosilane (OTS)	155°/1°	SHS formed had adhesiveness and COF reduced upto 90 and 75%, respectively	This method produced SHS with good adhesive and frictional characteristics
[34]	Chemical vapor deposition	Aluminum foil	(1) Al foil (2) SiC paper (3) NaOH solution (4) HCl solution (5) PDMS (6) Deionized water	(No coating material used)	158.7°	SHS showed excellent corrosion resistance in 3.5% NaOH solution	Facile method to fabricate corrosion-resistant SHS with good self-cleaning ability

Table 6. Other techniques.

anti-icing or anti-frosting which is the primary requirement for PV cells, airplanes, power lines, ships, etc. Moreover, drag reduction characteristics of SHS have also been discussed in this section, which are important for the submarine outer surfaces as well as internal surface of tubes during internal flow.

3.1. Boiling

Betz et al. [38] explained the effect of surface morphology on boiling heat transfer. The density of active nucleation sites and size of contact diameter of departing droplets are important parameters to obtain minimum heat flux (MHF) in boiling. For MHF, number of active nucleation sites should be maximum which can be obtained using hydrophobic surface and contact diameter should be smallest to maintain water transport to hot surface as in hydrophilic surface. Using a single type of surface could fulfill one criterion but not the second one, thereby reducing maximum heat flux. In this study, it was determined that surface having characteristics of both hydrophobic and hydrophilic surface showed CHF more than that obtained by using individual surfaces. This surface was named as biphilic surface for which maximum heat transfer coefficient was two and four times higher than hydrophilic and hydrophobic surface, respectively. Hydrophobic regions in biphilic surface increased the number of active nucleation sites while hydrophilic regions decreased the contact diameter, thereby aiding in liquid transport to hot surface.

Malavasi et al. [39] performed boiling heat transfer experiments on surfaces with various topographies to find the effect of roughness and wettability on boiling regimes. Six different types of surfaces (including two hydrophilic, one hydrophobic and three SHS) were prepared using stainless steel as substrate material. Shape of boiling curve for SHS was different from the typical boiling curve in sense that onset of nucleate boiling (ONB) was achieved near 0°C (<5°C) and Leidenfrost point was also reached earlier just after ONB. Due to more roughness of SHS, number of nucleation sites was also higher, so bubbles formation started just after T_{sat} leading to early ONB. Due to large number of vapors formed on surface, heat transfer to liquid started decreasing rapidly leading to early Leidenfrost point. It was also found that although all three SHS had different roughness values, but due to almost same WCA, all showed same boiling curves. Hence, wettability was dominant over roughness for any surface whose CA exceeded 135° which is the case of SHS.

Yamada et al. [40] created biphilic surfaces on copper substrate and compared it with plain copper surface during nucleate boiling process at pressure values ranging from atmospheric to 6.9 kPa. Difference of ONB between biphilic and plain surface increased with decreasing pressure. Higher HTC values were obtained for biphilic surface than plain surface at low pressure values showing enhanced efficiency of biphilic surfaces at sub-atmospheric pressure. However, this efficiency enhancement was not found below transition pressure at which continuous boiling converted into intermittent boiling. Effects of diameter and pitch of hydrophobic spots on biphilic surface were also investigated by testing three samples. At low heat flux, increasing hydrophobic spot diameter and reducing pitch led to more HTC while at low heat flux, smaller diameter with increased number of spots showed high HTC while pitch affected less.

Fan et al. [41] investigated the effect of four surfaces such as super-hydrophilic (0°), hydrophilic (23°), hydrophobic (119°) and super-hydrophobic (164°), on transient pool boiling by quenching methods. In quenching process, temperature was decreased and its decreasing rate was affected by the nature of surface. Cooling rate was found to be decreasing with increasing WCA. The reason for lowest cooling rate for SHS was the retention of vapor film on surface. On the other hand, super-hydrophilic surface showed maximum wettability leading to the collapse of water film. Boiling curves showed that CHF was found maximum for super-hydrophilic surface at higher values of superheat. Again the reason found was deterioration of vapor film at higher superheat.

3.2. Condensation

Kim et al. [42] investigated the effect of super-hydrophobicity of TFE-coated Pyrex glass tube and compared it with bare tube by conducting condensation experiment. Steam was produced by steam generator and then passed through 3 m long and 19 mm diameter glass tube inclined at 3–5° and finally collected as liquid at downstream. Air was cross-flowed through bare tube first and then through internally TFE-coated tube. Wall temperature distributions were taken for both cases. Flow regimes were also studied. In bare tube, condensation was film-wise and flow regimes were fully stratified while in TFE-coated tube, liquid droplets were formed on inside of tube. Averaged local heat flux values were higher for TFE-coated tube to 2 m length while lower or almost same in the remaining tube length. Temperature values at wall were higher for TFE-coated tube for the whole tube length due to the presence of droplets at inside wall instead of film. In the downstream region, local heat flux values were reduced due to increase in vapor quality. Overall, TFE-coated tube showed greater heat flux due to drop-wise condensation as compared to bare tube with film-wise condensation.

Lu et al. [43] fabricated super-hydrophobic Si nanowires through electroless etching and compared it with hydrophobic Si wafer during condensation process. Test samples were placed in experimental system where T_{sat} was kept constant at 80°C and pressure at 47 kPa. Heat transfer coefficient (HTC) and heat flux were determined for sub-cooling values to 20 K. Flooding was observed in case of hydrophobic Si but super-hydrophobic Si showed droplet jumping and efficient shedding without flooding. At low sub-cooling (1 K), the difference between HTC values of SHS and hydrophobic Si was very large as compared to the difference at high sub-cooling (15 K). This decrease was due to the increased surface coverage ratio of droplets at tube that acted as a resistant to heat transfer. Coverage ratio, departure diameter and departure frequency of water droplets were also found to be increasing with increase in sub-cooling.

Yin et al. [44] prepared three different types of super-hydrophobic coatings such as PFO, PA and RTC coatings with WCA 162, 161.7 and 158.3°, respectively, onto glass substrate through chemical etching method. All three types of coated substrates were put in different condensing environments. These included temperature values varying from 30°C to −10°C at each of the RH values 30, 60 and 90%. At 30% RH, all the coatings showed negligible changes in WCA values. But at 60% RH and −10°C, PFO, PA and RTC coating showed declined values of WCA as 148, 148 and 130°, respectively. A relatively more decreasing trend of WCA was found with 90% RH when all coatings lost their super-hydrophobicity with RTC coating showing the maximum decline in WCA to 114.7°. On the other hand, SA for each coating was found to be increasing with increase in RH values and decrease in temperature, showing considerable super-hydrophobicity loss.

Bisetto et al. [45] compared CHT on hydrophilic and super-hydrophobic copper substrates with WCA 86 and 159° values using thermosyphon two-phase test rig. SHS had micro-nanostructures which brought water molecules in Cassie-Baxter state due to entrapped air after condensation leading to the increased mobility of droplets. Higher droplets departure rate resulted in increased heat transfer for SHS. Effects of vapor velocity and wall sub-cooling on heat flux as well as heat transfer coefficient were also investigated. HTC and heat flux increased with increasing vapor velocity while the effect of wall sub-cooling was not much significant. Drop-wise condensation was observed on SHS for repeated experiments in 5 working days after which DWC transited into FWC with heat transfer performance even more worse than on hydrophilic copper substrate.

3.3. Anti-icing

Liu et al. [46] examined the anti-icing property or freezing delay of SHS. Freezing delay between bare (B) and SHS was also compared. 7075 Al alloy was used as a substrate material. Through laser processing and modification, SHS on substrate was fabricated. Three different types of microstructures, that is, round hump (R), square protuberance (S) and mountain-range-like structure (M) were generated with microstructure 'M' showing maximum WCA of $166 \pm 2°$. To check anti-icing property of all types of SHS as well as bare surface, an apparatus was developed. Testing samples were placed on experimental plate whose temperature was reduced from 16°C to −15°C at a rate of 0.2°C/s. Experimental plate was tilted at 5° and water was sprayed at testing sample. The freezing times of B, R, S and M were found to be as 319, 1153, 1165 and 1938 s, respectively. M surface showed maximum freezing delay indicating that with increasing WCA, freezing time of water droplet was also increased. So, type of micro-structure on surface played the key role in freezing delay. Model was also established in this study by which heat transfer between droplet and surface could be easily analyzed.

Kim et al. [47] applied stochastic approach for the determination of freezing delay of water droplet on bare and SHS. At first, SHS on Al 6061 was prepared by etching and surface modification. Then test sample was placed indirectly on thermoelectric element (TEC) with thermocouples and camera for observation. Temperature of TEC decreased from 25°C to −10°C at a rate of 10°C/min. Ice nuclei started growing near the surface, so thermocouples were also immersed in droplet near the surface for precise measurements. Temperature of every droplet on both bare and SHS was measured during cooling and super-cooling periods. For both bare and SHS, freezing delay points showed random distribution. By using better approach of stochastic data analysis instead of arithmetic average, it was found that freezing delay for SHS was three times more than that of bare surface showing anti-icing behavior of SHS.

Wang et al. [48] prepared three types of samples: sample 1 (bare Al, WCA = 86°), sample 2 (Al with nanostructures, WCA = 156°) and sample 3 (Al with micro-nanostructures, WCA = 160°) and compared their anti-icing characteristics by placing water droplet of 4 μL on each sample at −15°C. Freezing delay time for sample 1, 2 and 3 were 99, 1176.9 and 3942.1 s, respectively.

The presence of micro-nanostructures on sample 3 decreased solid/liquid area fraction which led to the maximum freezing delay. The results of frost formation measurements at 28°C and relative humidity 78% for all three samples showed frosting time of 81, 75 and 51 min, respectively. Again the roughness factor led to maximum frosting time for sample 3. Zuo et al. [49] prepared SHS on glass with ZnO nanorods and compared its anti-frosting characteristics with permanent room temperature vulcanized (PRTV)-coated glass and bare glass. Frosting process was investigated on all three surfaces using Peltier-based platform in which samples were maintained at three different temperatures: 0, −5 and −10°C. At −5°C, frost was formed on bare glass, PRTV-coated glass and SHS glass after 24, 43 and 153 min, respectively. This showed the prolonged delay for SHS compared to bare glass and PRTV-coated glass. At 0 and −10°C, SHS again showed frosting delay leading to the excellent anti-frosting performance of SHS. The reason for this delay was that nanorods in SHS increased the air cavities between solid surface and liquid droplets. This promoted self-transfer movement of condensed droplets and inhibited frost formation. Also, the SHS showed extreme durability after 30 cycles of frosting/defrosting process with WCA decreasing from 167 to 162° only.

Liu et al. [50] prepared SHS using fiberglass cloth (FC-coated SHS) on Al substrate and compared its anti-icing characteristics with bare Al surface (bare-Al). The testing device used consisted of container with low temperature environment (−35°C), air-blast atomizer to spray cold air, sample holding device tilted at 30° with electric heaters embedded. Without electrical heating, FC-coated SHS showed excellent ice formation delay with only 24% surface with accumulated ice at time when bare-Al was completely covered with ice. During electrical heating case, applied voltage varied as 5, 7, 7.4 and 10.8 V to find the power at which no ice could accumulate on both surfaces. At all voltages, ice accumulation on FC-coated SHS was always less than that on bare-Al and the voltages at which no ice accumulated on FC-coated SHS and bare-al were 7.4 and 10.8 V, respectively, showing more icing delay for FC-coated SHS. As a result, 51% energy saving to inhibit icing on FC-coated SHS was determined.

3.4. Drag reduction

Lv and Zhang [51] fabricated SHS on Al alloy and examined its drag reduction and heat transfer characteristics by turbulent flow of water in counter-current tube-in-tube heat exchanger. Bare Al and SHS tubes were used as inlet tubes separately. Inlet tubes of three different diameters: 4, 8 and 12 mm were investigated and their corresponding drag reduction values were found to be in range of 12.4–17.8%, 8.3–13.1% and 8.3–12.8%, respectively. Drag reduction increased with decreasing inlet tube diameter due to increasing area fraction of air-liquid interface. Effect of Reynolds number on drag reduction was also determined. At low Re, drag reduction was higher due to dominance of air-liquid interface causing slip flow while at high Re, turbulence replaced air in air-liquid interface with water, thereby decreasing drag reduction. Due to slip flow, friction factor as well as pressure drop were also less in SHS tube as compared to bare Al tube. Moreover, as air in air-liquid interface acted as insulator, HTC suppressed in SHS was compared to bare tube. To check overall performance of SHS, performance evaluation criterion (PEC) was determined for all three SHS tubes. For all values of Re, 8 and 12 mm tube showed PEC >1 while 4 mm tube showed PEC <1 after Re = 6800.

4. Conclusion

This chapter summarizes different techniques for the fabrication of super-hydrophobic surfaces. In each technique, micro-nanostructures are produced on the substrate surface due to which surface wettability is reduced to a great extent. The water droplets on such surface appear in Cassie-Baxter state in which they have almost spherical shape and very high mobility. This feature of super-hydrophobic surface has a great use in various heat transfer applications like boiling, condensation, anti-icing, drag reduction, etc. where high heat transfer coefficients, icing delay time, energy saving, friction reduction are the important parameters. The future challenge is to fabricate super-hydrophobic surfaces which have long-term stability and storability.

Nomenclature

SHS	Super-hydrophobic surface
WCA	Water contact angle
CAH	Contact angle hysteresis
FAS	Flouroalkyl silane
CA	Contact angle
SA	Sliding angle
RH	Relative humidity
PTES	1H,1H,2H,2H-perfluorooctyltriethoxysilane
PDMS	Polydimethyloxysilane
WSA	Water sliding angle
FNP	Fluorinated silica nanoparticles
CHT	Condensation heat transfer
PFTS	1H,1H,2H,2H-perfluorooctyltriethoxysilane
OTS	Octadecyltrichlorosilane
MHF	Minimum heat flux
CHF	Critical heat flux
TFE	Tetrafluoroethylene
DWC	Drop-wise condensation
FWC	Film-wise condensation
PEC	Performance evaluation criterion

Author details

Hafiz Muhammad Ali[1]*, Muhammad Arslan Qasim[1], Sullahuddin Malik[1] and Ghulam Murtaza[2]

*Address all correspondence to: h.m.ali@uettaxila.edu.pk

1 Mechanical Engineering Department, University of Engineering and Technology, Taxila, Pakistan

2 Mechanical Engineering Department, Government College University, Faisalabad, Pakistan

References

[1] https://en.wikipedia.org/wiki/Lotus_effect [Accessed: 22-11-2017]

[2] Darmanin T, Guittard F. Superhydrophobic and superoleophobic properties in nature. Biochemical Pharmacology. 2015;**18**:273-285. DOI: 10.1016/j.mattod.2015.01.001

[3] Esmaeilirad A, Rukosuyev MV, Jun MBG, van Veggel FCJM. A cost-effective method to create physically and thermally stable and storable super-hydrophobic aluminum alloy surfaces. Surface & Coatings Technology. 2016;**285**:227-234. DOI: 10.1016/j.surfcoat.2015.11.023

[4] Qi Y, Cui Z, Liang B, Parnas RS, Houfang L. A fast method to fabricate superhydrophobic surfaces on zinc substrate with ion assisted chemical etching. Applied Surface Science. 2014;**305**:716-724. DOI: 10.1016/j.apsusc.2014.03.183

[5] Chu F, Xiaomin W. Fabrication and condensation characteristics of metallic super-hydrophobic surface with hierarchical micro-nano structures. Applied Surface Science. 2016;**371**:322-328. DOI: 10.1016/j.apsusc.2016.02.208

[6] Yin L, Wang Y, Ding J, Wang Q, Chen Q. Water condensation on superhydrophobic aluminum surfaces with different low-surface-energy coatings. Applied Surface Science. 2011;**258**:4063-4068. DOI: 10.1016/j.apsusc.2011.12.100

[7] Liao R, Zuo Z, Guo C, Yuan Y, Zhuang A. Fabrication of superhydrophobic surface on aluminum by continuous chemical etching and its anti-icing property. Applied Surface Science. 2014;**317**:701-709. DOI: 10.1016/j.apsusc.2014.08.187

[8] Nguyen TPN, Dufour R, Thomy V, Senez V, Boukherroub R, Coffinier Y. Fabrication of superhydrophobic and highly oleophobic silicon-based surfaces via electroless etching method. Applied Surface Science. 2014;**295**:38-43. DOI: 10.1016/j.apsusc.2013.12.166

[9] Bo Y, Fang L, Tang A-q, Huang Q-l, Jia H, Mao J-h, Ge B, Bai H. Novel strategy in increasing stability and corrosion resistance for super-hydrophobic coating on aluminum alloy surfaces. Applied Surface Science. 2011;**258**:580-585. DOI: 10.1016/j.apsusc.2011.08.063

[10] Lin Y, Shen Y, Liu A, Zhu Y, Liu S, Jiang H. Bio-inspiredly fabricating the hierarchical 3D porous structure superhydrophobic surfaces for corrosion prevention. JMADE. 2016; **103**:300-307. DOI: 10.1016/j.matdes.2016.04.083

[11] Choi D, Yoo J, Park SM, Kim DS. Facile and cost-effective fabrication of patternable superhydrophobic surfaces via salt dissolution assisted etching. Applied Surface Science. 2017;**393**:449-456. DOI: 10.1016/j.apsusc.2016.10.057

[12] Jia X, Jinliang X, Cao Y, Ji X, Yan Y. Fabrication of non-flaking, superhydrophobic surfaces using a one-step solution-immersion process on copper foams. Applied Surface Science. 2013;**286**:220-227. DOI: 10.1016/j.apsusc.2013.09.051

[13] Zhao J, Li Z, Zhang M, Meng A. Super-hydrophobic surfaces of SiO_2-coated SiC nanowires: Fabrication, mechanism and ultraviolet-durable super-hydrophobicity. Journal of Colloid and Interface Science. 2015;**444**:33-37. DOI: 10.1016/j.jcis.2014.12.057

[14] Zheng Y, Yi H, Qing Y, Zhuo Z, Mo Q. Formation of SiO_2/polytetrafluoroethylene hybrid superhydrophobic coating. Applied Surface Science. 2012;**258**:9859-9863. DOI: 10.1016/j.apsusc.2012.06.043

[15] Fleming RA, Zo M. Silica nanoparticle-based films on titanium substrates with long-term superhydrophilic and superhydrophobic stability. Applied Surface Science. 2013;**280**:820-827. DOI: 10.1016/j.apsusc.2013.05.068

[16] Liu S, Liu X, Latthe SS, Li G, An S, Yoon SS, Liu B, Xing R. Self-cleaning transparent superhydrophobic coatings through simple sol-gel processing of fluoroalkylsilane. Applied Surface Science. 2015;**351**:897-903. DOI: 10.1016/j.apsusc.2015.06.016

[17] Mahadik SA, Parale V, Vhatkara RS, Mahadik DB, Kavale MS, Wagh PB, Gupta S, Gurav J. Superhydrophobic silica coating by dip coating method. Applied Surface Science. 2013;**277**:67-72.DOI:10.1016/j.apsusc.2013.04.001

[18] Huang W-H, Lin C-S. Robust superhydrophobic transparent coatings fabricated by a low-temperature sol–gel process. Applied Surface Science. 2014;**305**:702-709. DOI: 10.1016/j.apsusc.2014.03.179

[19] Ramezani M, Vaezi MR, Kazemzadeh A. Preparation of silane-functionalized silica films via two-step dip coating sol–gel and evaluation of their superhydrophobic properties. Applied Surface Science. 2014;**307**:147-153. DOI: 10.1016/j.apsusc.2014.08.095

[20] Chen C, Yang S, Liu L, Xie H, Liu H, Zhu L, Xiaoliang X. A green one-step fabrication of superhydrophobic metallic surfaces of aluminum and zinc. Journal of Alloys and Compounds. 2017;**711**:506-513. DOI: 10.1016/j.jallcom.2017.04.050

[21] Yiqiang W, Jia S, Wang S, Qing Y, Yan N, Wang Q, Meng T. A facile and novel emulsion for efficient and convenient fabrication of durable superhydrophobic materials. Chemical Engineering Journal. 2017;**328**:186-196. DOI: 10.1016/j.cej.2017.07.023

[22] Tang Y, Fu T, Liu Q, Luo W. Copper based superhydrophobic microchannels: Fabrication and its effect on friction reduction. Materials Science and Technology. 2015;**31**:730-736. DOI: 10.1179/1743284714Y.0000000658

[23] Li B-j, Huang L-j, Ren N-f, Kong X. Laser ablation processing of zinc sheets in hydrogen peroxide solution for preparing hydrophobic microstructured surfaces. Materials Letters. 2015;**164**:384-387. DOI: 10.1016/j.matlet.2015.11.035

[24] Chun D-M, Ngo C-V, Lee K-M. Fast fabrication of superhydrophobic metallic surface using nanosecond laser texturing and low-temperature annealing. CIRP Annals. 2016;**1**:519-522. DOI: 10.1016/j.cirp.2016.04.019

[25] Wang QC, Yang XD, Shang GR. Fabrication of copper-based Superhydrophobic surface through template deposition. Advanced Materials Research. 2014;**915-916**:799-802. DOI: 10.4028/www.scientific.net/AMR.915-916.799

[26] Bhagat SD, Gupta MC. Superhydrophobic microtextured polycarbonate surfaces. Surface & Coatings Technology. 2015;**270**:117-122. DOI: 10.1016/j.surfcoat.2015.03.013

[27] Liu X, Yang X, Ben K, Chen Z, Wang Y, Guan Z. Transparent, durable and thermally stable PDMS-derived superhydrophobic surfaces. Applied Surface Science. 2015;**339**:94-101. DOI: 10.1016/j.apsusc.2015.02.157

[28] Momen G, Farzaneh M. A ZnO-based nanocomposite coating with ultra water repellent properties. Applied Surface Science. 2012;**258**:5723-5728. DOI: 10.1016/j.apsusc.2012.02.074

[29] Ipekci HH, Harun Arkaz H, Serdar Onses M, Hancer M. Superhydrophobic coatings with improved mechanical robustness based on polymer brushes. Surface & Coatings Technology. 2016;**299**:162-168. DOI: 10.1016/j.surfcoat.2016.05.026

[30] Wang L, Yang J, Zhu Y, Li Z, Sheng T, Yang D-Q. An environment-friendly fabrication of superhydrophobic surfaces on steel and magnesium alloy. Materials Letters. 2016:297-299. DOI: 10.1016/j.matlet.2016.02.103

[31] Feng L, Yang M, Shi X, Liu Y, Wang Y, Qiang X. Copper-based superhydrophobic materials with long-term durability, stability, regenerability, and self-cleaning property. Colloids and Surfaces. 2016:39-47. DOI: 10.1016/j.colsurfa.2016.08.017

[32] Sun J, Wang L, Hu K, Song J, Liu X. Fabrication of superhydrophobic surfaces on copper substrates via flow plating. Micro and Nano Letters. 2015;**10**(2):88-92. DOI: 10.1049/mnl.2014.0440

[33] Huang D-J, Leu T-S. Condensation heat transfer enhancement by surface modification on a monolithic copper heat sink. Applied Thermal Engineering. 2015;**75**:908-917. DOI: 10.1016/j.applthermaleng.2014.10.019

[34] Zhang Z, Bo G, Men X, Li Y. Mechanically durable, superhydrophobic coatings prepared by dual-layer method for anti-corrosion and self-cleaning. Colloids and Surfaces. 2016;**490**:182-188. DOI: 10.1016/j.colsurfa.2015.11.049

[35] Stanton MM, Ducker RE, MacDonald JC, Lambert CR, Grant McGimpsey W. Superhydrophobic, highly adhesive, polydimethylsiloxane (PDMS) surfaces. Journal of Colloid and Interface Science. 2012;**367**:502-508. DOI: 10.1016/j.jcis.2011.07.053

[36] Song Y, Premachandran Nair R, Zou M, Wang YA. Adhesion and friction properties of micro/nano-engineered superhydrophobic/hydrophobic surfaces. Thin Solid Films. 2010;**518**:3801-3807. DOI: 10.1016/j.tsf.2010.01.009

[37] Sun W, Wang L, Yang Z, Li S, Tingting W, Liu G. Fabrication of polydimethylsiloxane-derived superhydrophobic surface on aluminium via chemical vapour deposition technique for corrosion protection. Corrosion Science. 2017;**128**:176-185. DOI: 10.1016/j.corsci.2017.09.005

[38] Betz AR, Jenkins J, Kim C-J, Attinger D. Boiling heat transfer on superhydrophilic, superhydrophobic, and superbiphilic surfaces. International Journal of Heat and Mass Transfer. 2013;**57**:733-741. DOI: 10.1016/j.ijheatmasstransfer.2012.10.080

[39] Malavasi I, Bourdon B, Di Marco P, de Coninck J, Marengo M. Appearance of a low superheat "quasi-Leidenfrost" regime for boiling on superhydrophobic surfaces. International Communications in Heat and Mass Transfer. 2015;**63**:1-7. DOI: 10.1016/j.icheatmasstransfer.2015.01.012

[40] Yamada M, Shen B, Imamura T, Hidaka S, Kohno M, Takahashi K, Takata Y. Enhancement of boiling heat transfer under sub-atmospheric pressures using biphilic surfaces. International Journal of Heat and Mass Transfer. 2017;**115**:753-762. DOI: 10.1016/j.ijheatmasstransfer.2017.08.078

[41] Fan L-W, Li J-Q, Li D-Y, Zhang L, Zi-Tao Y. Regulated transient pool boiling of water during quenching on nanostructured surfaces with modified wettability from superhydrophilic to superhydrophobic. International Journal of Heat and Mass Transfer. 2014;**76**:81-89. DOI: 10.1016/j.ijheatmasstransfer.2014.04.025

[42] Kim DE, Ho SA, Kwon T-S. Experimental investigation of filmwise and dropwise condensation inside transparent circular tubes. Applied Thermal Engineering. 2017;**110**:412-423. DOI: 10.1016/j.applthermaleng.2016.08.175

[43] Lu M-C, Lin C-C, Lo C-W, Huang C-W, Wang C-C. Superhydrophobic Si nanowires for enhanced condensation heat transfer. International Journal of Heat and Mass Transfer. 2017;**111**:614-623. DOI: 10.1016/j.ijheatmasstransfer.2017.04.021

[44] Yin L, Wang Y, Ding J, Wang Q, Chen Q. Water condensation on superhydrophobic aluminum surfaces with different low-surface-energy coatings. Applied Surface Science. 2012;**258**:4063-4068. DOI: 10.1016/j.apsusc.2011.12.100

[45] Bisetto A, Torresin D, Tiwari MK, Del Col D, Poulikakos D. Dropwise condensation on superhydrophobic nanostructured surfaces: Literature review and experimental analysis. In: IOP Publishing, editor. Heat Transfer Conference. Moltrasio (Como), Italy: IOP Publishing; 2014. DOI: 10.1088/1742-6596/501/1/012028

[46] Liu Y, Li X, Jin J, Liu J, Yan Y, Han Z, Ren L. Anti-icing property of bio-inspired microstructure superhydrophobic surfaces and heat transfer model. Applied Surface Science. 2017:498-505. DOI: 10.1016/j.apsusc.2016.12.219

[47] Kim M-H; Kim DR; Lee K-S. Stochastic approach to the anti-freezing behaviors of superhydrophobic surfaces. International Journal of Heat and Mass Transfer. 2017:841-846. DOI: 10.1016/j.ijheatmasstransfer.2016.10.015

[48] Wang G, Shen Y, Tao J, Luo X, Jin M, Xie Y, Li Z, Guo S. Facilely constructing micro-nanostructure superhydrophobic aluminum surface with robust ice-phobicity and corrosion resistance. Surface & Coatings Technology. 2017;**329**:224-231. DOI: 10.1016/j.surfcoat.2017.09.055

[49] Zuo Z, Liao R, Zhao X, Song X, Qiao Z, Guo C, Zhuang A, Yuan Y. Anti-frosting performance of superhydrophobic surface with ZnO nanorods. Applied Thermal Engineering. 2017;**110**:39-48. DOI: 10.1016/j.applthermaleng.2016.08.145

[50] Liu X, Chen H, Kou W, Zhang D. Robust anti-icing coatings via enhanced superhydrophobicity on fiberglass cloth. Cold Regions Science and Technology. 2017;**138**:18-23. DOI: 10.1016/j.coldregions.2017.03.004

[51] Lv FY, Zhang P. Drag reduction and heat transfer characteristics of water flow through the tubes with superhydrophobic surfaces. Energy Conversion and Management. 2016;**113**:165-176. DOI: 10.1016/j.enconman.2016.01.034

Chapter 15

Spouted Bed and Jet Impingement Fluidization in Food Industry

Dariusz Góral, Franciszek Kluza and
Katarzyna Kozłowicz

Additional information is available at the end of the chapter

http://dx.doi.org/10.5772/68105

Abstract

Spouted bed and jet impingement fluidization are the cases of classic fluidization modification obtained by proper distribution of the fluid and the construction of device. The condition for the realization of both fluidization types is strict determination of the apparatus dimensions and process parameters. The chapter presents the issues concerning the choice of optimal operating conditions and dimensions of working elements of devices for both spouted bed and jet impingement fluidization, as well as possibilities for analysis of heat transfer in the process. Furthermore, the examples of industrial application of spouted bed fluidization and jet impingement fluidization in food technology are presented.

Keywords: spouted bed, jet impingement fluidization, drying, freezing, roasting, food

1. Introduction

As a result of proper interaction between the gas-solid with the inner construction of fluidized bed apparatus and the mode of gas distribution, modifications of the classic fluidization, such as spouted beds and jet impingement fluidization, are developed. The term spouted bed fluidization was first used by Gishler and Mathur in 1954 who successfully applied this technique for drying of wheat. The first industrial device for drying peas, lentils, and flax was built and installed in Canada in 1962 [1].

2. Spouted bed fluidization

Spouted bed fluidization is the process in which the fluid is vertically introduced upwards at a suitable velocity through the center of the bottom granular material bed. The bed particles are carried up in the central jet by the fluid stream and on reaching the top medium layer they rain back onto the peripheral annular region. Next, they slowly move downward the column. The process cycle is repeated many times causing the bed's circulation (**Figure 1**) [2, 3].

Areas of spouted bed fluidization application overlap with that of classic fluidization use; however, the mechanisms of fluid flow and bed elements motion differ substantially in both cases. The motion and mixing of particles in the spouted bed are regular and cyclic, imposed by the constant upward-flowing fluid jet, opposed to a more random and complicated particle motion in the conventional fluidized bed [4]. There are at least five solutions of devices for spouted bed fluidization: rectangular, cylindrical flat base, cylindrical cone base, conical, and dilute jet spouted beds [5]. In view of the occurrence of bottom dead zones, the flat-base devices are used infrequently. Whereas equipment with conical beds is applied most often in numerous modifications, yet these are beds of low depth. Dilute jet-type spouted beds were studied extensively in the former Soviet Union over the 1950–1960 period [5]. They show a substantially higher fluid velocity than the minimal velocity of spouted bed fluidization that results in dilute phase occurrence in the entire device [2]. To overcome the constraints of conventional spouted beds (low aeration in the annular region, slow turnover of bed particles), there were developed several modifications concerning the process and devices. The most frequently applied are novel rotating jet spouted bed (RJSB) [6], rotating jet annular spouted bed (RJASB) [7], pulsating spouted bed [8], and draft tube spout-fluid bed (DTSFB) [9]. These

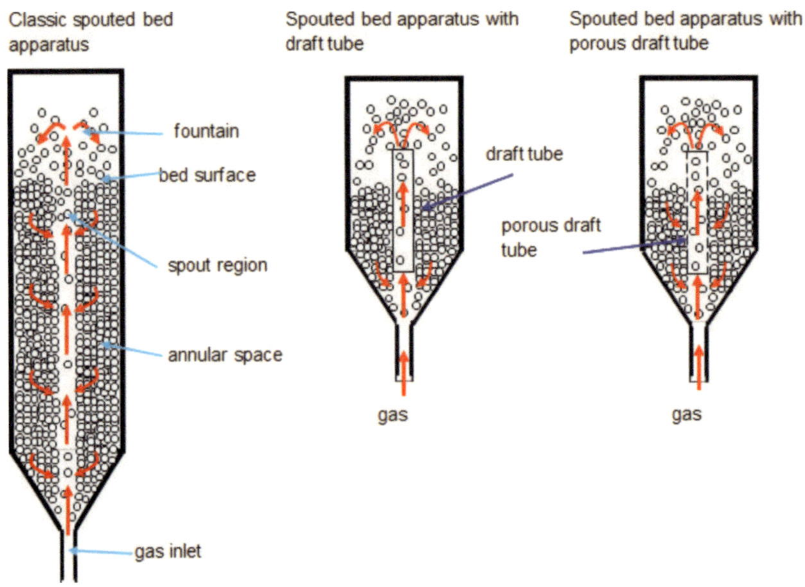

Figure 1. Scheme of gas motion in conventional spouted bed and spouted bed with draft tube.

http://dx.doi.org/10.5772/68105

modifications allow to obtain a higher heat transfer coefficient, more uniform distribution of particle residence time inside the device, and lower pressure drops as compared to conventional devices for spouted bed fluidization [10]. One of the most common solutions is using the apparatus with a draft tube where the fountain is formed over the tube top (**Figure 1**). Then neither bed height nor particle size homogeneity is limited and the ease of controlling the bed circulation in order to prevent particles from moving out of the peripheral annular zone to the central core is of importance [11, 12]. However, the application of this solution is restricted to the cases when the medium flowing out of the draft tube is not only the bed particles carrier but plays an active role in a process (drying, cooling, chemical reactions) as well. The reason for this is that the medium from the draft tube cannot percolate to the outer annular zone and therefore, the process is likely to be substantially slowed down through decreased efficiency at lowered heat transfer. To eliminate these shortcomings, a porous draft tube is used. The unique properties of the solution make it applicable, for example, for the thermal disinfestation process of wheat grain where both grain movement and proper conditions of gas-solid contacting with a concurrent slight pressure drop need to be controlled [11]. Another modification of spouted bed fluidization is provided by the Wurster apparatus—a spouted bed device with a draft tube and additional fluidizing air stream (spout-fluid bed). The additional gas stream is injected upwards through the flat or conical porous bottom. The aerodynamic characteristics of the spouted bed system as against conventional fluidization are not only dependent on fluid properties and solid particles making up the bed. The apparatus construction as well as bed height and width are of great importance. Subject to the fluid type and the kind of fluidizing solid particles and apparatus configuration, the types of bed structures/fluidization kinds are best represented by the dependence of bed height on the fluid velocity. **Figure 2** presents the exemplary ranges of the bed structures according to Mathur and Epstein [13].

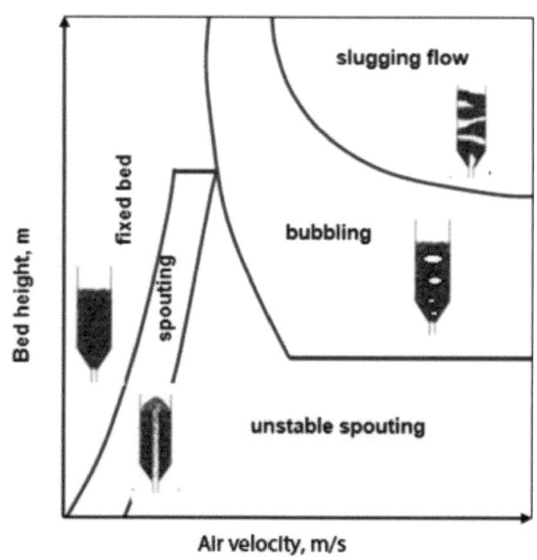

Figure 2. Ranges of the bed structures.

The line representing transition between the fixed bed and spouting bed area corresponds to minimum spouted bed velocity, whereas the horizontal line separating spouting from the bubbling area denotes the maximum bed depth. At the height of bed above some value, the transition from the packed bed to spouted bed is not possible. Increasing velocity in this range induces bubble bed formation followed by slugging flow. Working out the technology and spouted bed fluidization devices necessitates the determination of parameters at which the bed is converted into dynamic state as the parameters established for conventional fluidization cannot be used [2]. However, similar to classical fluidization, the dependence between bed pressure drop and fluid flow velocity can be presented (**Figure 3**).

Initially, particles remain stagnant (A-B state) but as the pressure drop increases linearly along with a fluid velocity increase, the so-called dilute zone forming at the bed base is observed to expand. At the B-point, the pressure drop reaches its maximum value while air velocity corresponds to the critical value. The B-C state is characterized by the moment of visible loosening of the bed with bubbling and channeling air flow. At the C-point, fluidization commences and the fluid velocity at the U_{ms} point is defined as the minimum spouting velocity. The C-D state corresponds to the process of ultimate breaking up of the bed dense phase by the fluid stream. Formation of the characteristic stable fountain of material is observed [14]. The increasing fluid flow rate makes the annular and fountain zones substantially loosened, whereas the further growing velocity results in a uniform bed without division into annular and fountain zones. Thus, the jet-type bed is created [2]. The analysis of the curves run in **Figure 4** shows a pressure drop reduction at increasing and decreasing gas velocity. Minimum spouted bed velocity indicates the onset of the bed stable work. The authors [15, 16] studied a difference in determining the velocity in the cases of increased and decreased gas flow rate. The minimum spouted bed velocity reported at the onset of the observed process is higher than that noted at decreased gas velocity. Some authors [17] established that the most repeatable velocity value of the spouting onset is the value obtained at increasing gas flow rate. Hence, the bottom line serves as the basis for minimum velocity

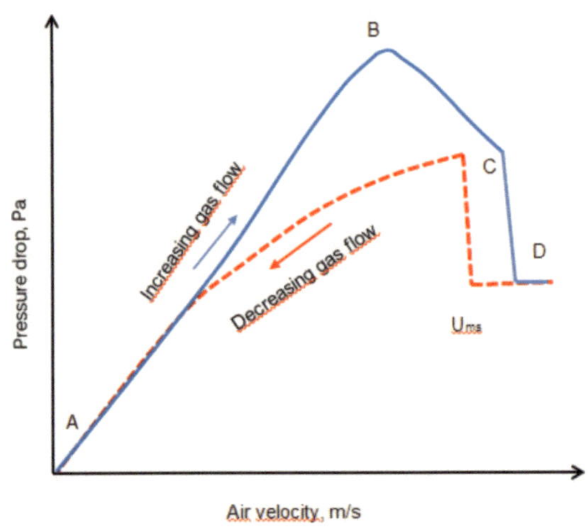

Figure 3. Pressure drop as affected by gas velocity during spouted bed process.

http://dx.doi.org/10.5772/68105

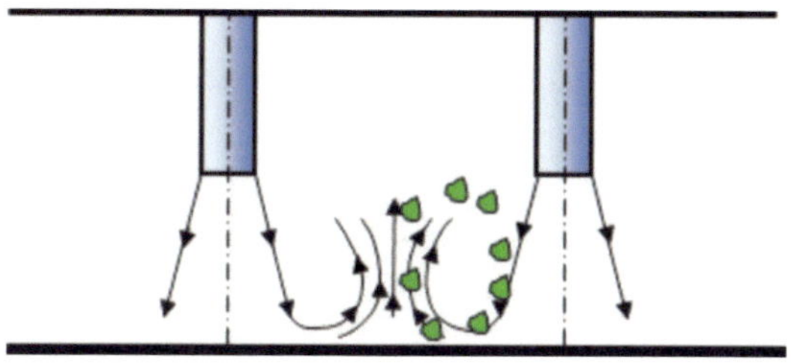

Figure 4. Motion of product in jet impingement fluidization.

determination in most papers. On the basis of the relationship between pressure drop and fluid flow rate in the bed, many authors developed the correlations enabling among others, determination of the minimum fluidization velocity [2, 18–20]. For cylindrical devices of up to 0.5 m diameter with/without conical base, the most generalized equation was proposed by Mathur and Gishler [21] to estimate minimum spouting velocity (prediction error within the range ±15%) for a wide variety of materials, bed dimensions, inlet diameter, and fluids from air to water (**Table 1**).

More accurate approximations of the minimum fluidization velocity value require knowing a number of process parameters. For instance, Zhong et al. [22] propose the correlation for estimation of minimum fluidization velocity of the bed particle of D dimension (according to Geldard classification) in the apparatus of cross section of 300 mm × 30 mm and 2000 mm height (Eq. (2)) and Olazar et al. [23] for materials such as glass or other materials of similar density (Eq. (3)) (**Table 1**). According to Bi [5], bed geometry, which determines the minimum spouting velocity, is sufficiently described by the following three parameters: bed height H_o,

Eq. no.	Correlation	Authors
(1)	$U_{ms} = \left(\frac{d_p}{D}\right)\left(\frac{D_o}{D}\right)^{\frac{1}{3}}\sqrt{\frac{2gH_o(\rho_s-\rho_g)}{\rho_g}}$	Mathur and Gishler [21]
(2)	$U_{ms} = 24.5\left(2gH_o\right)^{0.5}\left(\frac{d_p}{D_t}\right)^{0.472}\left(\frac{\lambda_i}{D_t}\right)^{0.183}\left(\frac{H_o}{D_t}\right)^{0.208}\left(1+\frac{U_f}{U_{mf}}\right)^{-0.284}\left(\frac{\rho_s-\rho_g}{\rho_g}\right)^{0.226}$	Zhong et al. [22]
(3)	$(Re_i)_{ms} = 0.126\sqrt{Ar}\left(\frac{D_b}{D_o}\right)^{1.68}\left(\tan\frac{\gamma}{2}\right)^{-0.57}$	Olazar et al. [23]
(4)	$(Re_i)_{ms} = 0.202\sqrt{Ar\left(\frac{D_t}{D_o}\right)\left[\left(\frac{D_t}{D_o}\right)^2+\left(\frac{D_t}{D_o}\right)+1\right]/3}$	Bi [5]

Table 1. Empirical equations for predicting minimum spouting velocity.

inlet nozzle diameter D_o, and diameter on bed surface D_t. Therefore, the minimum spouting velocity can be estimated from Eq. (4) (for $D_b/D_o < 1.66$). Sutkar and others [24] report that the minimum fluidization velocity increases along with increasing bed height, inlet tube diameter, and operating pressure. Pressure drop in the bed can be defined from the correlation [25]:

$$\frac{\Delta p_m}{H_o \rho_b g} = 1 + 0.0006 \left(\frac{D_t}{D_o}\right)^{5.04} \left(\frac{d_p}{D_t} - 1\right)^{-1.92} \left(tan\frac{\gamma}{2}\right)^{0.7} \quad (5)$$

This equation is valid for $\gamma = 30–60°$, $D_o = 0.5$ m, $H_o = 0.0541–0.28$ m. Pressure drop in the spouted bed increases along with the increasing bed height, inlet tube diameter, and particle density, while it decreases with elevating particle diameter and gas velocity [24]. Besides the minimum fluidization velocity and pressure drop across the bed, significant factors affecting the operating conditions of spouting beds are geometric parameters of the apparatus [2]. The most critical proved to be a fluid inlet nozzle diameter as well as the nozzle inlet diameter/column diameter ratio. Obviously, a nozzle inlet diameter is dependent on the dimensions of the bed elements. For most cases, the D_o/d_p ratio should range between 25 and 30 [26], while according to San Jose and others [27] for $\gamma > 60$ it should be 5:2 at $2 < D_o/d_p < 60$, for $\gamma < 60$ 5:2 at $2 < D_o/d_p < 80$ and for small diameter-solid particles $d_p = 1$ mm or >4 mm—the ratio should be equal to 2:1. The conical base angle is considered a less critical parameter. For a stable spouting state with mostly solid materials, the ratio should be over 40° [4]. Kmieć and others [2] analyzing the heat and mass transfer in the spouting bed concluded that the problem needs to be addressed separately for the central core zone and fountain zone. Many solutions to this problem are available in the literature, yet universal and generalized dependencies are still lacking. For instance, Rocha and others [25] present the dependence in a dimensionless form:

$$Nu = 0.9892 Re_p^{1.6421} Pr^{0.333} \left(\frac{H_o}{d_p}\right)^{-1.3363} \left(\frac{m_z}{m_g}\right)^{0.71} \left(\tan\frac{\gamma}{2}\right)^{0.1806} \quad (6)$$

which the authors developed studying the granular material drying process in a rectangular-based spout apparatus at the following parameters: $T = 70–80°C$, $75 < Re_p < 844$, $0.1065 < H_o < 0.2456$ m, $6.49 < d_p < 7.8039$ mm, $30° < \gamma < 60°$. Spouted bed fluidization has industrial application in agricultural engineering, chiefly in drying processes. Fountain driers are characterized by high efficiency and performance as well as operational reliability and simple structure. The products dried in such devices tend to maintain their physical properties, sensorial characteristics, and satisfy the microbiological safety requirements [14, 28].

3. Jet impingement fluidization

Jet impingement fluidization resembles spouted bed fluidization to some extent. During the jet impingement fluidization process, the bed particles displace in the fluid jet leaving the nozzle system to form characteristic fountains. Explanation of the operating principles of a jet impingement fluidization apparatus needs concerning the formation of a fluid jet flowing out of a single nozzle in the impingement phenomenon. The physical field of fluid jet flow from the nozzle includes three basic regions: free jet flow, stagnation, and horizontal flow or wall jet. The free jet flow area comprises a few regions whose size depends on the distance between the nozzle and rebound air jet area. This region is divided into two areas, that is, the core and mixing. In the region of the potential core, turbulences do not occur in the jet, whereas they are observed after fluid jet impinging on the stagnation area. Consequently, the flow is disturbed and the jet potential core becomes decentralized. This area is called the mixing region. A level of energy dissipation and length of the fluid jet central core rely mainly on the shape of nozzles and their exit profiling level as well as velocity at the nozzle exit [29, 30]. Vortices in the free jet region along with resulting energy dissipation are the factors of great weight as they contribute to intensive heat transfer in the stagnation area and the stream wall. The stream in the free jet region displays the free flow behavior pattern in the distance equal to 1–1.5 times the nozzle diameter, where upon it starts moving toward the stagnation region. In this region, fluid jet impingement occurs causing the fluid stream bouncing off the apparatus bottom and consequently, it experiences a velocity drop with a considerable increase in static pressure. A negative pressure gradient in the jet wall area induces the initial rapid flow speed increase followed by momentary stagnation and the final decrease [29]. The flow field of the jet leaving the multiple jet arrays is similar to that of a single impinging jet; it also comprises three regions. The differences, however, concern the jet wall area and depend on the nozzles spacing. The interaction mechanisms occurring in arrays of many jets can be observed in three types of cases [31]. First is related to interference between the neighboring jets in the mixing area. Second concerns the appropriate jet-to-jet spacing that promotes the rising fountain phenomenon. The fountain height is determined by, among others, the nozzle length and flow rate of fluid leaving the nozzles. Third pertains to the air outlet arrangement in the plane transversal to the nozzle direction and causes turbulent flow of fluid jet emerging from the working chamber [30]. Jet impingement fluidization is a method that utilizes the impingement phenomenon to induce fluidal bed boiling. The state of fluidal boiling occurs once the minimum fluidization velocity is achieved, that is the buoyancy force exceeds the gravitational force of bed solid particles, which are lifted toward the fountain top. Owing to the inter-particle force of adjacent bed particulates, single-bed elements are entrained to the fountain top to be afterwards pushed under the lower pressure zone or the air jet issuing from the nozzle. Consequently, solid particles fall back toward the bottom of the apparatus working chamber and the process cycle starts anew (**Figure 4**).

Kluza [32], Kluza and Stadnik [33], Góral and Kluza [34–36] thoroughly examined the effect of the apparatus parameters on attainability of fluidal beds of various products. The authors concluded that if the ratio of nozzle symmetry axis spacing to the inner diameters was $L/D \leq 1.8$, the bed boiled in a way atypical for jet impingement fluidization. The product tended to displace asymmetrically and was thrown outside the device. Besides, a change in the spacing ($1.8 \leq L/D \leq 4$) resulted in turbulent displacement of the bed solids followed by the occurrence of "bed loosening" phenomenon. Whereas if the $L/D > 4$ condition was satisfied, dead zones between the particular nozzles were found which were only slightly affected by the air jet emerging from the nozzles. A detailed analysis of the fluidized bed behavior in the jet impingement fluidization technology [33] showed that the fluidal boiling state occurred when the distance between the symmetry axes of nozzles was equal to four times the inner diameter of nozzle ($L/D = 4$). **Figure 5A–C** presents asymmetric displacement of the product in the case of multiple jet array at $L/D < 4$.

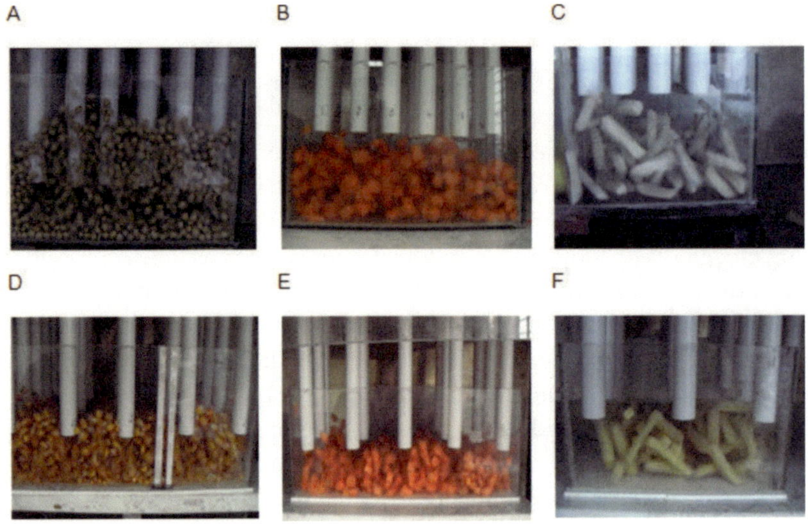

Figure 5. The view of impingement fluidization of different products [33].

There is a lack of visible typical curvilinear closing trajectories determined by the product. In the case of both, soya beans, maize, carrot cubes and French fries, the material was partly thrown outside the apparatus because of too small separation distances between the nozzles and thus, inability of appropriate return of the rebound air jet. **Figure 5D–F** shows typical trajectories of the product determined by the motion of the gas jet leaving the nozzles and rebound off the device bottom. The character of the bed behavior was obtained after setting and application of the optimum dependence between the nozzle diameter and the distance between the nozzles. On the basis of the bed development analysis in jet impingement fluidization, Kluza and Stadnik [33], Góral and Kluza [36] performed the assembly of 11 operating heads (nozzle diameter from 11 to 35 mm) with a nozzle pitch four times the nozzle inner diameter $L = 4D$. The heads were fitted with nozzles in a staggered or aligned configuration. To minimize the losses of velocity and pressure of air leaving the nozzles and the character of its flow, the authors assume that the ratio of nozzle length S to its diameter D should be higher than 6 ($S/D > 6$) that will ensure a turbulent character of the air flow [37]. According to the assumption, all the heads were equipped with nozzles of 230 mm length. Góral and Góral et al. [38, 39] reported that the maximum velocities of rebound air jets are obtained using the operating heads with nozzles of 23 and 20 mm diameter, while the lowest at 26 mm diameter and staggered configuration. As for the heads with 11, 14, and 16 mm nozzle diameter, the rebound air jet velocity was similar and ranged from 1 up to 5 m s^{-1}. The heads fitted with nozzles of 20 and 23 mm diameter enabled to achieve rebound air jet velocity as high as 18 m s^{-1} (**Figure 6**).

The objective of working toward the development of optimum processing conditions and dimensional dependencies of the operating elements of a device that provides appropriate conditions for jet impingement fluidization is to obtain proper bed boiling in its whole volume in a treatment process. For the purpose of full realization of the studied process, it is essential to strictly define the characteristic technical dimensions of the apparatus along with

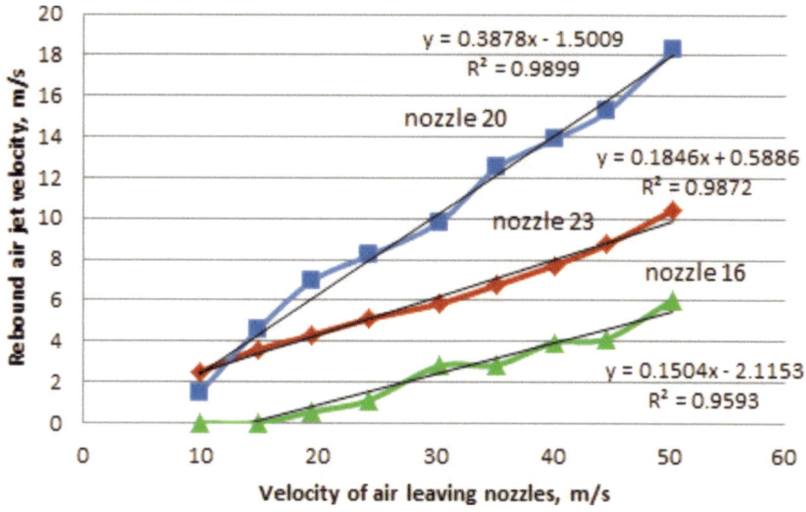

Figure 6. Rebound air jet velocity versus velocity of air leaving nozzles.

process parameters. Notably, for geometrically diversified bed particles, the parameters and dimensional dependencies of the device working space are universal only to some extent. Work on development of optimum conditions for bed fluidal boiling needs to equally consider treatment efficiency, energy saving, and quality of the product. Kluza and Stadnik [33] studied the boiling of carrot cubes beds and found that the greatest heights of beds at rest ensure the subsequent fluidal state with an operating head fitted with nozzles of 20 mm inner diameter in both aligned and staggered pattern (**Figure 7**).

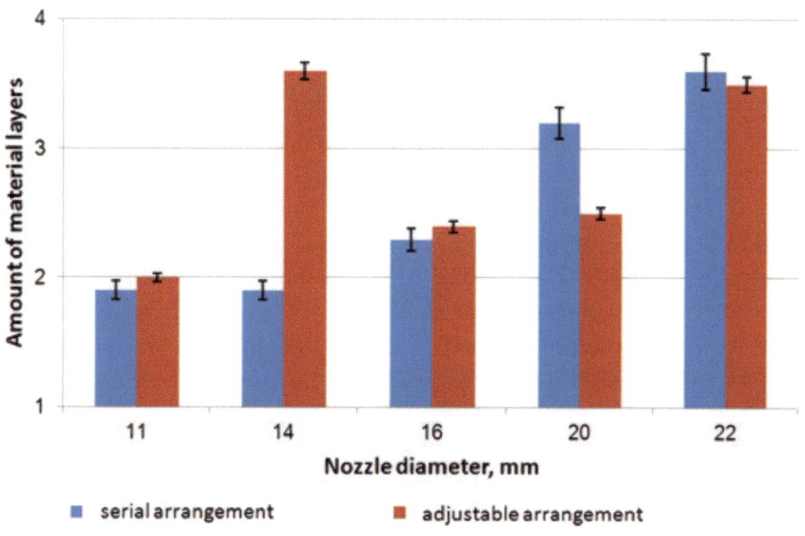

Figure 7. Dependence of bed layers of carrot on inner diameter and arrangement of nozzles.

It was also noted that using heads with nozzles in the in-line arrangement enables fluidization of the beds of the greatest heights. However, for both maize and carrot cubes bed, the best results were obtained with nozzles of 20 mm inner diameter. As for French fries, proper boiling bed phenomenon occurred when the operating heads with nozzles of inner diameter 20 and 35 mm were arranged by aligning. The papers of Góral and Kluza [36] and Góral et al. [39] investigated the problem of heat transfer at vegetable freezing with jet impingement fluidization technique. The authors assumed that the local heat transfer coefficient from the flat plate to air can be estimated from the Nusselt number. In most processing cases, a difference between the value of air temperature at the nozzle exit and of the impinging air jet is so small that the air jet temperature within the bed area can be assumed constant [40]. Prior to the development of a calculation model, Góral and Kluza [36] made the following basic assumptions: air jet flow is turbulent within the velocity range enabling reverse fluidization occurrence, air temperature range varies between −50°C and 0°C, material shape corresponds to infinite plate and non-stationary character of heat transfer. Assuming a turbulent fluid flow over the plate-shaped product, the estimation of the local Nusselt number characterizing heat transfer commonly involves using the criterial equation [41]:

$$Nu_x = CRe_x^m Pr^n \quad (7)$$

where constants *C*, *m*, and *n* are determined experimentally. In the investigated process case [36], the values of the Reynolds number were within the 8000–30,000 range and thus, also in the area described as transient flow. Considering the fact that a product moves within the rebound air jet, that is of a changed direction flow, then it can be assumed with high probability that the air flow around the studied product in form of the plate is turbulent. Regarding the scope of air temperature and humidity required in the industrial food-freezing systems, the Prandtl number changes only slightly. Therefore, the Prandtl number from the formula can be included into the constant *C* [42], and consequently, the dependence for the forced convection heat transfer can be expressed as:

$$Nu_x = CRe_x^m \quad (8)$$

Besides, as for the exponent m the value range of $0.5 < m < 0.8$ should be expected. For the purposes of the model, Góral and Kluza [36] established the *C* and *m* coefficients empirically during the freezing processes run using the operating head with nozzles of 20 mm inner diameter and the air velocity changing from the determined minimum fluidization velocity to rebound air jet rate equal to 10.5 m s^{-1}, essential to maintain a fluidization regime. The obtained Nusselt number values were subjected to regression analysis relative to the corresponding Reynolds number values to get the dependence Nu_x=0.353$Re_x^{0.608}$ at the determination coefficient 0.918 [36]. Góral and Kluza [36] compared the experimentally established heat transfer coefficients with those available in the literature. The mean heat transfer coefficients determined varied from 62 up to 200 W m^{-2} K^{-1} and were lower than 150–350 W m^{-2} K^{-1} values [43] obtained at impingement freezing of food, whereas they were considerably higher as against the mean heat transfer coefficients (35–100 W m^{-2} K^{-1}) determined at the freezing process of the zucchini slices in the fluidized tunnel at 10 m s^{-1} air velocity and −39°C temperature [44]. The paper of Góral and others [45] compares the effect of oval-shaped nozzle exit on uniformity of rebound air velocity fields. The heads fitted with oval nozzles were characterized by non-uniform distribution of rebound air jet velocity fields. Comparison of the obtained research results and those for the head with round nozzles showed an utterly different character of these fields as they distributed uniformly. However, the values of rebound air jet velocity obtained using all four operating heads were similar and ranged between 2 and 5.6 m s^{-1}. The authors assessed the influence of oval cross-sectioned nozzles on attainable boiling bed state studying the French fries freezing process by jet impingement fluidization technique. It was found that fluidal boiling state was achieved using the heads equipped with oval nozzles in the parallel and mixed staggered configuration and at 40 m s^{-1} velocity of air jets leaving the nozzles. In the other cases under study, the bed boiling state was not achieved or pneumatic transport of material occurred. Despite several advantages, jet impingement fluidization applicability in food industry is very limited. Industrial equipment based on this phenomenon is mainly used for toasting and puffing of snack pellets, toasting of cereals (corn flakes, bran flakes, rice flakes, etc.), drying of pet foods, toasting and drying of vegetables, toasting of pasta, roasting of coffee beans, peanuts, sunflower seeds, watermelon seeds, sesame seeds, soy beans, corns, grains, drying and baking of bread powder, crouton, and drying of cut meat products [46–48]. Application of the jet impingement fluidization method for freezing and thawing of foods has been still at the laboratory research stage [45].

4. Conclusion

A modification of conventional fluidization through a changed mode of fluid distribution and the construction of apparatus opens new potential application areas for this method. The work and studies on the modification have been carried out in two directions. The first is based on the vertical injection of fluid through the bottom of the device to induce spouted bed fluidization. The other concerns fluid distribution via nozzles as well long ones mounted above the bed. Both modifications have been intensively investigated and implemented. The fluid velocity in both modified methods should be appropriately chosen because of differences in bed particle size and structure and finally, substantial impact on economic conditioning of the realized processes. A too high fluidization velocity is likely to induce considerable worsening of product quality. Therefore, a large body of scientific papers was published and has been still presenting analytical research attempts to determine minimum fluidization velocity. The problem is also related to configuration and dimensioning of operating elements of devices for the realization of both fluidization modifications. These two methods enable better contact between the bed and fluid as compared to other types of fluidization systems. It pertains especially to spouting beds with low resistance of fluid permeability and is associated with enhanced intensity of heat and mass transfer. Introduction of jet fluidization technology for fruit and vegetable freezing allowed product treatment in a very short time. Compared to conventional fluidization, this method enables to obtain far higher heat transfer coefficients reaching 200 W m^{-2} K^{-1}. As against spouted bed fluidization problem, only few papers address the structure of devices for jet fluidization and they deal with a narrow range of products. Hence, it seems justifiable to state that despite several obvious advantages, applicability of both fluidization modifications still remains studied insufficiently, so working toward apparatus construction development needs further continuation.

Nomenclature

Ar	Archimedes number, $\frac{g\, d_p^3 (\rho_s - \rho_g)_{\rho_s}}{\mu_g^2}$
d_p	equivalent diameter of non-spherical particle calculated as the volume/area surface ratio, m
D	diameter of cylindrical column, m/nozzle inner diameters, m
D_o	diameter of fluid inlet, m
D_t	bed width , m
g	gravitational acceleration, m s^{-2}
h	heat transfer coefficient, W m^{-2} K^{-1}
H_o	bed depth, m
L	nozzle symmetry axis spacing
Nu	Nusselt number, $\frac{hx}{k}$
S	nozzle length, m
Re	Reynolds number, $\frac{d_p U_{ms} \rho_g}{\mu_g}$
T	temperature, °C
U_{ms}	minimum spouting velocity, m s^{-1}
U_f	fluidizing gas velocity, m s^{-1}
U_{mf}	minimum fluidizing velocity, m s^{-1}
x	characteristic dimension, m
Greek letters	
γ	cone angle, degree
λ_i	spout nozzle width, mm
μ	gas viscosity, Pa s
ρ_s	density of particles, kg m^{-3}
ρ_g	density of fluid, kg m^{-3}

Author details

Dariusz Góral*, Franciszek Kluza and Katarzyna Kozłowicz

*Address all correspondence to: dariusz.goral@up.lublin.pl

Department of Refrigeration and Food Industry Energetics, University of Life Sciences in Lublin, Lublin, Poland

References

[1] Peterson WS. Spouted bed drier. Can J Chem Eng. 1962;**40**(5):226–230. DOI: 10.1002/cjce.5450400512

[2] Kmieć A, Englart S, Ludwińska A. Theory and technique of fluidization [in Polish]. Monografie: Prace naukowe Instytutu Inżynierii Ochrony Środowiska Politechniki Wrocławskiej ed. Wrocław, Poland: Oficyna Wydawnicza Politechniki Wrocławskiej; 2007. 108 p.

[3] Altzibar H, Lopez G, Estiati I, Bilbao J, Olazar M. Particle cycle times and solid circulation rates in conical spouted beds with draft tubes of different configuration. Ind Eng Chem Res. 2013;**52**(45):15959–15967. DOI: 10.1021/ie401412j

[4] Epstein N, Grace JR. Spouting of particulate solids. In: Fayed ME, Otten L, editors. Handbook of powder science and technology. 2nd ed. Boston, MA USA: Springer US; 1997. pp. 532–567. DOI: 10.1007/978-1-4615-6373-0_10

[5] Bi HT. A discussion on minimum spout velocity and jet penetration length. Can J Chem Eng. 2004;**82**(1):4–10. DOI: 10.1002/cjce.5450820102

[6] Jumah RY, Mujumdar AS, Raghavan GSV. Aerodynamics of a novel rotating jet spouted bed. Chem Eng J. 1998;**70**(3):209–219. DOI: 10.1016/S1385-8947(98)00091-6

[7] Devahastin S. Flow and drying characteristics of a novel rotating jet annular spouted bed. Dry Technol. 1998;**16**(1–2):391–392. DOI: 10.1080/07373939808917414

[8] Niamnuy C, Kanthamool W, Devahastin S. Hydrodynamic characteristics of a pulsed spouted bed of food particulates. J Food Eng. 2011;**103**(3):299–307. DOI: 10.1016/j.jfoodeng.2010.10.028

[9] Xu J, Tang J, Wei W, Bao X. Minimum spouting velocity in a spout-fluid bed with a draft tube. Can J Chem Eng. 2009;**87**(2):274–278. DOI: 10.1002/cjce.20145

[10] Marmo L. Low temperature drying of pomace in spout and spout-fluid beds. J Food Eng. 2007;**79**(4):1179–1190. DOI: 10.1016/j.jfoodeng.2006.04.034

[11] Ishikura T, Nagashima H, Ide M. Hydrodynamics of a spouted bed with a porous draft tube containing a small amount of finer particles. Powder Technol. 2003;**131**(1):56–65. DOI: 10.1016/S0032-5910(02)00321-2

[12] Neto JV, Duarte CR, Murata VV, Barrozo MAS. Effect of a draft tube on the fluid dynamics of a spouted bed: Experimental and CFD studies. Dry Technol. 2008;**26**(3):299–307. DOI: 10.1080/07373930801897994

[13] Mathur KB, Epstein N. Spouted beds. New York, NY, USA: Academic Press Inc. Ltd.; 1974. 304 p.

[14] Peroń S, Surma M. Preliminary study on the hydrodynamics of spouted bed of corn grain [in Polish]. Inż Rol. 2005;**9**(7):215–221.

[15] Bi HT, Macchi A, Chaouki J, Legros R. Minimum spouting velocity of conical spouted beds. Can J Chem Eng. 1997 ;**75**(2):460–465. DOI: 10.1002/cjce.5450750221

[16] Nagahashi Y, Lee DH, Grace J R, Epstein N, Yokogawa A, Asako Y. Enhancement of large-particle gas-fluidization by adding liquid. AIChE J. 2003;**49**(3):675–681. DOI: 10.1002/aic.690490312

[17] Kmieć A. Hydrodynamics of flows and heat transfer in spouted beds. Chem Eng J. 1980;**19**(3):189–200. DOI: 10.1016/0300-9467(80)80029-3

[18] Grbavčíc ŽB, Vukovic DV, Zdanski FK, Littman H. Fluid flow pattern, minimum spouting velocity and pressure drop in spouted beds. Can J Chem Eng. 1976;**54**(1–2):33–42. DOI: 10.1002/cjce.5450540105

[19] Fane AG, Mitchell RA. Minimum spouting velocity of scaled up beds. Can J Chem Eng. 1984;**62**(3):437–439. DOI: 10.1002/cjce.5450620325

[20] Bi HT, Macchi A, Chaouki J, Legros R. Minimum spouting velocity of conical spouted beds. Can J Chem Eng. 1997;**75**(2):460–465. DOI: 10.1002/cjce.5450750221

[21] Mathur KB, Gisher PE. A technique for contacting gases with coarse solid particles. AIChE J. 1955;**1**(2):158–160. DOI: 10.1002/aic.690010205

[22] Zhong W, Chen X, Zhang M. Hydrodynamic characteristics of spout-fluid bed: Pressure drop and minimum spouting/spout-fluidizing velocity. Chem Eng J. 2006;**118**(1):37–46. DOI: 10.1016/j.cej.2006.01.008

[23] Olazar M, San Jose MJ, Aguayo AT, Arandes JM, Bilbao J. Stable operation conditions for gas-solid contact regimes in conical spouted beds. Ind Eng Chem Res. 1992;**31**(7):1784–1792. DOI: 10.1021/ie00007a025

[24] Sutkar VS, Deen NG, Kuipers JAM. Spout fluidized beds: Recent advances in experimental and numerical studies. Chem Eng Sci. 2013;**86**(4):124–136. DOI: 10.1016/j.ces.2012.06.022

[25] Rocha SCS, Taranto OP, Ayub GE. Aerodynamics and heat transfer during coating of tablets in two-dimensional spouted bed. Can J Chem Eng. 1995;**73**(3):308–312. DOI: 10.1002/cjce.5450730306

[26] Chandnani PP, Epstein N. Spoutability and spout destabilization of fine particles with a gas. In: Ostergaard K; Sorensen A, editors. 5. International Conference on Fluidization; 18–23 May 1986; Elsinore, Denmark. New York, NY, USA: Engineering Foundation; 1986. pp. 233–240.

[27] San José MJ, Olazar M, Aguado R, Bilbao J. Influence of the conical section geometry on the hydrodynamics of shallow spouted beds. Chem Eng J Biochen Eng. 1996;**62**(2):113–120. DOI: 10.1016/0923-0467(95)03064-6

[28] Markowski M, Białobrzewski I, Modrzewska A. Kinetics of spouted-bed drying of barley: diffusivities for sphere and ellipsoid. J Food Eng. 2010;**96**(3):380–387. DOI: 10.1016/j.jfoodeng.2009.08.011

[29] Narayanan V, Seyed-Yagoobi J, Page RH. An experimental study of fluid mechanics and heat transfer in an impinging slot jet flow. Int J Heat Mass Tran. 2004;**47**(8–9):1827–1845. DOI: 10.1016/j.ijheatmasstransfer.2003.10.029

[30] Sarkar A, Nitin N, Karwe MV, Singh RP. Fluid flow and heat transfer in air jet impingement in food processing. J Food Sci. 2004;**69**(4):R113–R122. DOI: 10.1111/j.1365-2621.2004.tb06315.x

[31] Geers LFG, Timmers JF, Hanjalić K. Experimental investigation of impinging jet arrays. Exp Fluids. 2004;**36**:946–958. DOI: 10.1007/s00348-004-0778-2

[32] Kluza F. Preliminary evaluation of impingement method utility for food freezing. In: Toldrá F, Ramón D, Navarro JL, editors. Proceedings of the International Congress "Improved traditional foods for the next century"; 28–29 October 1999; Valencia, Spain. DG XII-European Commission. Instituto de Agroquímica y Tecnología de Alimentos. CSIC; 1999. pp. 406–408.

[33] Kluza F, Stadnik Ł. Experimental formation of bed under jet tube fluidization conditions used for food cooling and freezing. EJPAU. 2009;**12**(3).

[34] Góral D, Kluza F. Impingement technique in thawing of food and agriculture products [in Polish]. Acta Sci Pol, Technica Agraria. 2003;**2**(1):23–31.

[35] Góral D, Kluza F. Experimental identification of refrigerating-system performance in the process of impingement [in Polish]. Inż Rol. 2004;**60**(5):117–125.

[36] Góral D, Kluza F. Heat transfer coefficient in impingement fluidization freezing of vegetables and its prediction. Int J Refrig. 2012;**35**(4):871–879. DOI: 10.1016/j.ijrefrig.2011.11.010

[37] Sarkar A, Singh RP. Air impingement technology for food processing: visualization studies. LWT-Food Sci Technol. 2004;**37**(8):873–879. DOI: 10.1016/j.lwt.2004.04.005

[38] Góral D. Technical and technological conditions of agricultural products refrigeration by using impingement method [in Polish]. Inż Rol. 2006;**80**(5):171–177.

[39] Góral D, Kluza F, Kozłowicz K. Analysis of heat transfer during the freezing of vegetables by impingement fluidization. Inż Rol. 2013;**141**(1):53–63.

[40] San JY, Lai MD. Optimum jet–to–jet spacing of heat transfer for staggered arrays of impinging air jets. Int J Heat Mass Tran. 2001;**44**(21):3997–4007. DOI: 10.1016/S0017-9310(01)00043-6

[41] Becker BR, Fricke BA. Heat transfer coefficients for forced–air cooling and freezing of selected foods. Int J Refrig. 2004;**27**(5):540–551. DOI: 10.1016/j.ijrefrig.2004.02.006

[42] Chan TL, Leung CW, Jambunathan K, Ashforth–Frost S, Zhou Y, Liu MH. Heat transfer characteristics of a slot jet impinging on a semi–circular convex surface. Int J Heat Mass Tran. 2002;**45**(5):993–1006. DOI: 10.1016/S0017-9310(01)00217-4

[43] Salvadori VO, Mascheroni RH. Analysis of impingement freezers performance. J Food Eng. 2002;**54**(2):133–140. DOI: 10.1016/S0260-8774(01)00198-4

[44] Amarante A, Lanoisellé JL. Heat transfer coefficients measurement in industrial freezing equipment by using heat flux sensors. J Food Eng. 2005;**66**(3):377–386. DOI: 10.1016/j.jfoodeng.2004.04.004

[45] Góral D, Kluza F, Kozłowicz K, Domin M. A preliminary assessment of using the operating head with oval nozzles for impingement fluidization of vegetables. Agric Eng. 2015;**153**(1):47–55. DOI: 10.14654/ir.2015.153.105

[46] Arakawa. THERMOZONE [Internet]. 2016. Available from: http://www.arakawa-mfg.co.jp/english/products_thermozone.html [Accessed: 2017-01-02]

[47] Buhler. AeroToast™ Fluid Bed Toaster [Internet]. 2016. Available from: http://www.buhlergroup.com/global/en/products/aerotoast-fluid-bed-toaster.htm#.WHfB5n2V72Y [Accessed: 2017-01-02]

[48] Volverine Proctor. Jet-Tube Fluidized Bed [Internet]. 2016. Available from: http://www.wolverineproctor.com/en-us/equipment/jetzone-fluidized-bed-dryerpuffertoaster/ [Accessed: 2017-01-02]

Chapter 16

High Thermal Conductivity Ceramics and Their Composites for Thermal Management of Integrated Electronic Packaging

Hyo Tae Kim

Additional information is available at the end of the chapter

http://dx.doi.org/10.5772/intechopen.75798

Abstract

Recently, ceramic substrates have been of great interest for use in light emitting diode (LED) packaging materials because of their excellent heat transfer capability. The thermal conductivities of ceramic-based substrates are usually one or two orders of magnitude higher than those of conventional epoxy-based substrates. The demand for ceramic substrates with high mechanical strength and thermal conductivity is also growing due to their use in thin and high-power device packaging substrates. Examples are direct bonded copper or aluminum or direct plated copper substrates for insulated gate bipolar transistors; thin and robust ceramic packages for image sensor modules that are used in mobile smart phones; ceramic packages for miniaturized chip-type supercapacitors; and high-power LED packages. This chapter will cover the development and application of ceramics and ceramic composites with high thermal conductivity for the thermal management of integrated electronic packaging substrates such as high-power LED packaging, power device packaging, etc.

Keywords: thermal conductivity, ceramics, composites, electronic, packaging

1. Introduction

Ceramic materials with high thermal conductivity are of great interest in the thermal management of integrated electronic device packaging such as high-power light emitting devices (LEDs), power semiconductor modules, micro and nano fluidics, thermoelectrics, solar cells, and wireless communication devices. These electronic devices and packages generate more heat than before as the system design goes into more integrated, miniaturized, and increasing data communication due to the multifunctional requirement in wireless communication and the Internet of things (IOT) environment. Moreover, a global movement for a clean environment has shifted the public interest from conventional cars with combustion engines to electric vehicles (EVs) and hybrid electric vehicles (HEVs), thus drastically increasing the use of integrated power modules with increased powers and operation frequencies. In this regard, the development of high thermal conductivity ceramics for packaging substrate and filler materials for composites is of great importance. This chapter briefly examines recent trends, development, and technical issues of selected high thermal conductivity ceramic materials and their composites.

2. High thermal conductivity ceramics and their composites

2.1. Aluminum nitride (AlN) base ceramics and composites

2.1.1. AlN ceramics with sintering additives

Aluminum nitride (AlN) has a highly covalent bonded wurtzite structure with a high thermal conductivity and a low thermal expansion coefficient (CTE) of 4.5 ppm/°C that matches well with silicon devices. Typical thermal conductivity of AlN is 140–180 W/mK but varies in the range 18–285 W/mK in polycrystalline AlN ceramics depending on the process condition, purity of starting materials, and microstructures [1]. AlN is stable at 700–1000°C in an oxygen atmosphere. It also has excellent dielectric properties: low dielectric constant (ε_r) = 9 and low loss (tan δ) = 0.0003 at 1 MHz. With these outstanding physical and thermal properties, AlN ceramic is frequently selected as a candidate material for insulating substrate (direct band gap energy ~6.015 eV) for power electronics device and package. However, sintering of AlN with high density for effective heat transfer and high mechanical strength is challenging due to its highly covalent and low diffusive nature that requires very high sintering temperatures over 1900°C in a reducing atmosphere with applied pressure even though it is stable near 1000°C in air. Also, degradation of thermal conductivity due to oxygen inclusion is another confronting issue in high thermal conductivity AlN substrate fabrication.

Many works to promote the densification of AlN ceramic bodies using different kinds of sintering additives such as CeO, Sm_2O_3, Y_2O_3, CaO, $CaZrO_3$, and their multiple co-additions have been investigated [2–5]. Some of the recent results showing the thermal conductivities obtained in polycrystalline bodies are 90–156 W/mK as summarized in **Table 1**, which is far below the theoretical value and has a wide span from each other. Ceria (CeO) doped AlN exhibited a stiff increase in thermal conductivity and decent increase in mechanical strength with a small amount of addition (~1.5 wt%), compared to the yttria (Y_2O_3) addition [2]. High energy sintering method like spark plasma sintering (SPS) was applied [3, 4] as well as conventional solid state reaction (SSR) method [2] for effective low temperature densification process. The two-step sintering technique was also conducted using different temperatures to minimize grain growth and purification of AlN grains [5] (**Figure 1**).

Chemical composition	Sintering additives	Thermal conductivity (W/mK)	Mechanical properties	Sintering temp/ method	Ref.
AlN	1.5 wt% CeO	156	72.46 kg/mm^2	1900°C/SSR	[2]
AlN	2 wt% Sm_2O_3	120	—	1700°C/SPS	[3]
AlN	Y_2O_3-CaO-B	90	—	1650°C/SPS	[4]
AlN	$CaZrO_3$-Y_2O_3	156	560 MPa	1550°C/Two-step	[5]

Table 1. Physical properties of AlN ceramics with the addition of sintering additives and densification methods.

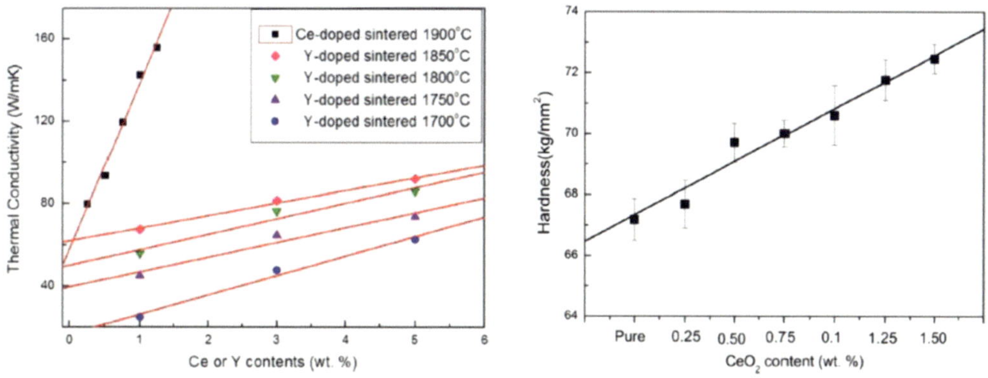

Figure 1. Effect of Ce and Y doping on the thermal conductivity and hardness of AlN ceramics [2].

2.1.2. AlN composites with GNP/GNS/rGO: electrically conducting

Carbon based nanostructured materials such as graphene nanoplatelets (GNP) or nanosheets (GNS) were added to AlN matrix to improve physical properties (**Table 2**). The electrical conductivity was increased with the addition of multilayer graphene as expected but the thermal conductivity was decreased with the addition in both in-plane and through-plane direction, which is adverse to other ceramic/graphene composites data. This sharp decline of thermal conductivity in both directions seems attributed to the large thermal resistance at the thin interaction zone existing in the interface between AlN and GNP [6]. The high directionality in the in-plane and through-plane of AlN/GNP composites, 74 W/mK for in-plane and 37 W/mK for through-plane, is ascribed to the thermal contact resistance existing in both phase interfaces that are severe in the perpendicular heat transfer direction of graphene nanoplatelets [7]. This strong directionality in heat transfer, thermal conductivity, can be easily found in the boron nitride (BN)/polymer composite system, in which 2D morphology BN filler materials are used [9–16]. In case of reduced graphene oxide (rGO) added AlN, the thermal conductivity decreased sharply from 92.5 to 37.4 W/mK when 2 wt% of rGO was added, though there are minor increases in flexural strength and fracture toughness at ≤1 wt% of rGO, which is due to the low crystallinity, high vacancy defects in rGO, and increased interfacial thermal resistance [8]. The declining thermal conductivity behaviors of AlN composites with these three carbon based 2D fillers exhibited almost similar results as GNP, GNS, and rGO basically

Chemical composition	Additives	Thermal conductivity (W/mK)	Mechanical properties	Sintering temp./ method	Ref.
AlN-2.9 wt%Y_2O_3-GNP~10 vol%	GNP 0.1 vol%	74 (in-plane) 37 (through-plane)	—	1700-1750/SPS	[6]
AlN-2wt%Y_2O_3	GNS 1.49 vol%	—	FS = 441 MPa FT = 4.09 MPam$^{1/2}$	1850/hot press	[7]
AlN	rGO 1 wt%	92.5 - > 37.4 (2 wt%)	FS = 375 MPa	1550/SPS	[8]

FS: Flexural strength; FT: Fracture toughness; SPS: Spark plasma sintering

Table 2. Physical properties of AlN with carbon nanomaterials addition.

had the same morphology and physical properties. High thermal conductivity materials with high directionality in heat transfer can be used in specific directional heat dissipation applications as thermal interface materials (TIM) (**Figures 2** and **3**).

2.1.3. Si_3N_4 base ceramics

Silicon nitride (Si_3N_4) ceramics has been drawing a lot of interest as a high thermal conductivity dielectric material used in insulated metal substrate (IMS) for power electronic circuit modules. Si_3N_4 have several benefits: high mechanical properties (flexural strength >800 MPa, Vickers' hardness >10 GPa), high electrical resistivity, and excellent thermal properties with thermal resistance, high thermal conductivity 70–180 W/mK. However, in reality, fabrication of Si_3N_4 with high thermal conductivity and high mechanical strength is not easy due to difficulties in densification and morphological control in microstructure. Typical approaches to get such a high performance Si_3N_4 are: (i) using raw materials with low oxygen to remove Si vacancies that cause phonon scattering, (ii) fabrication of Si_3N_4 ceramics with textured microstructure to utilize thermal anisotropy in Si_3N_4 crystals, (iii) using non-oxide sintering additives with low oxygen content to avoid oxygen content from the oxide phase, and (iv) selecting optimal additives that can minimize the Si vacancies [9]. Some of selected results based on these approaches

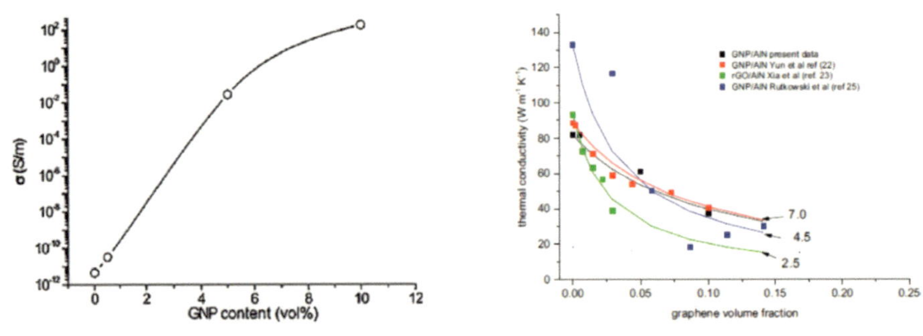

(a) (b)

Figure 2. AC electric conductivity (a) and thermal conductivity data (b) with GNP content in the AlN composites [6].

http://dx.doi.org/10.5772/intechopen.75798

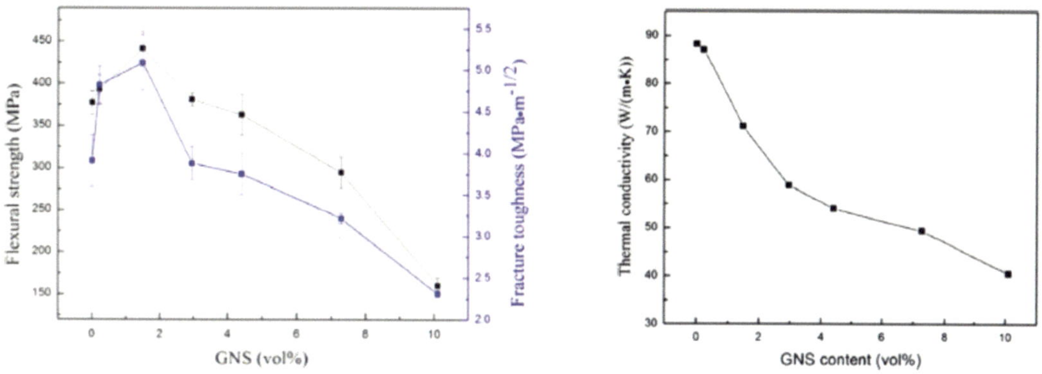

Figure 3. Fracture toughness and flexural strength of AlN/GNS composites [7].

are summarized in **Table 4**. When non-oxide additives like YF_3 was added to Si_3N_4 with MgO instead of Y_2O_3, the mechanical strength and thermal conductivities were improved from 40 to 52; it was further increased to 75 W/mK after the annealing treatment at 1850°C. However, Vickers' hardness was decreased slightly due to larger particle sizes than with the Y_2O_3 addition [9]. The thermal conductivity of Si_3N_4 increased up to 100 W/mK when Yb_2O_3/SiO_2 was added and the dielectric loss was decreased from 11.5×10^{-4} to 1.4×10^{-4} (2 GHz) by adjusting the ratio from 0.33 to 1.3 [10]. The influence of nitration and sintering conditions on the mechanical and thermal properties of the sintered reaction bonded Si_3N_4 (SRBSN) with Y_2O_3-MgO additives, and the coarsening of grain size and aspect ratio decreased the hardness and fracture toughness while increasing the thermal conductivity. Vickers' hardness of 17.32 GPa, fracture toughness of 8.36 $MPa{\cdot}m^{1/2}$, and thermal conductivity of 98.52 W/m·K were obtained by adjusting nitration and gas pressure sintering (GPS) [11]. A comparative study of the effects of oxide and non-oxide additives on the microstructure, lattice oxygen content, and thermal conductivity of Si_3N_4 ceramic was investigated. Non-oxide additives such as $MgSiN_2$, YF_3, YbF_3 induced a decrease in the amount of secondary phases and lattice oxygen contents, thus increasing thermal conductivity from 65 to 101.5 W/mK, while the flexural strength was not affected significantly [12].

2.1.4. AlN-BN base composites: electrically insulating

Boron nitride (BN) was introduced in the AlN matrix to realize low dielectric constant and moderate thermal conductivity [17]. Boron nitride (h-BN) has a hexagonal structure with good thermal shock resistance and high thermal conductivity together with directional preference in heat transfer, i.e., anisotropy in thermal conductivities at in-plane and through-plane of the substrate due to the 2D shape of the BN flake or BN nanosheet. The in-plane and through-plane thermal conductivities are about 300 and 30 W/mK, and the average apparent value is 33 W/mK. The h-BN has low dielectric constant and loss tangent, $\varepsilon_r = 4$–4.6, $\tan \delta =$ 0.0012–0.0017 at 8.8 GHz, and dielectric strength at AC = 67–95 kV, which varies depending on the purity. BN has been used in many applications due to these excellent properties, for example, in microelectronic packaging especially in thermal management parts such as heat sinks and power electronic substrates, etc. The addition of BN to AlN also can improve chemical resistance and moisture resistance since the AlN can be hydrolyzed slowly in water (**Table 3**).

Chemical composition	Additives	Thermal conductivity (W/mK)	Sintering temp/method	Ref.
AlN–BN	Sm_2O_3-CaF_2	40–85	1800/SPS	[17]
AlN–BN(15%)	Y_2O_3	141	1800/SPS	[18]
AlN–BN	CaF_2	110	1850/hot-press	[19]
(Al-O-N)–BN		⊥14, //42.5	1900/hot-press	[20]

Table 3. Thermal conductivities of AlN–BN composites.

In the AlN–BN composites with 8 wt% Sm_2O_3-CaF_2 as sintering aid, the highest thermal conductivity of 85 W/mK and lowest loss tangent of 4×10^{-3} were achieved at the sintering temperature of 1800°C by the SPS method. The obtained thermal conductivity was lower than that of pure AlN because the platelet BN particles randomly distributed along the AlN matrix hinders direct contact of AlN so that phonon scattering is inhibited [17]. When yttrium oxide (Y_2O_3) from 3 to 8 wt% is added to the AlN–15%BN composites, the thermal conductivity was increased from 110 to 140 W/mK, which is attributed to the significant decrease in residual grain boundary phase containing yttrium by using SPS method [18]. The addition of CaF_2 and increasing temperature also improved the densification, thermal conductivity, and grain boundary purification at the AlN–BN system. As a result, high thermal conductivity of 110 W/mK was obtained when 3 wt% of CaF_2 was added and sintered at 1850°C [19]. In the aluminum oxynitride (γ-AlON)–BN system that was prepared by the self-propagating high-temperature synthesis (SHS) process, platelet shaped h-BN grains are re-oriented during the hot-pressing process resulting in the anisotropy of thermal conductivities. The thermal conductivity of (γ-AlON)–BN composites were 14 W/mK for through-plane direction of BN grains and 42.5 W/mK for in-plane direction which is perpendicular to the hot-pressing force [20] (**Figure 4**).

2.2. Polymer matrix composites with high thermal conductivity ceramic fillers

Polymer matrix composites for thermal management packaging are usually filled with high thermal conductivity ceramics such as AlN, h-BN, and carbon based fillers like carbon nano

Main element	Sintering additives	Thermal conductivity (W/mK)	Mechanical properties	Sintering condition	Ref.
Si_3N_4	YF_3, MgO	52–75	87 MPa/14.8 GPa	1750–1850°C/20 MPa	[9]
Si_3N_4	Yb_2O_3, SiO_2	46–100	—	1900°C/0.9 MPa	[10]
Si_3N_4	Y_2O_3, MgO	98.52	—/17.52 GPa	1950°C/GPS	[11]
Si_3N_4	MgO-Y_2O_3, $MgSiN_2$-Y_2O_3, $MgSiN_2$-YF_3, MgO-Yb_2O_3, $MgSiN_2$-Yb_2O_3, and $MgSiN_2$-YbF_3	101.5	862 MPa	1800°C/250 MPa	[12]

Table 4. Physical properties of Si_3N_4 ceramics with the addition of sintering additives and densification methods.

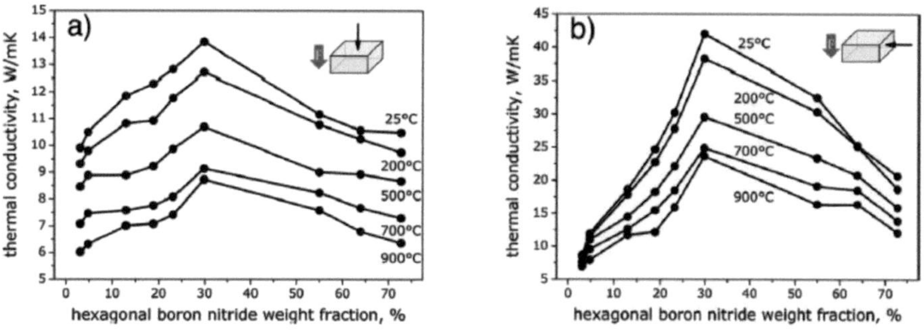

Figure 4. Compositional dependence of thermal conductivity of the composites in the (γ-AlON)–BN system determined in (a) perpendicular and (b) parallel to the pressing force [20].

fibers (CNTs), graphite or graphene nanosheets (GNSs), and reduced graphene oxide (rGO). Polymers with AlN and h-BN ceramic filler systems are mostly preferred for high thermal conductivity with electrically insulating heat transfer substrates or thermal interface materials (TIM) due to the high thermal conductivity, low dielectric constant and low loss characteristics of AlN and BN ceramics. On the other hand, carbon based fillers are preferred in TIMs where electrically conducting characteristics are allowed.

2.2.1. Polymer: BN composites

In the hexagonal-boron nitride (h-BN) filled polymer composites, the major issues to enhance heat transfer property are surface treatment of h-BN platelet particles to improve the dispersion of the filler particles in the polymer matrix; to lower the interface thermal resistance; and to increase the alignment of h-BN particles to the preferred orientation in order to achieve high directional thermal conductivity in composites. **Table 5** summarizes several technical efforts to enhance the heat transfer properties of polymer–BN composites [13–16, 21–24].

A combined technique that uses mechanical vibration and rotating magnetic field induced high degree of alignment of 10% filler loaded composite exhibited 74% improvement in thermal conductivity compared to the unaligned sample by the formation of conductive network and the reduction of the thermal boundary resistance. The reduction in the thermal boundary resistance between h-BN and bisphenol-A based resin was induced by a high degree of alignment of h-BN platelets via the combined process [13].

The effect of AC and DC electric fields on the anisotropically aligned microstructure in the h-BN filled silicone rubber composites was studied. It was found that the degree of re-orientation of h-BN was more effective under the AC than the DC field during the curing process of the h-BN-silicone composite (**Figure 5**), and the thermal conductivity of the e-field assisted curing composite was about 250% higher than that cured without *E*-field [14]. In a recent study, the largest total number of linear densely packed BN nanosheets (LDPBNs) was formed by applying AC field, and thickening of LDPBNs and narrowing of interparticle gaps were achieved by applying a switching DC field (**Figure 6**). As a result, the thermal conductivity was increased four times that of the composite without LDPBN structure [15].

Composition	BN %	Thermal conductivity (W/mK)	BN align method	Ref.
BN–epoxy	10–20	74% ↑	Mechanical and magnetic	[13]
BN–silicone	20	250% ↑	E-field(AC, DC)	[14]
BN–polysiloxane	15 vol%	400%↑	AC/switching DC	[15]
BN–PVA	27 vol%	1.63 W/mK ⊥/8.44 W/mK//	PVA solution infiltration	[16]
BN–polyurethane acrylate (PUA)	30 vol%	190%↑(in-plane) 72%↓(through-plane)	e-field (TiO_2-coated BN)	[21]
BN–PEN(poly(arylene ether nitride))	30 wt%	140%	Magnetic + mussel inspired co-modification (Fe_3O_4/PDA + KH550)	[22]
BN		3.49 W/mK(45.4%)	Silane coupling agents with different carbon chain	[23]
BN–PVA	50 vol%	1.1 W/mK ⊥/13 W/mK//	Mechanical exfoliation, compression	[24]

Table 5. Examples of thermal conductivities of polymer–BN composites.

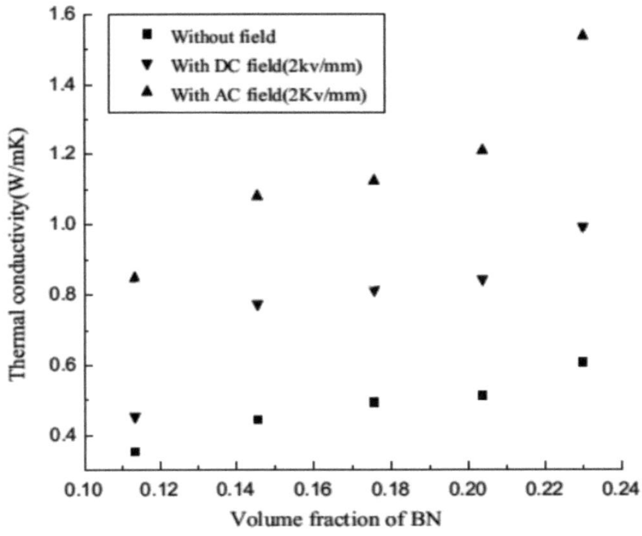

Figure 5. Thermal conductivity of BN/silicone composites at different volume fractions [14].

A flexible h-BN/poly(vinyl alcohol) composite tape was fabricated by the infiltration of poly(vinyl alcohol) (PVA) solution into the h-BN stack with vacuum filtration to reduce the gap between the h-BN particles and to increase the degree of alignment of h-BN platelets. The in-plane and through-plane thermal conductivities of h-BN/PVA composites thus obtained were 1.63 and 8.44 W/mK, respectively [16]. In the same h-BN/PVA system, the degree of the orientation of h-BN platelet particles can be improved by pressure assisted casting [24], where the degree of orientation of the h-BN particle can be observed by the characteristic peaks in the X-ray diffraction (XRD) data. In order to boost the alignment of h-BN particles, the coating of electric or magnetic field sensitive materials such as TiO_2 or Fe_3O_4 ceramics on to the h-BN

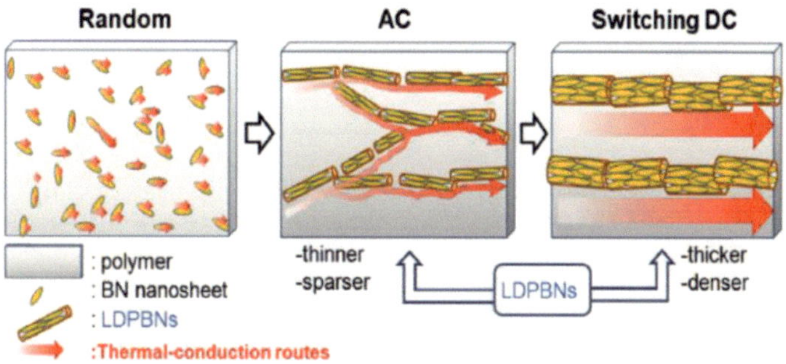

Figure 6. Schematic model of the generation of higher conduction routes through LDPBNs using various applications of electric fields [15].

particles are also explored to enhance the thermal conductivity [21, 23]. The h-BN particles coated with TiO_2 by the sol–gel process were aligned in a vertical direction to the applied field such that the through-plane thermal conductivity of h-BN/polyurethane acrylate (PUA) composite was increased by 190%, while the in-plane thermal conductivity of the composite was decreased by 72% compared to the untreated h-BN composite [21]. High dielectric constant and high thermal conductivity h-BN/poly(arylene ether nitrile) (PEN) composites were developed by magnetic alignment of h-BN through the coating of magnetic Fe_3O_4 particles together with an additional surface modification by polydopamine (PDA) and functional monomer KH550: (3-aminopropyl) triethoxy-silan). These surface modifications improved the dispersion of h-BN fillers in PEN matrix and the interfacial adhesion. In the BN/Fe_3O_4/PDA + KH550/PEN composite system, both dielectric constant and loss tangent were increased significantly with the amount of BN/Fe_3O_4/PDA + KH550, and the thermal conductivity was increased by 140% compared to the neat PEN film [22]. The enhancement of thermal conductivity in h-BN/epoxy composites through the surface modification of h-BN particles via silane coupling agents with different carbon chain has been investigated. The thermal conductivity of h-BN/epoxy composite was improved by 45.4% due to better dispersion of h-BN in epoxy resin than untreated h-BN, which attributed to the higher interfacial affinity of the composite obtained by using longer carbon chain of silane on the h-BN surface [23] (**Figures 7**, **8**, and **9**).

The effects of h-BN particle sizes, exfoliation of BN particles, and compression of h-BN/PVA composites on the thermal conductivity behavior were investigated [24]. **Figure 10(a)** shows that the thermal conductivity of h-BN/PVA composites was increased to almost two times when the as-received h-BN flakes are exfoliated into a thin h-BN nanosheet. Also, h-BN particles with smaller size exhibited a higher thermal conductivity in the h-BN/PVA composites, as shown in **Figure 10(b)**. Further increase in the thermal conductivity of h-BN/PVA composites were realized by uniaxial thermal compression after solution casting of h-BN/PVA composite film at 90°C which is above the glass transition temperature of the PVA polymer. **Figure 11** shows FE-SEM micrographs of the cross-sectional views of 30 vol% h-BN/PVA composites before (a) and after compression, which clearly show the h-BN particles alignment perpendicular to the pressing direction (b). The through (transverse)-plane and in-plane thermal conductivities measured by laser flash method are plotted together with theoretical modeling in **Figure 12**. Two models, arithmetic and Wiener models are used for the calculation of the two-phase composite system with the following equations.

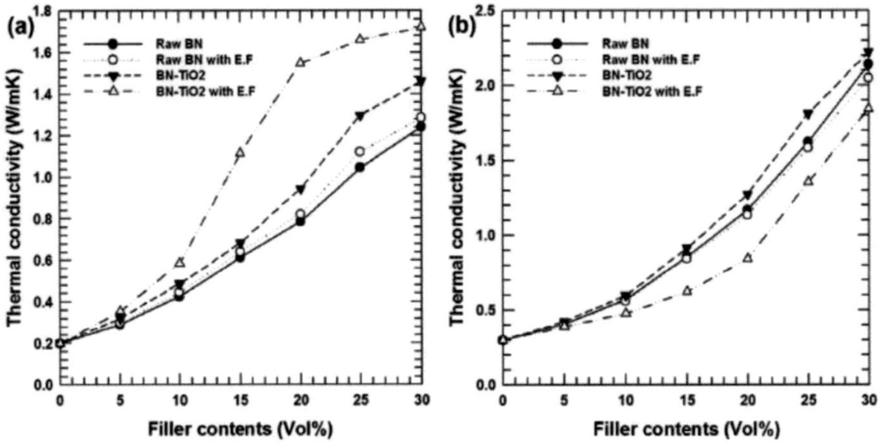

Figure 7. Thermal conductivity of h-BN/PUA composite and TiO_2 coated h-BN/PUA composite before and after electric field alignment: (a) through-plane and (b) in-plane thermal conductivity [21].

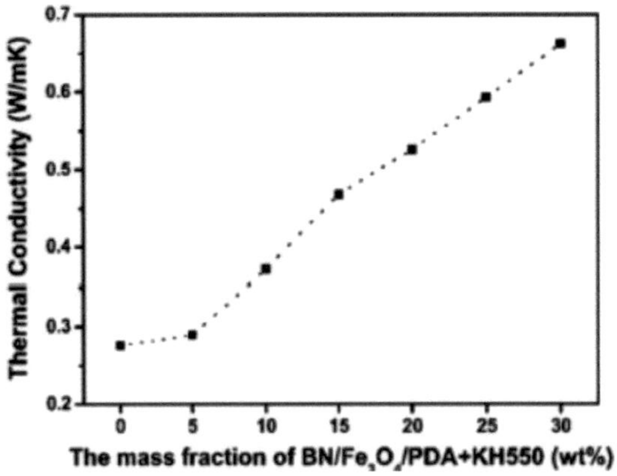

Figure 8. Thermal conductivities of neat PEN and BN/Fe_3O_4/PDA + KH550/PEN composite films with various amounts of filler loading content [22].

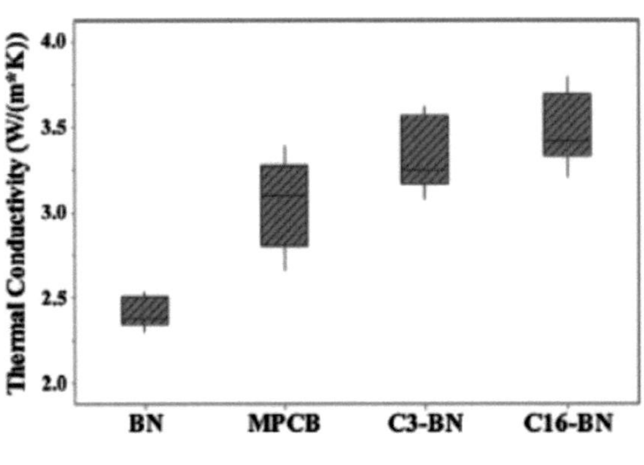

Figure 9. Thermal conductivities of h-BN/epoxy composites with untreated BN, MPCB (Al_2O_3/epoxy), silane (C3)/BN, and silane (C16)/BN [23].

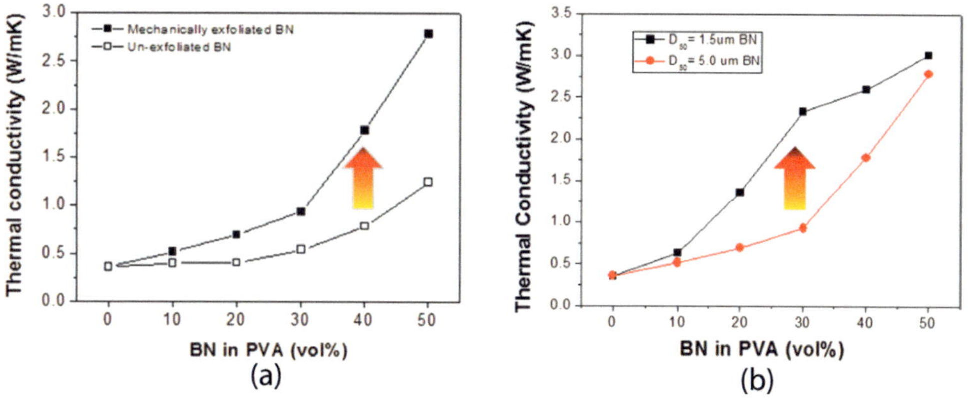

Figure 10. Effect of mechanical exfoliation (a) and h-BN particle sizes (b) on the through-plane thermal conductivities of h-BN/PVA composites [24].

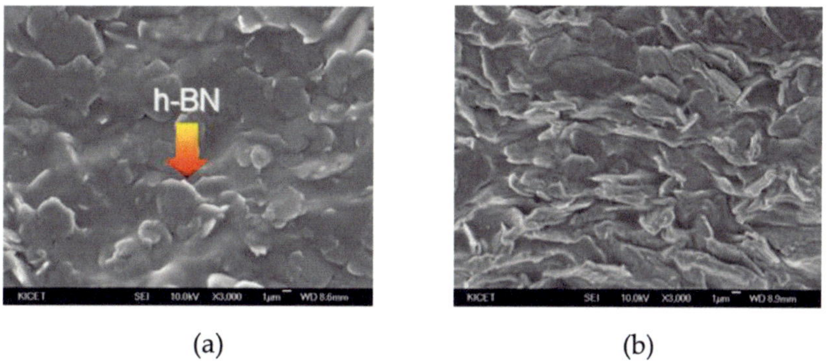

Figure 11. Cross-sectional views of 30 vol% h-BN/PVA composite films observed by FE-SEM: (a) uncompressed and (b) compressed [24].

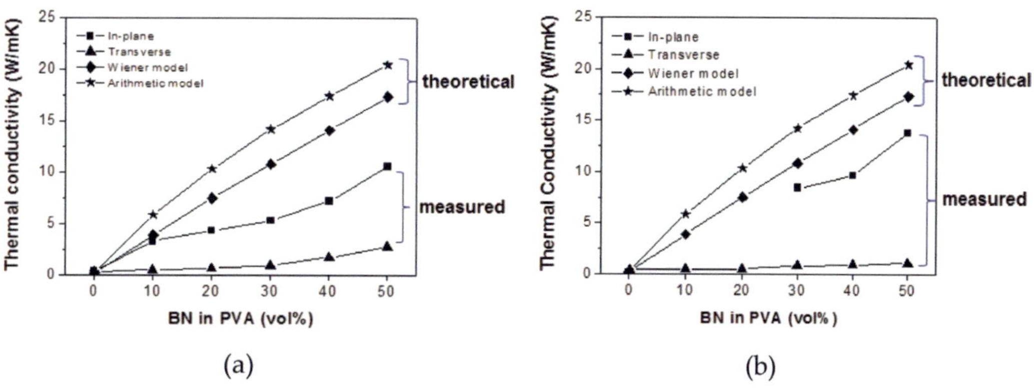

Figure 12. Effect of uniaxial compression on the transverse and in-plane mode thermal conductivities of h-BN/PVA composite films: (a) uncompressed and (b) compressed samples [24].

Arithmetic model:

$$\lambda = c\lambda_1 + (1 - c)\lambda_2 \tag{1}$$

where, c = proportion of component 1.

$1–c$ = proportion of component 2

λ_1, λ_2 = thermal conductivity of component 1, 2.

Wiener model:

$$\lambda/\lambda_2 = \left(1 - c\left((1 - \lambda_1/\lambda_2)/(1 + \alpha\,\lambda_1/\lambda_2)\right)\right) / \left(1 + \alpha c\left((1 - \lambda_1/\lambda_2)/(1 + \alpha\,\lambda_1/\lambda_2)\right)\right) \tag{2}$$

where, $\alpha = 0.5$ for dispersion model, $\lambda_1 < \lambda_2$.

In this calculation, the thermal conductivity of PVA and h-BN was assigned as 0.2 and 33 W/mK, respectively. The in-plane, perpendicular to the compressing direction, thermal conductivities were 5 times higher than those of through-plane at the un-compressed samples, and had 10 times higher values at the compressed samples. The highest thermal conductivity obtained at 50 vol% h-BN loaded PVA composite was 13 W/mK in the in-plane mode, while that of through-plane mode was about 1.1 W/mK, which is lower than that the uncompressed sample [24]. The decrease in the through-plane mode thermal conductivity at the h-BN/PVA composites is due to the reduction in through-plane particle contact by the alignment of h-BN platelets parallel to the in-plane direction, which results in the increasing in-plane thermal conducting paths and decreasing through-plane thermal conducting paths. The Wiener model seems more close to the experimental data than the arithmetic model but the gap between these theoretical and experimental data is wide indicating that there still remain many factors that should be improved to reach an optimum condition in the fabrication of polymer/ceramic composites with high thermal conductivity.

In summary, several methods were explored to increase the thermal conductivity of h-BN/polymer composites. Examples are (i) surface modification of h-BN particles with functional organics to improve the affinity and dispersion of the h-BN/PVA solution, (ii) coating dielectric and ferrous ceramic materials on the h-BN particle to increase the alignment performance with the application of electric and magnetic field, (iii) exfoliation of h-BN flakes into thin h-BN nanosheets for better particle connection in a given filler loading, and (iv) mechanical compressing to promote particle alignment and inter-particle contact.

2.2.2. Polymer: AlN composites

Aluminum nitride (AlN) is a priority choice for filler material in high thermal conductivity polymer/ceramic composites since it has high thermal conductivity (320 W/mK, theoretical), good insulating (electrical resistivity > 10^{14} Ω·cm) characteristic, low dielectric constant (2.2–3.7 at 1 MHz), and low CTE (4.4 ppm/°C) which is close to silicon. In hot-pressed AlN/PMMA (polymethyl methacrylate) composites, thermal conductivity of 1.87 W/mK was obtained at 70 vol% AlN loaded composite which is about 10 times higher than the PMMA resin (0.18 W/mK) as shown in **Figure 13(a)**. The dielectric constant and loss of the composite were 4.4 (**Figure 13(b)**) and

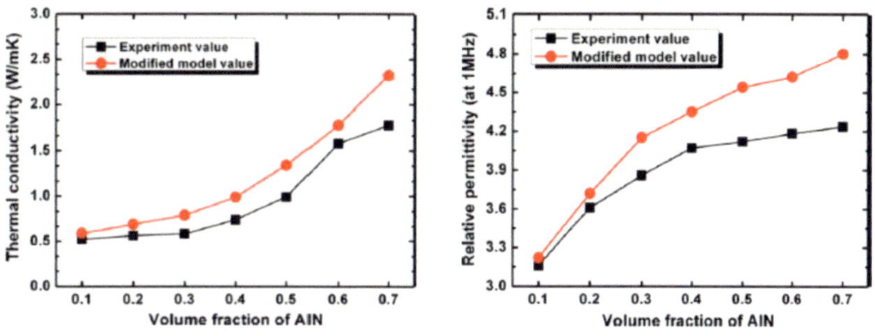

Figure 13. Experimental and calculated thermal conductivities data of PMMA/AlN composites with variation of AlN fillers [25].

0.017 at 1 MHz [25]. In the polypropylene (PP)/AlN composites with 3-D segregated structure made by mechanical grinding of PP and AlN mixture followed by hot-pressing at 190°C, core–shell structured PP/AlN composites were obtained. The comparative results of this 3D core–shell structure composite with conventional solution and melt mixed composites revealed that this mechanically ground composite with 10 vol% AlN showed 23% higher thermal conductivity than the others [26]. Thermal conductivity of aluminum nitride loaded poly(propylene glycol) (PPG) fluidic solution was also studied and the results have shown that the thermal conductivities of AlN/PPG fluids were dependent on the AlN solid loading and molecular weight of PPG [27].

In summary, in spite of efforts to increase thermal conductivity in polymer/ceramic composites, the thermal conductivities obtained in polymer matrix ceramic filled composites are still far below that of the fully ceramic base materials due to the low thermal conductivities in the polymer matrix which is limited in improving thermal conductivity when they are electrically insulating. So, further elaboration is needed in the development of high thermal conductivity polymers with electrical insulation to get the utmost high thermal conductivity in polymer/ceramic composites, since the thermal conductivities of filler ceramics are more likely to depend on their own intrinsic nature and are hardly changed by material science and engineering manipulation.

2.3. LTCC ceramics with high thermal conductivity

Low temperature co-fired ceramics (LTCC) have several benefits in microelectronic packaging. Typical sintering temperature of LTCC is below 1000°C, so they can be co-fired with highly conductive electrodes such as silver (Ag) or copper (Cu) metal. Most of the current LTCC materials are composed of low temperature melting glass matrix and ceramic fillers for functional adjustment such as electrical, mechanical, and thermal properties depending on the requirement of the application. Hence, LTCC is sometimes called glass–ceramics but technically, LTCC is part of glass–ceramic composites. For example, in the field of high frequency (RF, microwave, and mm wave) devices and packaging substrates, low loss and low dielectric constant ceramic powder such as alumina (Al_2O_3) powder is added to the low loss glass matrix. As a result, secondary phases are evolved during the heat treatment process due to part of the alumina filler particles being subjected to react with glass matrix. Interestingly, these secondary phases contribute to the improvement of dielectric properties and mechanical strength when proper filler particle and matrix composition are selected. LTCC has been used for many applications due to energy saving in the low temperature sintering process, excellent dielectric properties, and ease in 3D integration and miniaturization. Examples are radio frequency-system in-a-package (RF-SiP) module, LED packages, high temperature sensors, microelectromechanical system (MEMS) package, micro-heaters, microfluidics etc. [28–38].

2.3.1. *Glass-ceramic base LTCC system*

Recently, the use of high thermal conductivity with insulating ceramic substrate is rapidly increasing to enhance the heat transfer property of integrated electronic device and package. Since the conventional LTCCs are based on the glass matrix ceramic composite (GMC) system with low thermal conductivity of the glass phase (1–2 W/mK), the thermal conductivity of most LTCCs are as low as 2–5 W/mK as shown in **Table 6** [38–44], which is still higher than FR-4 (typically 0.1–0.2 W/mK) substrate but far lower than high thermal conductivity bulk ceramics such as alumina, AlN, or Si_3N_4.

Increasing the thermal conductivity in the glass–ceramic system can be realized by recrystallization, addition of high thermal conductivity fillers such as Al_2O_3, AlN, BN, and Si_3N_4 particles.

Category	Maker code/ Chemical composition	Sintering temp (°C)	Thermal conductivity (W/mK)	CTE (ppm/°C)	Dielectric constant	Dielectric loss	Ref.
Commercial products	ESL41110-70C	—	2.5–3	6.4	4.3	0.004	[37, 38]
	Heraeus CT2000	850	3	5.6	9.1	0.002	[37, 39]
	DuPont 951	850	3.2–3.9	5.8	7.8	0.006	[37, 40]
	DuPont 9K7	850	4.6	4.4	7.1	0.001	[37, 41]
	DuPont 943	850	4.4	6	7.4	0.002	[37, 42]
	Ferro A6	850	2	7	5.9	0.002	[37, 43]
	Murata	850	2.5	5.5	7.7	—	[37, 44]
Research works	$MgO–CaO–Al_2O_3–SiO_2$	900	1.95–2.6	8	—	—	[33, 45]
	$Al2O_3$–BBSZ glass	850	7.2	6.9	10.9	0.009	[46]
	$ZnTiO_3–B_2O_3$	900	6.6	—	20	0.001	[47]
	ABS–MWCNT	—	2.2	—	—	—	[48]
	CAS + β-Si_3N_4	850	7.9	—	7.1	0.006	[49]
	LZT–LMZBS	900	5.8		11.97	—	[50]
	Glass–diamond	750	9.01	4.35	—	—	[51]
	Al_2O_3/glass+ AlN whisker/carbon fiber/ copper fiber	850	38.9	4.6–6.1	6.6–7.4	—	[52]
	Borosilicate glass–AlN–β–Si_3N_4	850	18.8	4.2–4.4	6.5	0.0016	[53]

BBSZ: $Bi_2O_3–ZnO–B_2O_3–SiO_2$, ABS: $Al_2O_3–B_2O_3–SiO_2$, CAS: $CaO–Al_2O_3–SiO_2$, LZT: $Li_2ZnTi_3O_8$, LMZBS: $Li_2O–MgO–ZnO–B_2O_3–SiO_2$

Table 6. Physical properties of commercial LTCC systems and some of recent research works on high thermal conductivity glass–ceramics.

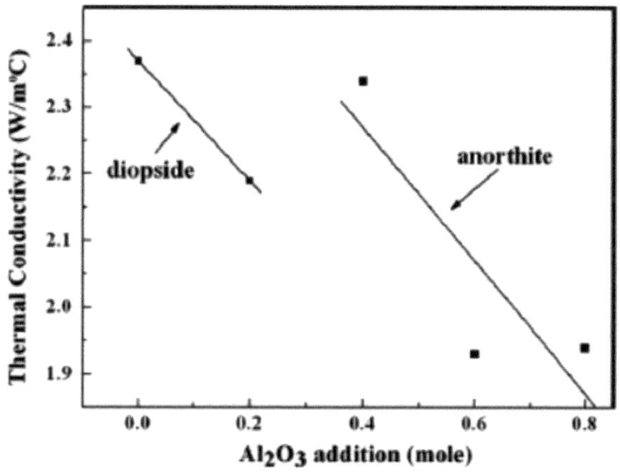

Figure 14. Dependences of thermal conductivity for MgO–CaO–SiO_2–Al_2O_3 glass–ceramic system on Al_2O_3 additions [33, 45].

Various types of filler particle morphologies such as platelet, fibrous types are addressed to improve the inter-particle contact to lower interfacial heat resistance and tailor heat transfer directionality in the substrate. Microstructural manipulations such as filler particle re-orientation, low thermal conductivity, secondary phase removal, and grain boundary control are also explored. Among them, some of key research works are summarized in **Table 6** [33, 45–53].

In the alumina (Al_2O_3) filled glass–ceramic system, MgO–CaO–Al_2O_3–SiO_2, the addition of alumina decreased the thermal conductivity of the glass–ceramic, where diopside or anorthite phase is a major re-crystallized secondary phase as shown in **Figure 14**. The thermal conductivities of diopside-based and anorthite-based glass–ceramics that sintered at <1000°C are 2.37 and 2.35 W/mK, respectively. It was found that the crystallinity is a more important factor than the ratio of dioposide and anorthite such that the highest bending strength and thermal conductivity were obtained at the samples with high crystallinity, since the main peak intensities in the XRD patterns of glass–ceramics is linearly proportional [33, 45]. In the 40 wt% alumina–60 wt% BSSZ (Bi_2O_3–ZnO–B_2O_3–SiO_2) glass system [46], the reported thermal conductivity of 7.2 W/mK is unusually high compared to that of previously known glass–ceramics [38–44] with a high dielectric constant of 10.9 and low loss of 0.009 at the sintered tape.

Aluminoborosilicate (ABS) glass-ceramics containing ≤15 wt% of multiwalled carbon nanotubes (MWCNTs) exhibited an improvement of electrical conductivity by $\sim 10^6$ and a thermal conductivity by ~70%. The maximum electrical conductivity of 2.1 S/cm was obtained when 15 wt% MWCNTs was added to an ABS base LTCC, while that of pure ABS was only $\sim 10^{-6}$ S/cm. The percolation threshold exists at the 2.5–5 wt% MWCNTs added region owing to the uniform dispersion of MWCNTs up to 10 wt%, which is a relatively higher loading rate than others [48] (**Figure 15**).

In the calcium aluminosilicate (CAS) glass system, the thermal conductivity was increased from 1.6 to 7.9 W/mK when 35 vol% of β-Si_3N_4 whiskers added to CAS–Si_3N_4 composites were sintered at 775–850°C in air (**Figure 16**). This thermal conductivity is much higher than other LTCC systems reported [38–44, 46, 47, 50]. However, the thermal conductivity was decreased when the β-Si_3N_4 whiskers loading exceed 35 wt%. The dielectric constant and loss measured at 1 MHz were 7.1 and 0.006, respectively [49].

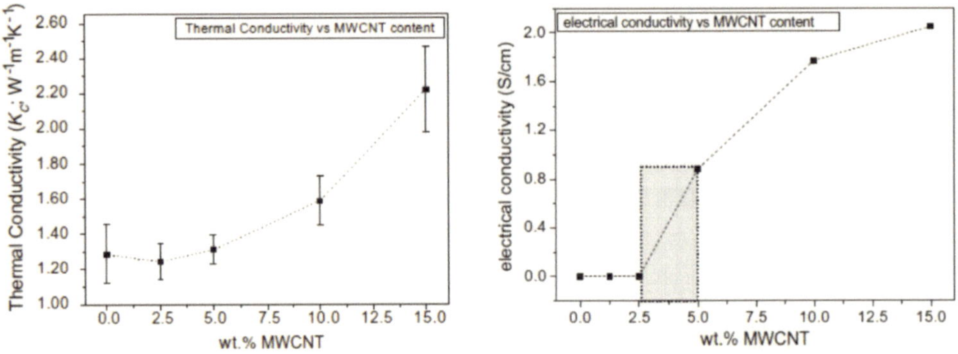

Figure 15. Thermal and electrical conductivity with MWCNT content for the ABS–MWCNT nanocomposites [48].

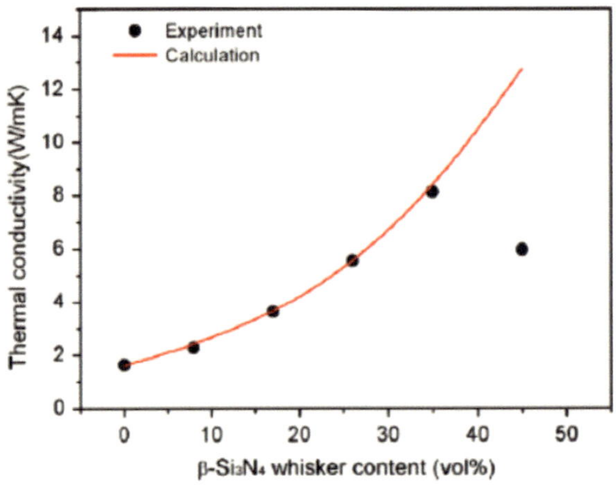

Figure 16. Experimental data and theoretical curve of the thermal conductivity of CAS–Si_3N_4 composites as a function of β-Si_3N_4 whisker contents [49].

The $Li_2ZnTi_3O_8$ (LZT) system with 1 wt% of LMZBS (Li_2O: MgO: ZnO: B_2O_3: SiO_2 = 1: 1: 1: 1: 1) as a sintering aid, thermal conductivity of 5.8 W/mK, and CTE of 11.97 ppm/°C were obtained at samples sintered at 875°C. High dielectric constant and loss of the LZT–LMZBS system at 1 MHz were 24.14 and 5.1×10^{-4}, respectively. Microwave dielectric properties of the sintered tape measured by split post dielectric resonator (SPDR) technique were $\varepsilon_r = 21.9$, tan $\delta = 6 \times 10^{-4}$ at 5 GHz, and τ_ε of −29 ppm/°C [50] (**Figure 17**).

Glass–ceramics filled with 3–40 μm size monocrystal diamond particles were studied, and the results showed highest thermal conductivity at the glass–diamond composites with 30 μm size diamonds that sintered at 750°C and revealed the lowest CTE, the highest thermal conductivity and bending strength: 4.35 ppm/°C, 9.01 W/mK, and 108.25 MPa [51] (**Figure 18**).

The addition of 1D materials such as AlN whiskers, carbon fibers, and copper fibers to the alumina/30 vol% glass composites was studied. The addition of AlN whiskers did not improve the thermal conductivity compared with AlN powder addition and fibrous fillers was more effective in increasing thermal conductivity of the composites. The highest thermal conductivity obtained in these composites was Al_2O_3/glass with 30 vol% copper fibers that sintered at 850°C as shown in

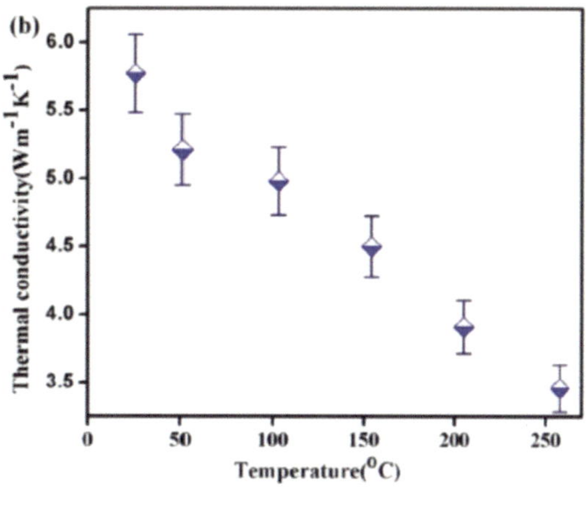

Figure 17. Thermal conductivity measurement data of LZT + LMZBS bulk sintered at 900°C [50].

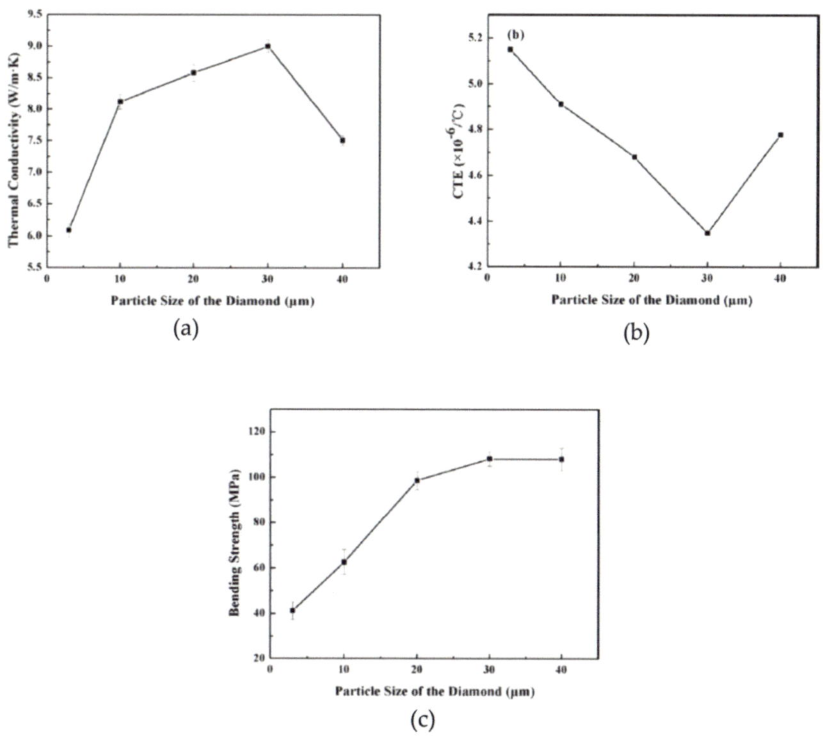

Figure 18. Thermal conductivity (a), CTE (b), and bending strength (c) of glass–diamond LTCCs sintered at 750°C [51].

Figure 19 [52]. The specific resistivity of Al_2O_3/30% glass and Al_2O_3/30% glass/30%AlN-whiskers are 2.0–2.1 × 10^{12} Ω·cm while those of 30% carbon fiber and 30% copper fiber added composites were 8.7 × 10^{-3} and 3.4 × 10^{-6} Ω cm, respectively. Therefore, these two electrically conductive composites are not applicable to electrically insulating thermal management material, or they can be coated with a insulating glass layer on the surface to improve electrical resistivity [52].

In the borosilicate glass–AlN composite LTCC, the thermal conductivity was increased from 11.9 to 18.8 W/mK by the addition of 14 vol% β-Si_3N_4 whiskers as shown in **Figure 20**. This enhanced

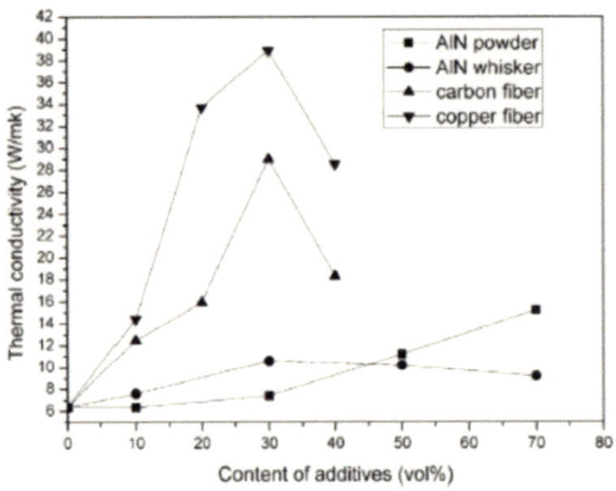

Figure 19. Variation of thermal conductivities of Al_2O_3/glass composites with the addition of 1D filler content [52].

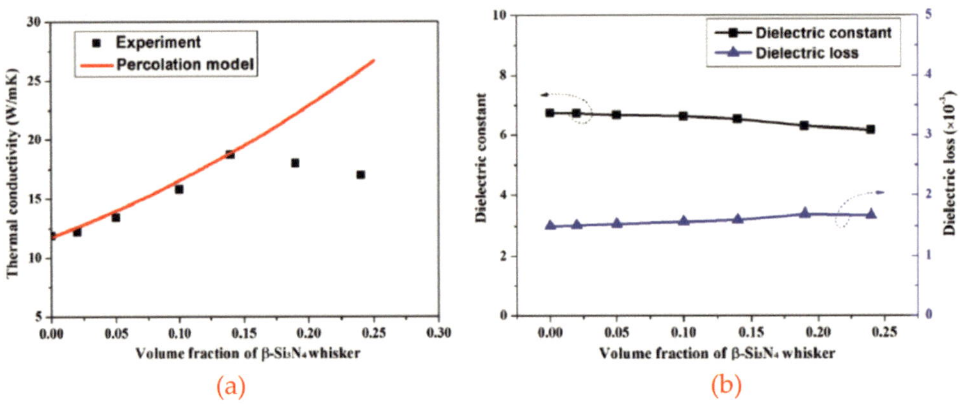

Figure 20. (a) Experimental data and theoretical curve of the thermal conductivity of the CMBS–AlN–Si_3N_4 ceramic composites as a function of β-Si_3N_4 whisker volume fraction and (b) relative dielectric constant and dielectric loss of CMBS–AlN–Si_3N_4 ceramic composites as a function of β-Si_3N_4 whisker volume fraction [53].

thermal conductivity may be due to the formation of a thermal conducting path by bridging the isolated AlN particles through β-Si_3N_4 whiskers. The dielectric properties of this composites are $\varepsilon_r = 6.5$ and $\tan\delta = 0.0016$ at 1 MHz, and the values are not significantly changed with the amount of whiskers due to their similarity in dielectric properties between AlN and Si_3N_4 [53].

2.3.2. Glass-free LTCC system

Glass-free or non-glass base LTCC systems have been investigated to reduce the complexity of the LTCC systems due to the multiple phases included such as glass, filler particles, and additional sintering additives. Due to their complexity, several problems occurred during the preparation of LTCC circuits and devices in the integrated electronic module. To overcome this complexity in chemical interaction and inhomogeneous dielectric properties and difficulties in slurry dispersion, LTCC systems with simple phase components were developed [54–60] (**Table 7**).

In a conventional LTCC system, glass was used as a matrix phase to lower the sintering temperatures below 1000°C because the functional dielectric ceramic materials were mostly fully densified at the high sintering temperatures, over 1200°C, where high electrically conductive

Chemical composition	Sintering temp (°C)	Thermal conductivity (W/mK)	CTE (ppm/°C)	Dielectric constant	tan δ/ Q factor	Reactivity with electrode	Ref.
CaO–(GeO_2, SiO_2, TeO_2)	780–1200	—	—	6.5~19.3	2.4×10^{-3} ~1×10^{-4} (MHz)	N(Ag: germanates, silicate) Y(Ag: tellurate)	[54]
$AMPO_2O_7$ (A = Ca, Sr., M = Zn, Cu)	950	—	—	7.06	52,871 (GHz)	Y(Ag) N(Cu)	[55]
$LiMgPO_4$	950	7.1	10.5	6.4	0.0002	—	[56]
$Bi_4(SiO_4)_3$	900	2.82	7.09	13.3	0.0007 (15GHz)	—	[57]
$Li_2Mg_2Ti_3O_8$	925	—	0.45(τf)	27	58,480 (5.8 GHz)	N(Ag)	[58]
$Li_{2.08}TiO_3$-LiF	900	4.75	22.4	22.4	35,490 (GHz)	N(Ag)	[59]
11ZnO-10MoO$_3$	850	1.3	4.7	11.1	—	—	[60]

Table 7. Compositions and physical properties of glass-free LTCC systems.

metals such as Ag or Cu cannot be used as a matching electrode. LTCCs containing glass phase matrix generally exhibited low thermal conductivity as we have seen in **Table 6**. In the glass-free LTCCs, lowering the sintering temperature below 1000°C is a primary requirement without introducing secondary phases except the minor content of the sintering agent. Finding a low temperature synthesis and low temperature melting crystalline phase ceramic compound is a crucial point to develop glass-free LTCCs. Glass-free LTCC compositions applicable in the industry with proven mechanical properties and reliabilities are hardly found, even though several primary research results showing excellent dielectric properties were reported [54–60]. The substantial problems exposed in the previous glass-free LTCC systems are weak mechanical strength, reactive with matching electrode materials during heat treatment, and vulnerable in moisture environment.

In the calcium germinates and silicates system, the dielectric constants were 6.5–10.8, quality factor (Q × f) = 16,000–39,000 (@10 GHz), temperature coefficients of dielectric constant were 70–140 ppm/°C for samples sintered at 1180–1200°C, which are slightly higher temperatures for LTCC processing. These systems did not show any chemical reaction with Ag electrode. On the other hand, in the calcium tellurates system, the dielectric constants were 15.5–23.6, Q × f = 13,400–49,300 (~10 GHz), temperature coefficients of dielectric constant were 130–140 ppm/°C for samples sintered at 780–840°C; but this system was vulnerable at Ag electrode. The high temperature coefficients of dielectric constant were suppressed by the addition of 10 mol% of $CaTiO_3$ [54].

For an AMP_2O_7 (A = Ca, Sr.; B = Zn, Cu) system, all of the compounds reacted with Ag but can be co-fired with Cu under reduced atmosphere. Among them, $SrZnP_2O_7$ sintered at 950°C exhibited a dielectric constant of 7.06, Q × f = 52,781 GHz, and temperature coefficient of resonance frequency (τ_f) = −70 ppm/°C; therefore, this compound can be modified into a temperature stable composition if a proper counter dielectric material with negative temperature coefficient is mixed with it. The thermal conductivity of this system was not provided [55].

In the $LiMgPO_4$ tape sintered at 950°C, the microwave dielectric properties were, $\varepsilon_r = 6.4$, tan $\delta = 0.0002$, CTE = 10.5 ppm/°C, and the thermal conductivity was 7.1 W/mK, which is twice as high as that of conventional glass–ceramic base LTCCs. The microwave dielectric properties of sintered tape were measured by using split post dielectric resonator (SPDR) method connected with a vector network analyzer [56].

$Bi_4(SiO_4)_3$ glass-free LTCC tape system has shown dielectric constant of 13.3, loss (tan δ) of 0.0007 at 15GHz, and thermal conductivity of 2.82 W/mK [57]. The $Li_2MgTi_3O_8$ glass-free ceramics sintered at 925°C exhibited a dielectric constant of 27, Q × f value of 58,480 GHz (@5.8 GHz), and very stable temperature coefficient of resonance frequency $\tau_f = 0.45$ ppm/°C. This system is compatible with silver electrode [58]. Another Li-base glass-free LTCC is $Li_{2.08}TiO_3$–LiF system, where the microwave dielectric properties of $\varepsilon_r = 22.4$, Q × f = 35,490 GHz were obtained at the 900°C sintered tape. The CTE and thermal conductivity of the system were 22.4 ppm/°C and 4.75 W/mK, respectively. The system also is compatible with silver electrode and has a high insulating rate of 50 kV/mm that has a potential in high power application. For the aforementioned glass-free system, the thermal conductivities obtained are 2.28–7.1 W/mK, which is well above that of the most of conventional glass–ceramic base LTCCs [59]. Zinc molybdate with 1% B_2O_3 that sintered at 850–900°C exhibited a dielectric constant of 11.1 CTE of 4.7 ppm/°C, and a break down voltage of 17.6 kV/mm. However, the thermal conductivity was relatively low, 1.4 W/mK, compared to the other glass-free LTCCs. This system may be applicable to high temperature insulating dielectrics due to low CTE and high break down voltage [60].

However, regardless of excellent dielectric and thermal properties, some of the glass-free LTCC compounds containing lithium element have a water soluble problem that limits the application. Therefore, they might need a protective layer coating to resist under weathering conditions.

3. Summary and future prospects

In this chapter, recent research and development works on high thermal conductivity ceramics and their composites for thermal management of integrated electronic packages are briefly explored. Key lessons drawn from these prior works can be summarized as follows:

3.1. High thermal conductivity bulk ceramics

Most frequently found HTCC base high thermal conductivity ceramics are alumina and nitride ceramics such as AlN, BN, and Si_3N_4 materials. Among them, silicon nitride ceramic seems the most frequently used in the power electronic applications these days. In nitrides and nitride based ceramic matrix composites, key parameters that control the thermal property are densification including pore removal, grain size and grain boundary control, impurity, and secondary phase control. Among them, densification is the primary factor to achieve high thermal conductivity due to high thermal resistance of pores. These nitride ceramics are difficult to sinter with high density so that spark plasma sintering and two-step sintering methods together with the addition of small amount of sintering aids should be applied to realize high densification. In nitride ceramics, controlling the oxygen content is a very important factor in addition to the parameters required in oxide materials.

In the LTCC-base materials, there are many research works on the development of high mechanical strength LTCC materials but few works are found in the improvement of thermal conductivity of LTCCs. Some works found in the literature are mainly on re-crystallization and phase control in the matrix and show only minor improvement in thermal properties compared with the noticeable enhancement in mechanical properties. The main reason for this minor change in thermal property in the conventional glass–ceramic type LTCC comes from the glass matrix which comprises more than half the volume of the sintered body. The volume fraction of newly evolved nano-crystalline phases through heat treatment process are so small that the overall apparent thermal conductivity might not change significantly while the mechanical strength can be easily boosted by the inclusion of the nano-crystalline phase in the matrix.

In thermally conductive and electrically conducting ceramics composites, the electrical conductivity was improved with the addition of 2D carbon allotropes like graphene nano sheet/platelet; however, the thermal conductivity was decreased with the involvement of graphene in the AlN matrix, which may be due to the thin interaction layer at the AlN–GNS/GNP interface that would cause thermal resistance. Large difference was observed between the in-plane and through-plane thermal conductivity of the composites as can be observed in the polymer/graphene composites.

3.2. Polymer matrix composites with high thermal conductivity

Polymer matrix composites with high thermal conductivity and electrically insulating ceramic filler materials are mostly used for dielectric insulation layers in LED packaging substrate for effective heat dissipation to the metallic heat spreading panels. Most frequently used insulating ceramic filler materials are alumina, BN, and AlN powders. Among them, BN platelet powders are preferred due to the anisotropic thermal conductivity behavior in the 2D structure of the BN crystal. Lots of researches have been focused on the tailored re-orientation of BN nanosheets in the BN/polymer solution into in-plane or through-plane of the BN/polymer composite tape using magnetic and electric field during casting process. Also, there are some efforts to coat ferrous or dielectric nano particles on BN platelet particles to promote the easy alignment of BN platelet particles into the intended direction. Surface modification of BN particles with functional organics such as silane coupling agent, dopamine, and secondary functional monomers are applied to enhance the thermal conductivity of the composites by improving the BN/polymer affinity and interfacial adhesion, thereby lowering interfacial heat resistance.

Polymer matrix composites with high thermal conductivity inorganic fillers such as CNT, graphite flake, and graphene nanosheets exhibited a great improvement in thermal conductivity with a little amount of additions. However, they are mostly electrically conducting so that they cannot be used for electrical circuit substrates. Instead, these composites are mainly used for thermal interface materials. These polymer/carbon allotropes base composites can also be used for flexible device application as well as rigid substrates since the morphology of carbonates filler particles are 1D or 2D.

3.3. High thermal conductivity ceramics for LED and IGBT packages

Ceramic materials used for applications in LED packages are typically of two types: dielectric insulating substrates for circuit forming bed and high thermal conductivity fillers for thermal interface material. The insulating ceramic substrates for IGBT modules mostly use alumina, AlN ceramics, but recent development moves to Si_3N_4 and LTCC for high power device due to reliability or low cost.

3.4. Future prospects

Regardless of the aforementioned progresses in the development and commercialization of ceramics, there are several challenges in the high thermal conductivity ceramic based heat transfer materials:

Continuous efforts in lowering costs and cost-effective processing of high temperature sintering HTCCs with high thermal conductivity are required in materials chemistry and innovative processing techniques. Compared to the HTCC based high thermal conductivity ceramics, LTCCs still require further enhancement in both thermal and mechanical characteristics in order to be adopted in thermal management applications. Since the major part of the conventional LTCC formulation consists of glass, the utmost thermal conductivity of the glass–ceramic filler composites thus obtained is limited and far below that of HTCC based high thermal conductivity ceramics. Therefore, first, we need to investigate ways to improve thermal conductivity of the glass phase itself as it is done in many polymer matrix composites. Second, the mechanical strength of LTCC should be further improved even though some results demonstrated enhanced mechanical strength via recrystallization process through the interfacial reaction and nucleation between glass phase and crystalline filler phase. Other challenges in high thermal conductivity LTCCs may be the development of non-glass base LTCCs, which have already been attempted earlier as applications in RF and microwave dielectric materials. The non-glass based LTCCs are exempted from the usage of low thermal conductivity glass matrix phase; they will exhibit higher thermal conductivity than the conventional type.

There are many reports that state the achievement of high thermal conductivities in polymer matrix composites using high thermal conductivity ceramic fillers. As a practical point of view, however, simply increasing thermal conductivity of ceramic filled polymer composites does not ensure the potential use in thermal management application, especially when they are used as thermal interface materials. Other factors, such as adhesion strength to the substrates or heat sink materials for TIM application and tensile strength of thermal tapes or flexible device substrates, also should be considered in addition to thermal and electrical properties since the more filler loading, the less adhesion and tensile strength is provided.

For insulated metal substrates (IMS) using high thermal conductivity ceramics, a reliable solution for CTE mismatch between ceramic and metal joining inducing delamination and crack generation failure in harsh conditions such as cyclic temperature environment is still required. In addition, for highly effective heat transfer performance IMS, we may need an ultra-thin insulation layer with high dielectric breakdown voltage together with high mechanical strength, which enables both low thermal resistance and low package profile of high power device and module.

Acknowledgements

This work was supported by the R&D convergence program of MSIP (Ministry of Science, ICT and Future Planning) and NST (National Research Council of Science & Technology) of the Republic of Korea (Grant No. CAP-13-02-ETRI).

Conflict of interest

None.

Author details

Hyo Tae Kim

Address all correspondence to: hytek@kicet.re.kr

Korea Institute of Ceramic Engineering and Technology, Jinju-si, Gyeongsangnam-do, Republic of Korea

References

[1] Franco Júnior A, Shanafield DJ. Thermal conductivity of polycrystalline aluminum nitride (AlN) ceramics. Cerâmica. 2004;**50**(315). DOI: 10.1590/S0366-69132004000300012

[2] Choi HS, Im HN, Kim YM, Chavan A, Song SJ. Structural, thermal and mechanical properties of aluminum nitride ceramics with CeO_2 as as sintering aid. Ceramics International. 2016;**42**:11519-11524. DOI: 10.1016/j.ceramint.2016.04.028

[3] He X, Yea F, Zhanga H, Liu L. Effect of Sm_2O_3 content on microstructure and thermal conductivity of spark plasma sintered AlN ceramics. Journal of Alloys and Compounds. 2009;**482**:345-348. DOI: 10.1016/j.jallcom.2009.04.013

[4] Kobayashi R, Ohishi K, Tu R, Goto T. Sintering behavior, microstructure, and thermal conductivity of dense AlN ceramics processed by spark plasma sintering with Y_2O_3–CaO–B additives. Ceramics International. 2015;**41**:1897-1901. DOI: 10.1016/j.ceramint.2014.09.040

[5] Lee HM, Kim DK. High-strength AlN ceramics by low-temperature sintering with $CaZrO_3$–Y_2O_3 co-additives. Journal of the European Ceramic Society. 2014;**34**:3627-3633. DOI: 10.1016/j.jeurceramsoc.2014.05.008

[6] Simsek ING, Nistal A, García E, Pérez-Coll D, Miranzo P, Osendi MI. The effect of graphene nanoplatelets on the thermal and electrical properties of aluminum nitride ceramics. Journal of the European Ceramic Society. 2017;**37**:3721-3729. DOI: 10.1016/j.jeurceramsoc.2016.12.044

[7] Yun C, Feng Y, Qiu T, Yang J, Li X, Yu L. Mechanical, electrical and thermal properties of graphene nanosheet/aluminum nitride composites. Ceramics International. 2015;**41**(7):8643-8649. DOI: 10.1016/j.ceramint.2015.03.075

[8] Xia H, Zhang X, Shi Z, Zhao C, Li Y, Wang J. Mechanical and thermal properties of reduced graphene oxide reinforced aluminum nitride ceramic composites. Materials Science and Engineering A. 2015;**639**:29-36. DOI: 10.1016/j.msea.2015.04.091

[9] Liang H, Zengn Y, Zuo K, Xia Y, Yao D, Yin J. Mechanical properties and thermal conductivity of Si_3N_4 ceramics with YF_3 and MgO as sintering additives. Ceramics International. 2016;**42**:15679-15686. DOI: 10.1016/j.ceramint.2016.07.024

[10] Miyazaki H, Yoshizawa Y, Hirao K. Fabrication of high thermal-conductive silicon nitride ceramics with low dielectric loss. Materials Science and Engineering B. 2009;**161**:198-201. DOI: 10.1016/j.mseb.2008.11.029

[11] Golla BR, Ko JW, Kim HD. Processing and characterization of sintered reaction bonded Si_3N_4 ceramics. International Journal of Refractory Metals & Hard Materials. 2017;**68**:75-83. DOI: 10.1016/j.ijrmhm.2017.07.005

[12] Lee HM, Lee EB, Kim DL, Kim DK. Comparative study of oxide and non-oxide additives in high thermal conductive and high strength Si_3N_4 ceramics. Ceramics International. 2016;**42**:17466-17471. DOI: 10.1016/j.ceramint.2016.08.051

[13] Yuan C, Xie B, Huang M, Wu R, Luo X. Thermal conductivity enhancement of platelets aligned composites with volume fraction from 10% to 20%. International Journal of Heat and Mass Transfer. 2016;**94**:20-28. DOI: 10.1016/j.ijheatmasstransfer.2015.11.045

[14] Han Y, Lva S, Haob C, Dinga F, Zhanga Y. Thermal conductivity enhancement of BN/silicone composites cured under electric field: Stacking of shape, thermal conductivity, and particle packing structure anisotropies. Thermochimica Acta. 2012;**529**:68-73. DOI: 10.1016/j.tca.2011.11.029

[15] Cho HB, Nakayama T, Suematsu H, Suzuki T, Jiang W, Niihara K, Song E, Eom NA, Kim S, Choa YH. Insulating polymer nanocomposites with high-thermal-conduction routes via linear densely packed boron nitride nanosheets. Composites Science and Technology. 2016;**129**:205-213. DOI: 10.1016/j.compscitech.2016.04.033

[16] Zhang J, Wang X, Yu C, Li Q, Li Z, Li C, Lu H, Zhang Q, Zhao J, Hu M, Yao Y. A facile method to prepare flexible boron nitride/poly(vinyl alcohol) composites with enhanced thermal conductivity. Composites Science and Technology. 2017;**149**:41-47. DOI: 10.1016/j.compscitech.2017.06.008

[17] He X, Gong Q, Guo Y, Liu J. Microstructure and properties of AlN-BN composites prepared by sparking plasma sintering method. Journal of Alloys and Compounds. 2016;**675**:168-173. DOI: 10.1016/j.jallcom.2016.03.058

[18] Li YL, Zhang J, Zhang JX. Fabrication and thermal conductivity of AlN/BN ceramics by spark plasma sintering. Ceramics International. 2009;**35**:2219-2224. DOI: 10 .1016/j.ceramint.2008.12.003

[19] Zhao H, Wang W, Fu Z, Wang H. Thermal conductivity and dielectric property of hot-pressing sintered AlN–BN ceramic composites. Ceramics International. 2009;**35**:105-109. DOI: 10.1016/j.ceramint.2007.09.111

[20] Wilk A, Rutkowski P, Zientara D, Bu'cko MM. Aluminium oxynitride–hexagonal boron nitride composites with anisotropic properties. Journal of the European Ceramic Society. 2016;**36**:2087-2092. DOI: 10.1016/j.jeurceramsoc.2016.02.029

[21] Kim K, Ju H, Kim J. Filler orientation of boron nitride composite via external electric field for thermal conductivity enhancement. Ceramics International. 2016;**42**:8657-8663. DOI: 10.1016/j.ceramint.2016.02.098

[22] Zhan Y, Long Z, Wan X, Zhan C, Zhang J, He Y. Enhanced dielectric permittivity and thermal conductivity of hexagonal boron nitride/poly(arylene ether nitrile) composites through magnetic alignment and mussel inspired co-modification. Ceramics International. 2017;**43**:12109-12119. DOI: 10.1016/j.ceramint.2017.06.068

[23] Jang I, Shin KH, Yang I, Kim H, Kim J, Kim WH, Jeon SW, Kim JP. Enhancement of thermal conductivity of BN/epoxy compositethrough surface modification with silane coupling agents. Colloids and Surfaces A: Physicochemical and Engineering Aspects. 2017;**518**:64-72. DOI: 10.1016/j.colsurfa.2017.01.011

[24] Lee ST, Kim HT, Nahm S, Lee SH, Lee SG. Fabrication of high thermal conductivity ceramic hybrid meatrials for power electronics and integrated packages. In: Proceedings of HEFAT; 20-23 July 2015

[25] Zhou Y, Wang H, Wang L, Yu K, Lin Z, He L, Bai Y. Fabrication and characterization of aluminum nitride polymer matrix composites with high thermal conductivity and low dielectric constant for electronic packaging. Materials Science and Engineering B. 2012;**177**:892-896. DOI: 10.1016/j.mseb.2012.03.056

[26] Hu M, Feng J, Ng KM. Thermally conductive PP/AlN composites with a 3-D segregated structure. Composites Science and Technology. 2015;**110**:26-34. DOI: 10.1016/j.compscitech.2015.01.019

[27] Wozniak M, Anna Danelska A, Rutkowski P, Kata D. Thermal conductivity of highly loaded aluminium nitride–poly(propylene glycol) dispersions. International Journal of Heat and Mass Transfer. 2013;**65**:592-598. DOI: 10.1016/j.ijheatmasstransfer.2013.06.048

[28] Chen C, Hou F, Liu F, She Q, Cao L, Wan L. Thermo-mechanical reliability analysis of a RF SiP module based on LTCC substrate. Microelectronics Reliability. 2017;**79**:38-47. DOI: 10.1016/j.microrel.2017.10.003

[29] Nowak D, Dziedzic A. LTCC package for high temperature applications. Microelectronics Reliability. 2011;**51**:1241-1244

[30] Chutani RK, Galliou S, Passilly N, Gorecki C, Sitomaniemi A, Heikkinen M, Kautio K, Keranen A, Jorno A. Thermal management of fully LTCC-packaged Cs vapour cell for MEMS atomic clock. Sensors and Actuators A. 2012;**174**:58-68

[31] Jiang B, Muralt P, Maeder T. Meso-scale ceramic hotplates – A playground for high temperature microsystems. Sensors and Actuators B. 2015;**221**:823-834

[32] Vasudev A, Kaushik A, Tomizawa Y, Norena N, Bhansali S. An LTCC-based microfluidic system for label-free, electrochemical detection of cortisol. Sensors and Actuators B. 2013;**182**:139-146. DOI: 10.1016/j.snb.2013.02.096

[33] Kang M, Kang S. Influence of Al_2O_3 additions on the crystallization mechanism and properties of diopside/anorthite hybrid glass-ceramics for LED packaging materials. Journal of Crystal Growth. 2011;**326**:124-127. DOI: 10.106/j.jcrysgro.2011.01.081

[34] Liu S, Li X, Yu X, Chang Z, Che P, Zhou J. A route for white LED package using luminescent low-temperature co-fired ceramics. The Journal of Alloys and Compounds. 2016;**655**:203-207. DOI: 10.1016/j.jallcom.2015.09.177

[35] Ding Y, Liu YS, Li X, Wang R, Zhou J. Luminescent low temperature co-fired ceramics for high power LED package. The Journal of Alloys and Compounds. 2012;**521**:35-38. DOI: 10.1016/j.jallcom.2011.12.143

[36] Bienert C, Roosen A. Characterization and improvement of LTCC composite materials for application at elevated temperatures. Journal of the European Ceramic Society. 2010;**30**:369-374. DOI: 10.1016/j.jeurceramsoc.2009.05.023

[37] Sebastian MT, Jantunen H. Low loss dielectric materials for LTCC applications: A review. International Materials Reviews. 2018;**53**:57-90. DOI: 10.1179/174328008X277524

[38] Golonka LJ. Technology and applications of low temperature co-fired ceramic (LTCC) based sensors and microsystems. Bulletin of the Polish Academy of Sciences, Technical Sciences. 2006;**54**:221-231

[39] http://natelems.com/wp-content/uploads/2014/10/Natel-LTCC-Quick Ref.pdf. Heratape™ CT2000

[40] http://www.dupont.com, DataSheetofDupontTM Green TapeTM 951

[41] http://www.dupont.com, DataSheetofDupontTM Green TapeTM 9K7

[42] http://www.ltcc-consulting.com/LTCC_technology_materials

[43] http://www.dupont.com/content/dam/assets/products-and-services/electronic-electrical materials/ assets/datasheets/prodlib/943LowLossTape.pdf

[44] http://www.murata.com-Murata's LTCCsubstratetechnology: LFC Series

[45] Jang S, Kang S. Influence of MgO/CaO ratio on the properties of MgO–CaO–Al_2O_3–SiO_2 glass–ceramics for LED packages. Ceramics International. 2012;**38S**:S543-S546. DOI: 10.1016/j.ceramint.2011.05.073

[46] Induja IJ, Abhilash P, Arun S, Surendran KP, Sebastian MT. LTCC tapes based on Al_2O_3–BBSZ glass with improved thermal conductivity. Ceramics International. 2015;**41**:13572-13581. DOI: 10.1016/j.ceramint.2015.07.152

[47] Kim HT, Kim SH, Nahm S, Byun JD. Low-temperature sintering and microwave dielectric propertiesof zinc metatitanate-rutile mixtures using boron. The Journal of the American Ceramic Society. 1999;**82**(11):3043-3048. DOI: 10.1111/j.1151-2916.1999.tb02200.x

[48] Mukhopadhyay A, Otieno G, Chu BTT, Wallwork A, Green MLH, Todd RI. Thermal and electrical properties of aluminoborosilicate glass–ceramics containing multiwalled carbon nanotubes. Scripta Materialia. 2011;**65**:408-411. DOI: 10.1016/j.scriptamat.2011.05.023

[49] Ma M, Liu Z, Li Y, Zeng Y, Yao D. Thermal conductivity of low-temperature sintered calcium aluminosilicate glass–silicon nitride whisker composites. Ceramics International. 2013;**39**:4683-4687. DOI: 10.1016/j.ceramint.2012.11.056

[50] Arun S, Sebastian MT, Surendran KP. $Li_2ZnTi_3O_8$ based high κ LTCC tapes for improved thermal management in hybrid circuit applications. Ceramics International. 2017;**43**:5509-5516. DOI: 10.106/j.ceramint.2017.01.073

[51] Feng D, Li Z, Zhu Y, Ji H. Influence of diamond particle size on the thermal and mechanical properties of glass-diamond composites. Materials Science and Engineering B. 2018;**227**:122-128. DOI: 10.1016/j.mseb.2017.10.017

[52] Wang S, Zhang D, Ouyang X, Wang Y, Liu G. Effect of one-dimensional materials on the thermal conductivity of Al_2O_3/glass composite. Journal of Alloys and Compounds. 2016;**667**:23-28. DOI: 10.1016/j.jallcom.2016.01.120

[53] Ma M, Liu Z, Li Y, Zeng Y, Yao D. Enhanced thermal conductivity of low-temperature sintered borosilicate glass–AlN composites with β-Si_3N_4 whiskers. Journal of the European Ceramic Society. 2013;**33**:833-839. DOI: 10.1016/j.jeurceramsoc.2012.09.030

[54] Valant M, Suvorov D. Glass-free low-temperature cofired ceramics: Calcium germanates, silicates and tellurates. Journal of the European Ceramic Society. 2004;**24**:1715-1719. DOI: 10.1016/S0955-2219(03)00483-7

[55] Bian JJ, Kim DW, Hong KS. Glass-free LTCC microwave dielectric ceramics. Materials Research Bulletin. 2005;**40**:2120-2129. DOI: 10.1016/j. materresbull.2005.07.003

[56] Thomas D, Abhilash P, Sebastian MT. Casting and characterization of $LiMgPO_4$ glass free LTCC tape for microwave applications. Journal of the European Ceramic Society. 2013;**33**:87-93. DOI: 10.1016/j.jeurceramsoc.2012.08.002

[57] Abhilash P, Sebastian MT, Surendran KP. Glass free, non-aqueous LTCC tapes of $Bi_4(SiO_4)_3$ with high solid loading. Journal of the European Ceramic Society. 2015;**35**:2313-2320. DOI: 10.106/j.jeurceramsoc.2015.02.002

[58] Zhou H, Wang N, Gong J, Fan G, Chen X. Processing of low-fired glass-free $Li_2MgTi_3O_8$ microwave dielectric ceramics. Journal of Alloys and Compounds. 2016;**688**:8-13. DOI: 10.1016/j.jallcom.2016.07.214

[59] Bian JJ, Yu Q, He JJ. Tape casting and characterization of $Li_{2.08}TiO_3$-LiF glass free LTCC for microwave applications. Journal of the European Ceramic Society. 2017;**37**:647-653. DOI: 10.1016/j.jeurceramsoc.2016.09.022

[60] Wang Z, Freer R. Low firing temperature zinc molybdate ceramics for dielectric and insulation applications. Journal of the European Ceramic Society. 2015;**35**:3033-3042. DOI: 10.1016/j.jeurceramsoc.2015.04.020